McKinsey, J. C. C. (John Charles Chenoweth), 1908-

Introduction to the theory of games

INTRODUCTION TO THE THEORY OF GAMES

The RAND Series

The RAND Series

This is one of a series of publications presenting the results of research undertaken by The RAND Corporation, a nonprofit organization, chartered "to further and promote scientific, educational, and charitable purposes, all for the public welfare and security of the United States of America."

ALREADY PUBLISHED

THE OPERATIONAL CODE OF THE POLITBURO
Nathan Leites

AIR WAR AND EMOTIONAL STRESS
Psychological Studies of Bombing and Civilian Defense
Irving L. Janis

SOVIET ATTITUDES TOWARD AUTHORITY
An Interdisciplinary Approach to Problems of Soviet Character
Margaret Mead

THE ORGANIZATIONAL WEAPON
A Study of Bolshevik Strategy and Tactics
Philip Selznick

INTRODUCTION TO THE THEORY OF GAMES
J. C. C. McKinsey

WEIGHT-STRENGTH ANALYSIS OF AIRCRAFT STRUCTURES
F. R. Shanley

INTRODUCTION TO THE THEORY OF GAMES

J. C. C. McKinsey

The RAND Corporation

NEW YORK · TORONTO · LONDON
McGRAW-HILL BOOK COMPANY, INC.
1952

INTRODUCTION TO THE THEORY OF GAMES

Printed in the United States of America

Library of Congress Catalog Card Number: 52-11846

III

Foreword

THIS book is intended to be used as a college textbook for a course for upper division and graduate students.

It is assumed that the student will have at least the knowledge of analysis that is ordinarily acquired from a course in advanced calculus. I have therefore employed without explanation such notions as: convergence, continuity, derivatives, (Riemann) integrals, greatest lower and least upper bounds, and maxima and minima. And I have availed myself, without specific reference, of the more familiar theorems involving these notions.

Some familiarity with classical algebra and matrix theory would be an advantage for the understanding of Chap. 3; but for the sake of students who are untrained in this branch of mathematics, I have included in that chapter a brief survey of the elementary operations on matrices. It should be noticed that nothing in the later parts of the book depends in any essential way on the results of Chap. 3; so the instructor can feel at liberty simply to omit this chapter.

In order to make the book accessible to a wider class of students, I have introduced some less familiar notions in a detailed way. This applies, in particular, to distribution functions and Stieltjes integrals, to which I have devoted separate chapters, and to some elementary topological notions which are explained in Sec. 2 of Chap. 3.

I have attempted to assign credit for the various results formulated here by some indications in the Historical and Bibliographical Remarks at the end of the several chapters. Besides these debts of a general scientific nature, I want to express my personal gratitude to a number of friends, without whose assistance the book could hardly have been written. Mr. Oliver Gross, of the RAND Corporation, has supplied several examples for Chap. 10; and Mr. J. D. Williams, also of the RAND Corporation, has kindly made available several examples of games which he had collected for a forthcoming book of his own. Dr. A. V. Martin, of the University of California, and Dr. J. G. Wendel, of the RAND Corporation, have made some valuable suggestions in connection with Chaps. 9 and 10; and Professor David Blackwell, of Howard University, has helped to formulate the discussion of statistical inference in Chap. 13. Dr. Norman Dalkey and Dr. F. M. Thompson, both of the RAND Corporation, have assisted with Chaps. 5 and 6; and Mr. L. S. Shapley, of Princeton University, has made a careful examination of the entire manuscript, eliminating many absurdities and errors. My most special thanks, finally, are due to Dr. Melvin Dresher and Dr. Olaf Helmer, both of the RAND Corporation, who have often taken time from their own work to help me with mine: their assistance has been invaluable.

J. C. C. McKinsey

Stanford University
Stanford, California
January, 1952.

Contents

CHAPTER 12

CHAPTER 13

CHAPTER 14

CHAPTER 15

CHAPTER 16

CHAPTER 17

CHAPTER 18

CHAPTER 1

RECTANGULAR GAMES

1. Introduction. In this book we shall be concerned with the mathematical theory of games of strategy. Examples of parlor games of strategy are such games as chess, bridge, and poker, where the various players can make use of their ingenuity in order to outwit each other. Aside from this application within the sphere of social amusement, the theory of games is gaining importance because of its general applicability to situations which involve conflicting interests, and in which the outcome is controlled partly by one side and partly by the opposing side of the conflict. Many conflict situations which form the subject of economic, social, political, or military discourse are of this kind.

Although many real-life conflicts as well as parlor games involve elements of chance (as in the cards dealt in bridge or the weather encountered in a military operation), we shall ordinarily exclude from our discussion games in which the outcome depends entirely on chance and cannot be affected by the cleverness of the players.

The essential difference between games of strategy and games of (pure) chance lies in the circumstance that intelligence and skill are useful in playing the former but not the latter. Thus an amateur would be extremely unwise to play chess for even money and high stakes against a master: he would face almost certain ruin. But, contrary to the stories occasionally heard (stories which are most likely fabricated by the proprietors of gaming houses), there is no "system" for playing roulette on an unbiased wheel: an idiot has as good a chance of winning at this game as has a man of sense. (This is not to say that there do not remain difficult unsolved mathematical problems in connection with games of chance; but there exist, at least, standard methods for attacking such problems, and we shall not treat of them here.)

Although our attention will be devoted almost entirely to the purely mathematical aspects of the theory of games of strategy, it is perhaps well to begin with some brief remarks about the history of economics. These remarks may serve to convince the student that our theory is not altogether frivolous; for buying and selling are customarily regarded as more serious and respectable occupations than is playing poker—or even chess, for that matter.

For many decades economists tended to take as a standard model for their science the situation of Robinson Crusoe, marooned on an uninhabited island and concerned with behaving in such a way as to maximize the goods he could obtain from nature. It was generally felt that it would be possible to get an insight into the behavior of groups of individuals by starting with a detailed analysis of the behavior proper in this simplest possible case: the case of a single individual all alone and struggling against nature.

This line of attack on economic problems, however, suffers from the defect that in going from a one-man society to even a two-man society, qualitatively different situations arise which could hardly have been foreseen from the one-man case.[1] In a society which contains two members, it may happen that each desires a certain commodity (the supply of which is not sufficient for both) and that each member has control of some, but not all, of the factors which determine how the commodity is to be distributed. The behavior of each, then, if it is to be rational, must take into account the expected behavior of the other. No such situation as this can arise in the one-man case, where the one member of society is concerned simply with maximizing the amount of the commodity he is to receive from nature. For, though we often personify nature (by capitalizing the word and even by treating it as being of feminine gender) and though we sometimes poetically speak of the "perverseness" of nature, no one seriously believes that nature is really a conscious being, who takes thought about what we are to do and adjusts her own behavior accordingly.

In a society with two or more members, entirely new problems appear which are radically different from anything found in a one-man society. For this reason it is not possible to determine the properties of ordinary society by simply extrapolating from the case of a Robinson Crusoe society.

It was through considerations of this sort that the mathematician John von Neumann was led, some twenty years ago, to believe that economics could more profitably be viewed under the analogy to parlor games (of strategy) than under the simpler analogy to the analytic problem of finding maxima and minima. This approach to economics has by now been explored rather thoroughly by mathematicians as well as by economists. References to the relevant books and papers can be found in the Bibliography at the end of this book.

---

[1] It is as though we were to try to get an insight into the nature of circles by studying points, which, after all, are a kind of circles—circles with zero radius. But ordinary circles differ too radically from point-circles for this type of approach to be helpful.

(In connection with the question of guessing at the economic laws of an n-man society by using the laws of a one-man society, it should be remarked that, rather paradoxically, a better approximation is obtained in this way to the case that n is large, than to the case that n is small but greater than one. For if Smith has only one competitor, he must take account of the very real possibility that his antagonist will behave in a rational way—that he will even, indeed, attempt to guess what Smith is going to do, and to adjust his own behavior accordingly. And if Smith has but a few competitors, he should not neglect the possibility that they may all behave rationally, or that they may combine together in a coalition against him and thus, in essence, behave as if they were but one. But as n becomes very large, the probability that a large proportion of Smith's opponents will behave rationally becomes small, and the advantage they could gain by combining against him becomes negligible. Therefore, it becomes increasingly plausible for Smith to assume that the average behavior of the rest of the population is determined by prevalent superstitions and fallacies, for instance, or by the average level of intelligence; and it becomes reasonable for Smith to feel confidence in his predictions of the behavior of his competitors—e.g., predictions based on their past behavior—and to treat the rest of mankind like a part of nature. But the advocate of Robinson Crusoe economics, before he becomes too complacent from considering this little paradox, should reflect that in modern society men tend to combine into a few large coalitions—corporations, cooperatives, labor unions, and the like—which, for many practical purposes, behave like individual human beings.)

It should be mentioned, finally, that the theory of games of strategy can be expected to find practical application in domains which would not ordinarily be regarded as economic: to the problems arising in connection with courtship and marriage, for instance, where the end in view is not necessarily monetary; or to the problems which confront a politician trying to get elected to office in a country which allows more than one candidate to have his name on the ballot. It is possible that this theory will throw light on all kinds of situations in which various people have opposing goals and in which each of them, although he may exert some influence on the outcome, cannot completely dominate the course of events.

2. Terminology, and Classification of Games. The word "game," as used in everyday English, is ambiguous. Sometimes, as when we say "Chess is a more difficult game than checkers," we use the word "game" to refer to a set of rules and conventions for playing; at other times, as when we say "I played three games of chess last night," we use the word to refer to a particular possible realization of the rules. For our purposes, it is convenient to distinguish these two notions: accordingly, we shall use the word "game" only for the first meaning and shall employ the word "play" for the second meaning. Thus we shall still say "Chess is a more difficult game than checkers," but now we shall say "I played three plays of chess last night."

In a similar way, we shall use the word "move" to mean a point in a game at which one of the players (or chance, in some cases) picks out an alternative from some set of alternatives, and we shall use the word "choice" to mean the alternative picked out. (In ordinary speech, the word "move" is used, ambiguously, for both notions.) Thus we shall say "Black won by a clever choice in his tenth move."

The number, and variety, of games of strategy is enormous. We shall now indicate some modes of classification.

We first distinguish a game according to the *number of players*: one-person games, two-person games, and so on. Solitaire, for example, is a one-person game and chess is a two-person game. When we call a game n-person, however, we do not necessarily mean that in every play of it exactly n people participate, but, rather, merely that the rules of the game are such that the players fall into n mutually exclusive sets in such a way that the people within each set have identical interests. These n sets of people with identical interests are referred to as "persons" (just as, in law, a corporation is referred to as a person). For example, although chess is ordinarily played by just two people, it could also be played by two "teams," each consisting, say, of three people; and even if this were done, the game would still be chess and would still be a two-person game, not a six-person game. In the same way, bridge is to be regarded as a two-person game, not a four-person game, since North and South have identical interests and are therefore considered as one person, and East and West are similarly considered as one person.

When people play social games, they sometimes decide that at the end of the play they will make monetary *payments* among themselves in a manner determined by the rules of the game. This is almost always done, for example, in games of pure chance such as craps (since otherwise these games would hardly be interesting to play), and it is usually done in poker and often in bridge. In other cases, the players keep track of the "score," so that at the end of the play numbers are calculated which measure the relative skill with which the participants have played, but no money is exchanged; this is often done in bridge, for example. Finally, in some cases, no attempt is made even to calculate any kind of scores, but it is simply announced who has "won" and who has "lost"; this is usually done, for example, in ticktacktoe, checkers, and chess. For our purposes, however, it turns out to be convenient to neglect the second and third of the above alternatives and to speak as though all games were played for money; thus we shall usually speak of the "payments" made among the players at the end of a play and shall think of these payments as sums of money. (The assumption that there are money payments may appear to constitute simply a limitation of our field of inquiry; but it is also possible to argue that, even when no money changes hands, the players derive pleasure or pain from their relative scores and to maintain that they would be able, if questioned, to set a monetary value to their experiencing the emotions in question—so that the game could just as well be conceived as being played for these equivalent sums of money. But we do not want to enter into these knotty problems about value, which lie in the province of economics or philosophy rather than in that of mathematics.)

Suppose, now, that we consider a play of an n-person game with players $P_1, P_2, \cdots, P_n$ and let p_i (for $i = 1, \cdots, n$) be the payment made to P_i at the end of the play (if P_i has to pay, p_i is negative). Then if

$$\sum_{i=1}^{n} p_i = 0,$$

we call the play *zero-sum*. If every possible play of a game is zero-sum, we call the game itself zero-sum.[2] It is clear that all the ordinary parlor games which are played for money are zero-sum, since wealth is neither created nor destroyed in the course of playing them. But non-zero-sum games are nevertheless very important; for if we wish to find models for economic processes in the theory of games, then we shall be forced to consider non-zero-sum games, since economic processes usually create (or destroy) wealth. It can happen that an economic process increases (or decreases) the wealth of each of the participants.

We can also classify games according to *how many moves* they have. Thus ticktacktoe, when played to the bitter end, has nine moves, five of which are made by one player and four by the other. Some games do not have all of their plays of the same length—a play of chess may be short or long, depending on the relative skill of the two players.

A *finite* game has a finite number of moves, each of which involves only a finite number of alternatives; other games are called *infinite*.

Finally, games can be classified according to the *amount of information* available to the players regarding the past choices. In checkers and chess, for example, players are kept informed at all times as to what the previous choices have been; but in bridge a player does not know what cards have been dealt to the other people and is therefore in partial ignorance. It is clear that, starting out with a given game, it is possible to get an entirely different game by altering the rules regarding the information to be given to the

---

[2] We shall give the term "zero-sum game" a somewhat wider meaning in Chap. 6, after we have introduced the notion of a strategy.

players. Thus bridge would become a quite different game if everyone had to expose his cards at the beginning of the play. And one obtains from chess a completely new game (called *Kriegsspiel*) by denying to the players information about the choices of their opponents.

3. Definition of Rectangular Games. A zero-sum one-person game presents no intellectual difficulties; for regardless of what the one player does, he gets zero, and he may as well do one thing as another. In playing a non-zero-sum one-person game, the player has merely to solve an ordinary maximization problem: he must simply pick out from among the various courses of action open to him the one which will maximize his gain, or, in case the game involves also some chance moves, the one which will maximize his mathematical expectation of gain. Thus, in order to study the characteristic properties of games of strategy, it is necessary to go to games which involve more than one player.

We shall begin our studies with the case of two-person zero-sum games where each player has but one move. The first player chooses a number from the first m positive integers, and the second player, without being informed what choice the first player has made, chooses a number from the first n positive integers. The two numbers are then compared, and one of the players pays the other an amount, depending on the choices made, which is specified by the rules of the game. In order to have a name for such games, we shall call them, rather arbitrarily, *rectangular games.* (We are going to see later that rectangular games do not constitute such an extremely special kind of games as might appear at first glance; a very wide variety of other games can be put into the form of rectangular games.)

An example of a rectangular game is the following. Player P_1 chooses a number from the set $\{1, 2, 3\}$ and player P_2, without having been informed what choice P_1 has made, chooses a number from the set $\{1, 2, 3, 4\}$. After the two choices have been made, P_2 pays P_1 an amount given by the following table:

	1	2	3	4
1	2	1	10	11
2	0	−1	1	2
3	−3	−5	−1	1

That is to say, if, for example, P_1 chooses 1 and P_2 chooses 3, then P_2 pays P_1 ten dollars (or ten cents, or ten of whatever has been taken as the unit of money). If P_1 chooses 3 and P_2 chooses 2, then P_2 pays P_1 minus five dollars; i.e., P_1 pays P_2 five dollars. For the sake of brevity, we shall henceforth describe such a game as this by giving merely the *payoff matrix*:

$$\begin{Vmatrix} 2 & 1 & 10 & 11 \\ 0 & -1 & 1 & 2 \\ -3 & -5 & -1 & 1 \end{Vmatrix}.$$

An additional example of a rectangular game is "Two-finger Morra," which has been played in Italy since classical antiquity. This game is played by two people, each of whom shows one or two fingers and simultaneously calls his guess as to the number of fingers his opponent will show. If just one player guesses correctly, he wins an amount equal to the sum of the fingers shown by himself and his opponent; otherwise the game is a draw. If, by $\|1 \quad 2\|$, we indicate that a player shows one finger and guesses that his opponent will show two fingers, then the payoff matrix for this game is given by

	$\|1 \quad 1\|$	$\|1 \quad 2\|$	$\|2 \quad 1\|$	$\|2 \quad 2\|$
$\|1 \quad 1\|$	0	2	−3	0
$\|1 \quad 2\|$	−2	0	0	3
$\|2 \quad 1\|$	3	0	0	−4
$\|2 \quad 2\|$	0	−3	4	0

The most important question which can be asked regarding a rectangular game (indeed, regarding any game at all) is whether there is any optimal way of playing it. That is to say, whether one can give rational arguments in favor of playing one way rather than another.

In the case of the first game described above (but not of Morra), it so happens that this question is very easily answered. For we notice that each element of the first row is greater than the corresponding element of the second row and is also greater than the corresponding element of the third row. Hence, regardless of what number P_2 chooses, P_1 will do better by choosing 1 than by choosing 2 or 3, so the optimal way for P_1 to play is to choose 1. Similarly, each element of the second column is less than the corresponding element of each of the other columns; hence, since P_2 wants to play in such a way as to make the payoff as small as possible, the optimal way for P_2 to play is to choose 2.

This argument, however, has rested on a very special property of the payoff matrix: the fact that each element of a certain row (or column) is greater than the corresponding element of another row (or column). In order to give an analysis of rectangular games which will apply to a wider range of cases, we shall have to introduce some new notions.

4. Rectangular Games with Saddle-points. Let us consider now the rectangular game whose $m \times n$ matrix is

$$A = \begin{Vmatrix} a_{11} & a_{12} & \cdots & a_{1n} \\ a_{21} & a_{22} & \cdots & a_{2n} \\ \vdots & \vdots & & \vdots \\ a_{m1} & a_{m2} & \cdots & a_{mn} \end{Vmatrix}.$$

If player P_1 chooses the number 1 in a given play of this game, then he is certain to get paid at least the minimum of the elements of the first row, i.e., at least

$$\min_j a_{1j}.$$

And, in general, if he chooses the number i, then he is sure to get paid at least

$$\min_j a_{ij}.$$

Since he can choose i at will, however, he can in particular choose it so as to make

$$\min_j a_{ij}$$

as large as possible. Thus there is a choice for P_1 which will ensure that he gets at least

$$\max_i \min_j a_{ij}.$$

In an analogous way, remembering that the payments to P_2 are the negatives of the elements of A, we see that there is a choice for P_2 which will ensure that *he* gets at least

$$\max_j \min_i - a_{ij}.$$

We now recall the rather elementary fact about maxima and minima, that if f is any real-valued function, and if the indicated maxima and minima exist, then

$$\max_x - f(x) = - \min_x f(x),$$

and

$$\min_x - f(x) = - \max_x f(x).$$

And since, in the case under consideration, the ranges of variation of i and

j are finite, and hence all the maxima and minima exist, we conclude that

$$\max_j \min_i - a_{ij} = \max_j - [\max_i a_{ij}] = - \min_j \max_i a_{ij}.$$

Thus P_2 can play in such a way that he will be sure to get at least

$$- \min_j \max_i a_{ij}$$

and hence such that P_1 will get at most

$$\min_j \max_i a_{ij}.$$

In summary, then, P_1 can ensure that he will get at least

$$\max_i \min_j a_{ij},$$

and P_2 can keep him from getting more than

$$\min_j \max_i a_{ij}.$$

If it happens that

$$\max_i \min_j a_{ij} = \min_j \max_i a_{ij} = v, \tag{1}$$

then P_1 must realize, if he gives the matter sufficient thought, that he can get v and that he can be prevented by his opponent from getting more than v. Thus, unless he has some sound reason for believing that P_2 is going to do something wild (and this reason would have to be based on something cxtraneous to the game itself—such as, for instance, a knowledge that P_2 has a superstition which makes him play always in a certain way), P_1 might as well settle for v and play in such a way as to get it. And, similarly, P_2 might as well settle for $-v$ and play in such a way as to get it.

If (1) were true for every matrix A, then, in view of the above considerations, the search for an optimal way of playing rectangular games would be at an end. But, unfortunately, the situation is not quite so simple; it is easy, indeed, to give examples of matrices which make (1) false. Suppose, for instance, that we consider the matrix

$$\begin{Vmatrix} a_{11} & a_{12} \\ a_{21} & a_{22} \end{Vmatrix},$$

where $a_{11} = a_{22} = +1$ and $a_{12} = a_{21} = -1$; then

$$\max_i \min_j a_{ij} = \max [\min_j a_{1j}, \min_j a_{2j}] = \max [-1, -1] = -1,$$

and

$$\min_j \max_i a_{ij} = \min\,[\max_i a_{i1}, \max_i a_{i2}] = \min\,[+1, +1] = +1,$$

so that

$$\max_i \min_j a_{ij} \neq \min_j \max_i a_{ij}.$$

In view of the importance of (1) for our subject, it is natural that we should seek a simple necessary and sufficient condition that this equation hold. Since we shall need this result later, however, in a more general form, we shall here establish it for arbitrary real-valued functions, deducing the result for matrices only as a corollary. We show first that (as in the above example) the maximum of the minima is never greater than the minimum of the maxima.

THEOREM 1.1. Let A and B be sets, let f be a function of two variables such that $f(x, y)$ is a real number whenever $x \in \mathsf{A}$ and $y \in \mathsf{B}$, and suppose that

$$\max_{x \in \mathsf{A}} \min_{y \in \mathsf{B}} f(x, y)$$

and

$$\min_{y \in \mathsf{B}} \max_{x \in \mathsf{A}} f(x, y)$$

both exist. Then

$$\max_{x \in \mathsf{A}} \min_{y \in \mathsf{B}} f(x, y) \leq \min_{y \in \mathsf{B}} \max_{x \in \mathsf{A}} f(x, y).$$

PROOF. For any fixed x and y, we have, by the definition of a minimum,

$$\min_{y \in \mathsf{B}} f(x, y) \leq f(x, y),$$

and, by the definition of a maximum,

$$f(x, y) \leq \max_{x \in \mathsf{A}} f(x, y);$$

hence

$$\min_{y \in \mathsf{B}} f(x, y) \leq \max_{x \in \mathsf{A}} f(x, y). \tag{2}$$

Since the left-hand member of (2) is independent of y, we conclude that

$$\min_{y \in \mathsf{B}} f(x, y) \leq \min_{y \in \mathsf{B}} \max_{x \in \mathsf{A}} f(x, y). \tag{3}$$

Since the right-hand member of (3) is independent of x, we conclude that

$$\max_{x \in A} \min_{y \in B} f(x, y) \leq \min_{y \in B} \max_{x \in A} f(x, y),$$

as was to be shown.

REMARK 1.2. The application of the above result to matrices rests on the fact that a matrix,

$$\begin{Vmatrix} a_{11} & a_{12} & \cdots & a_{1n} \\ a_{21} & a_{22} & \cdots & a_{2n} \\ \vdots & \vdots & & \vdots \\ a_{m1} & a_{m2} & \cdots & a_{mn} \end{Vmatrix},$$

can be regarded as a real-valued function f of two variables, such that $f(i, j)$ is defined (for $i = 1, 2, \cdots, m$ and $j = 1, 2, \cdots, n$) by the equation

$$f(i, j) = a_{ij}.$$

COROLLARY 1.3. Let

$$\begin{Vmatrix} a_{11} & \cdots & a_{1n} \\ \vdots & & \vdots \\ a_{m1} & \cdots & a_{mn} \end{Vmatrix}$$

be an arbitrary $m \times n$ matrix. Then

$$\max_i \min_j a_{ij} \leq \min_j \max_i a_{ij}.$$

PROOF. This follows from Theorem 1.1, by taking **A** to be the set of the first m positive integers and **B** to be the set of the first n positive integers.

In order to formulate a necessary and sufficient condition that (1) hold, it is convenient to introduce a new notion regarding real-valued functions of two variables.

DEFINITION 1.4. Suppose that f is a real-valued function such that $f(x, y)$ is defined whenever $x \in A$ and $y \in B$; then a point $\| x_0 \quad y_0 \|$, where $x_0 \in A$ and $y_0 \in B$, is called a *saddle-point* of f if the following conditions are satisfied:

(i) $\quad f(x, y_0) \leq f(x_0, y_0) \quad$ for all x in **A**,

(ii) $\quad f(x_0, y_0) \leq f(x_0, y) \quad$ for all y in **B**.

Thus the function $y^2 - x^2$ has the point $\| 0 \quad 0 \|$ as a saddle-point, since, for all real x and y,

$$0^2 - x^2 \leq 0^2 - 0^2 \leq y^2 - 0^2.$$

(This example is, of course, in no way surprising, since the hyperboloid of one sheet,

$$z = y^2 - x^2,$$

is ordinarily called a saddle-shaped surface. It should be remarked, however, that our definition of a saddle-point by no means coincides with the notion as used in differential geometry; e.g., according to our definition, the function $x^2 - y^2$ has no saddle-point.)

THEOREM 1.5. Let f be a real-valued function such that $f(x, y)$ is defined whenever $x \in \mathbf{A}$ and $y \in \mathbf{B}$, and suppose, moreover, that

$$\max_{x \in \mathbf{A}} \min_{y \in \mathbf{B}} f(x, y)$$

and

$$\min_{y \in \mathbf{B}} \max_{x \in \mathbf{A}} f(x, y)$$

both exist. Then a necessary and sufficient condition that

$$\max_{x \in \mathbf{A}} \min_{y \in \mathbf{B}} f(x, y) = \min_{y \in \mathbf{B}} \max_{x \in \mathbf{A}} f(x, y)$$

is that f possess a saddle-point. If $\| x_0 \quad y_0 \|$ is any saddle-point of f, then

$$f(x_0, y_0) = \max_{x \in \mathbf{A}} \min_{y \in \mathbf{B}} f(x, y) = \min_{y \in \mathbf{B}} \max_{x \in \mathbf{A}} f(x, y).$$

PROOF. To see that the condition is sufficient, suppose that $\| x_0 \quad y_0 \|$ is a saddle-point of f. Then we have, for all x in $\mathbf{A}$ and y in $\mathbf{B}$,

$$f(x, y_0) \leq f(x_0, y_0), \tag{4}$$

$$f(x_0, y_0) \leq f(x_0, y). \tag{5}$$

From (4) we conclude that

$$\max_{x \in \mathbf{A}} f(x, y_0) \leq f(x_0, y_0),$$

and from (5), that

$$f(x_0, y_0) \leq \min_{y \in \mathbf{B}} f(x_0, y),$$

so that

$$\max_{x \in A} f(x, y_0) \leq f(x_0, y_0) \leq \min_{y \in B} f(x_0, y) . \tag{6}$$

Since

$$\min_{y \in B} \max_{x \in A} f(x, y) \leq \max_{x \in A} f(x, y_0)$$

and

$$\min_{y \in B} f(x_0, y) \leq \max_{x \in A} \min_{y \in B} f(x, y) ,$$

we conclude from (6) that

$$\min_{y \in B} \max_{x \in A} f(x, y) \leq f(x_0, y_0) \leq \max_{x \in A} \min_{y \in B} f(x, y) . \tag{7}$$

But, by Theorem 1.1, the first term of (7) is not less than the third; hence we conclude that all three members are equal, as was to be shown.

To see that the condition is also necessary, let x_0 be a member of A which makes

$$\min_{y \in B} f(x, y)$$

a maximum, and let y_0 be a member of B which makes

$$\max_{x \in A} f(x, y)$$

a minimum; i.e., let x_0 and y_0 be members of A and B, respectively, which satisfy the conditions

$$\left.\begin{aligned} \min_{y \in B} f(x_0, y) &= \max_{x \in A} \min_{y \in B} f(x, y) , \\ \max_{x \in A} f(x, y_0) &= \min_{y \in B} \max_{x \in A} f(x, y) . \end{aligned}\right\} \tag{8}$$

We shall show that $\| x_0 \quad y_0 \|$ is a saddle-point of f.

Since we are supposing that

$$\max_{x \in A} \min_{y \in B} f(x, y) = \min_{y \in B} \max_{x \in A} f(x, y) ,$$

we see from (8) that

$$\min_{y \in B} f(x_0, y) = \max_{x \in A} f(x, y_0) . \tag{9}$$

From the definition of a minimum, we have

$$\min_{y \in B} f(x_0, y) \leq f(x_0, y_0) ,$$

and hence from (9) we have

$$\max_{x \in A} f(x, y_0) \leq f(x_0, y_0).$$

From the last inequality, together with the definition of a maximum, we conclude that, for all x in A,

$$f(x, y_0) \leq f(x_0, y_0),$$

which is condition (i) of Definition 1.4. In a similar fashion, we show that condition (ii) of Definition 1.4 is satisfied, which completes the proof.

REMARK 1.6. In view of the interpretation of a matrix as being a real-valued function (indicated in Remark 1.2), we see that a saddle-point of a matrix is a pair of integers $\| i \quad j \|$ such that a_{ij} is at the same time the minimum of its row and the maximum of its column. Thus the matrix

$$\begin{Vmatrix} 21 & 11 & 31 \\ 32 & 0 & 4 \end{Vmatrix}$$

has a saddle-point at $\| 1 \quad 2 \|$, since 11 is the smallest element in the first row and the largest element in the second column. The matrix

$$\begin{Vmatrix} 12 & 13 & 12 \\ 10 & 31 & 9 \end{Vmatrix}$$

has two saddle-points: one at $\| 1 \quad 1 \|$ and one at $\| 1 \quad 3 \|$. But the matrix

$$\begin{Vmatrix} 12 & 13 & 12 \\ 10 & 31 & 13 \end{Vmatrix}$$

has only one saddle-point, since 12 is not the maximum of the third column.

Using this notion of the saddle-point of a matrix, we now derive from Theorem 1.5 the following corollary.

COROLLARY 1.7. If

$$A = \begin{Vmatrix} a_{11} & \cdots & a_{1n} \\ \vdots & & \vdots \\ a_{m1} & \cdots & a_{mn} \end{Vmatrix}$$

is a matrix, then a necessary and sufficient condition that

$$\max_i \min_j a_{ij} = \min_j \max_i a_{ij}$$

is that A possess a saddle-point, i.e., that there be a pair of integers $\| i_0 \quad j_0 \|$ such that $a_{i_0 j_0}$ is at the same time the minimum of its row and the maximum of its column. If $\| i_0 \quad j_0 \|$ is any saddle-point of A, then

$$a_{i_0 j_0} = \max_i \min_j a_{ij} = \min_j \max_i a_{ij}.$$

From Corollary 1.7 we see that (1) holds if, and only if, the matrix A has a saddle-point. Therefore, if the matrix of a rectangular game has a saddle-point $\| x_0 \quad y_0 \|$ (in which case, simply for brevity, we shall sometimes say that the game itself has a saddle-point), then it is ordinarily best for P_1 to choose x_0 and for P_2 to choose y_0. For this reason we call x_0 and y_0 *optimal choices* for P_1 and P_2, respectively. We call $a_{x_0 y_0}$ the *value* of the game (to P_1).

Thus, for example, the matrix

$$\begin{Vmatrix} -5 & 3 & 1 & 20 \\ 5 & 5 & 4 & 6 \\ -4 & -2 & 0 & -5 \end{Vmatrix}$$

has a saddle-point at $\| 2 \quad 3 \|$, since 4 is the minimum of the second row and the maximum of the third column. Hence, in playing the rectangular game of which this is the matrix, the optimal choice for P_1 is 2 and the optimal choice for P_2 is 3. The value of the game is 4. By choosing 2, P_1 can make sure that he will receive at least 4 and, by choosing 3, P_2 can keep P_1 from getting more than 4.

It should be noticed that when we say the optimal choice for P_1 is 2, we do not mean that it would be wisest for him to choose 2 under all conceivable circumstances. For example, suppose that P_1 has information which makes him absolutely sure that P_2 will choose 4 (for instance, suppose that P_1 knows P_2 always follows the advice of a certain sorcerer, and that P_1 has bribed the latter to tell P_2 to choose 4); then, of course, it would be best for P_1 to choose 1 instead of 2, since this would give him 20 instead of merely 6. But it is only in unusual cases that a player can have such knowledge of his opponent's intentions; so, in general, it is wise to play in the (technically) optimal way.

Thus, in case the matrix of a rectangular game has a saddle-point, we are provided with an adequate theory of how best to play it. We are still left, however, with the problem of how to play a game with a matrix such as

$$\begin{Vmatrix} 1 & -7 \\ -1 & 2 \end{Vmatrix},$$

which has no saddle-point. We shall consider this problem in Chap. 2.

HISTORICAL AND BIBLIOGRAPHICAL REMARKS

The earliest publication dealing with the theory of games and the application of that theory to economics is by von Neumann [1];[3] see also Kalmar [1]. The theory was brought to a very high state of development in von Neumann and Morgenstern [1], from which much of the material of this book has been taken. A general survey of the subject, as of 1949, can be found in Paxson [1], and a more popular account is contained in McDonald [1], [2], and [3].

EXERCISES

1. Do you think that Robinson Crusoe economics would be a better approximation to the British economic system of 1900 or of 1952? Why?

2. If A is a set of real numbers, then by a *lower bound* of A we mean a number y such that, for every x in A, $y \leq x$; by a *greatest lower bound* of A, we mean a lower bound which is not less than any other lower bound. Give an example of a set which has no lower bound. Show that a set cannot have more than one greatest lower bound.

3. Define upper bounds and least upper bounds, analogous to lower bounds and greatest lower bounds, and prove analogues of the statements made in Exercise 2.

4. If f is a real-valued function, defined over a set A, then by

$$\inf_{x \in \mathsf{A}} f(x),$$

we mean the greatest lower bound of the set B of all values of $f(x)$, i.e., of the set B of all numbers y such that, for some x in A,

$$y = f(x).$$

Similarly, by

$$\sup_{x \in \mathsf{A}} f(x),$$

we mean the least upper bound of the set B of all values of $f(x)$. Show that if the indicated bounds exist, then

$$\sup_{x \in \mathsf{A}} - f(x) = - \inf_{x \in \mathsf{A}} f(x),$$

and

[3] The numbers in square brackets refer to items in the Bibliography at the end of the book.

$$\inf_{x \in A} - f(x) = - \sup_{x \in A} f(x).$$

5. Show that if the indicated bounds exist, then

$$\sup_x [f(x) + g(x)] \leq \sup_x f(x) + \sup_x g(x),$$

and

$$\inf_x f(x) + \inf_x g(x) \leq \inf_x [f(x) + g(x)].$$

6. Show that the "$\leq$" signs in Exercise 5 cannot be replaced by the "$=$" signs.

7. By the *maximum* of a set A of real numbers, we mean a member of A which is at the same time an upper bound of A. Show that the maximum of a set, if it exists at all, is unique. Show that the maximum of a set, if it exists, is the least upper bound of the set. Show that a set can have a least upper bound without having a maximum.

8. Define the minimum of a set, analogous to the definition of a maximum in Exercise 7, and prove analogues of the statements made in Exercise 7.

9. If f is a real-valued function, defined over a set A, then by

$$\max_{x \in A} f(x)$$

we mean the maximum of the set B of all values of $f(x)$ for $x \in A$. Similarly,

$$\min_{x \in A} f(x)$$

is the minimum of the set of all values of $f(x)$ for $x \in A$. Show that if the indicated maxima and minima exist, then

$$\max_{x \in A} - f(x) = - \min_{x \in A} f(x),$$

and

$$\min_{x \in A} - f(x) = - \max_{x \in A} f(x).$$

10. Show that if f and g are real-valued functions, defined whenever $x \in A$, and if the indicated maxima and minima exist, then

$$\max_{x \in A} [f(x) + g(x)] \leq \max_{x \in A} f(x) + \max_{x \in A} g(x),$$

and

$$\min_{x \in A} f(x) + \min_{x \in A} g(x) \leq \min_{x \in A} [f(x) + g(x)].$$

11. Show that if f is a real-valued function of two variables, which is defined whenever $x \in \mathsf{A}$ and $y \in \mathsf{B}$, and if the indicated maxima and minima exist, then

$$\max_{x \in \mathsf{A}} \max_{y \in \mathsf{B}} f(x, y) = \max_{y \in \mathsf{B}} \max_{x \in \mathsf{A}} f(x, y),$$

and

$$\min_{x \in \mathsf{A}} \min_{y \in \mathsf{B}} f(x, y) = \min_{y \in \mathsf{B}} \min_{x \in \mathsf{A}} f(x, y).$$

12. Prove the following: If f is a real-valued function of two real variables, which is defined whenever $x \in \mathsf{A}$ and $y \in \mathsf{B}$, and if

$$\sup_{x \in \mathsf{A}} \inf_{y \in \mathsf{B}} f(x, y)$$

and

$$\inf_{y \in \mathsf{B}} \sup_{x \in \mathsf{A}} f(x, y)$$

both exist, then

$$\sup_{x \in \mathsf{A}} \inf_{y \in \mathsf{B}} f(x, y) \leq \inf_{y \in \mathsf{B}} \sup_{x \in \mathsf{A}} f(x, y).$$

13. Find the saddle-points of the following matrices:

(a) $$\begin{Vmatrix} 1 & 3 \\ -2 & 10 \end{Vmatrix},$$

(b) $$\begin{Vmatrix} 3 & 5 & 2 & 4 \\ 2 & 6 & 1 & 1 \end{Vmatrix},$$

(c) $$\begin{Vmatrix} 2 & 2 & 2 & 2 \\ 1 & 2 & 3 & 4 \end{Vmatrix},$$

(d) $$\begin{Vmatrix} 2 & 2 & 2 & 2 \\ 2 & 2 & 3 & 4 \end{Vmatrix}.$$

14. Find the min max and the max min for the following matrix:

$$\begin{Vmatrix} 1 & 3 & 6 \\ 2 & 1 & 3 \\ 6 & 2 & 1 \end{Vmatrix}.$$

15. Show that if $\| x_1 \quad y_1 \|$ and $\| x_2 \quad y_2 \|$ are saddle-points of a real-valued function, then so are $\| x_1 \quad y_2 \|$ and $\| x_2 \quad y_1 \|$. What does this mean as applied to matrices?

16. Find the max min and the min max of two-finger Morra.

17. In three-finger Morra, each player shows one, two, or three fingers and simultaneously guesses the number of fingers shown by his opponent. The rest of the rules are the same as in two-finger Morra. Compute the payoff matrix and show that the game does not have a saddle-point.

18. Two players own n dollars each and, between them, an object of value $c > 0$. Each player makes a sealed bid, offering i dollars (where i is one of the integers from 0 to n) for sole possession of the object. The higher bidder gets the object and pays the other player whatever amount he had offered. If both bid the same amount, the object is assigned, without a compensating side-payment, to one of the players by tossing a coin, so that each in this case has an expected share of $\frac{1}{2}c$ in the object. Write out the payoff matrix and determine whether it has a saddle-point.

CHAPTER 2

THE FUNDAMENTAL THEOREM FOR RECTANGULAR GAMES

1. Mixed Strategies. Consider now the rectangular game whose matrix is

$$\begin{Vmatrix} 1 & -1 \\ -1 & 1 \end{Vmatrix}.$$

Since the matrix has no saddle-point, our previous methods do not suffice to enable us to determine optimal ways for P_1 and P_2 to play. Moreover, it appears to make little difference whether P_1 chooses 1 or 2, for in either case he will receive 1 or -1, according as P_2 makes the same or the opposite choice. On the other hand, if P_2 knows what choice P_1 will make, then P_2 can ensure that P_1 will have to pay him 1 (merely by making the opposite choice); thus it seems to be of the greatest importance to P_1 that he make it difficult for P_2 to guess what choice he is going to make. One way for P_1 to guarantee this is to decide what he will do by means of some chance device.

Suppose, for example, that P_1 decides to make his choice by tossing a (true) coin—choosing 1 if the coin shows heads and 2 if it shows tails. In this case, since the probability that P_1 will choose 1 is ½, and the probability that he will choose 2 is the same, we see that the mathematical expectation of P_1, in the event that P_2 chooses 1, is

$$1 \cdot \left(\frac{1}{2}\right) + (-1) \cdot \left(\frac{1}{2}\right) = 0,$$

and his expectation is the same if P_2 chooses 2. Therefore, if P_1 chooses in this way, then his expectation will be 0, regardless of what P_2 does.

As a matter of fact, this is the *only* way P_1 can play the game in question without running the risk of losing if P_2 discovers what he is going to do. For suppose that P_1 uses a chance device which assigns the probability x to 1 and the probability $1 - x$ to 2, and suppose that P_2 discovers what chance device P_1 is using. Then the expectation of P_1, if P_2 chooses 1, is

$$(1)(x) + (-1)(1 - x) = 2x - 1;$$

while, in the event that P_2 chooses 2, the expectation of P_1 is

$$(-1)x + (1)(1 - x) = 1 - 2x.$$

If $x > 1/2$, then $1 - 2x < 0$, so that P_1 will have an expectation less than 0 in case P_2 chooses 2; and if $x < 1/2$, then $2x - 1 < 0$, so that P_1 will have an expectation less than 0 in case P_2 chooses 1.

It therefore appears that the optimal way for P_1 to play this game is to choose 1 and 2 each with the probability $1/2$, and that the optimal way for P_2 to play is, by a similar argument, the same. The value of the game to P_1 (i.e., his expectation if he plays in the optimal way) is 0.

In the above discussion we have always talked about using chance devices which assign probabilities to the various choices. It is sometimes intuitively simpler, however, to speak as though the game were played many times in succession, and to speak of the relative frequencies with which the various choices are made. In the following pages we shall often employ this less exact manner of speaking.

Let us now consider a slightly more difficult example: the rectangular game whose matrix is

$$\begin{Vmatrix} 1 & 3 \\ 4 & 2 \end{Vmatrix}.$$

Since the matrix has no saddle-point, it would again seem desirable for P_1 and P_2 to play only with certain frequencies. Suppose that P_1 plays 1 with frequency x and plays 2 with frequency $1 - x$ (so that neither x nor $1 - x$ is negative), and suppose that P_2 plays 1 with frequency y and plays 2 with frequency $1 - y$. Then the mathematical expectation of P_1 is

$$E(x, y) = 1xy + 3x(1 - y) + 4(1 - x)y + 2(1 - x)(1 - y).$$

We are presented with the problem of giving a precise mathematical meaning to the intuitive notion of an optimal choice (for P_1) of x and of an optimal choice (for P_2) of y.

By elementary algebra, however, we have

$$\begin{aligned} E(x, y) &= -4xy + x + 2y + 2 \\ &= -4\left(x - \frac{1}{2}\right)\left(y - \frac{1}{4}\right) + \frac{5}{2}. \end{aligned} \tag{1}$$

When $E(x, y)$ is written in the form (1), we see that if P_1 takes $x = 1/2$, he can ensure that his expectation will be at least $5/2$. Moreover, he cannot be sure of more than $5/2$; for, by taking $y = 1/4$, P_2 can ensure that the expectation of P_1 will be exactly $5/2$ and not greater than $5/2$. Thus P_1 might as well settle for 5/2 and play $x = 1/2$ so as to obtain this amount. And, similarly, P_2 might as well reconcile himself to getting $-5/2$ and play $y = 1/4$ so as to get it. Therefore, in this particular game, it appears reasonable to say that an optimal way for P_1 to play is to choose 1 and 2 equally often and that an optimal way for P_2 to play is to choose 1 with probability 1/4 and to choose 2 with probability 3/4. It also seems reasonable to call 5/2 the value of the game.

Expressing this in a different way, we see from (1) that, for all x and y between 0 and 1,

$$E\left(x, \frac{1}{4}\right) \leq E\left(\frac{1}{2}, \frac{1}{4}\right) \leq E\left(\frac{1}{2}, y\right); \tag{2}$$

thus the point $\| 1/2 \quad 1/4 \|$ is a saddle-point of the function E. It seems reasonable to take (2), in general, as the definition of optimal frequencies for any 2×2 rectangular game. Thus, if $E(x, y)$ is the expectation of the first player in such a game (when P_1 plays 1 and 2 with relative frequencies x and $1 - x$, and when P_2 plays 1 and 2 with relative frequencies y and $1 - y$), then we say that x^* is an optimal frequency for P_1 and that y^* is an optimal frequency for P_2 if, for all x and y between 0 and 1,

$$E(x, y^*) \leq E(x^*, y^*) \leq E(x^*, y).$$

We shall now extend this definition so as to cover arbitrary rectangular games (i.e., rectangular games whose matrices have arbitrary numbers of rows and columns).

Let us consider the rectangular game whose matrix is

$$A = \left\| \begin{matrix} a_{11} & a_{12} & \cdots & a_{1n} \\ a_{21} & a_{22} & \cdots & a_{2n} \\ \vdots & \vdots & & \vdots \\ a_{m1} & a_{m2} & \cdots & a_{mn} \end{matrix} \right\|.$$

By a *mixed strategy* for P_1 we shall mean an ordered m-tuple $\| x_1 \quad \cdots \quad x_m \|$ of non-negative real numbers satisfying the condition

$$\sum_{i=1}^{m} x_i = 1;$$

these numbers are, of course, to be thought of as the frequencies with which P_1 chooses the numbers $1, 2, \cdots, m$. We shall henceforth use the symbol $\mathbf{S}_m$ to stand for this set of m-tuples. Similarly, by a mixed strategy for P_2 we shall mean a member of $\mathbf{S}_n$, i.e., an ordered n-tuple $\| y_1 \quad \cdots \quad y_n \|$

of non-negative real numbers satisfying the condition

$$\sum_{j=1}^{n} y_j = 1.$$

We sometimes call the numbers $1, \cdots, m$ themselves *pure strategies for* P_1, and the numbers $1, \cdots, n$ *pure strategies for* P_2. It is evident that for P_1 to play the pure strategy k is equivalent to his playing the mixed strategy $\| x_1 \quad \cdots \quad x_m \|$, where $x_k = 1$, and $x_i = 0$ for $i \neq k$.

If P_1 uses the mixed strategy $X = \| x_1 \quad \cdots \quad x_m \|$, and if P_2 uses the mixed strategy $Y = \| y_1 \quad \cdots \quad y_n \|$, then the mathematical expectation of P_1 is given by the formula

$$E(X, Y) = \sum_{j=1}^{n} \sum_{i=1}^{m} a_{ij} x_i y_j.$$

If it happens that, for some X^* in $\mathbf{S}_m$ and some Y^* in $\mathbf{S}_n$, we have

$$E(X, Y^*) \leq E(X^*, Y^*) \leq E(X^*, Y), \tag{3}$$

for all X in $\mathbf{S}_m$ and all Y in $\mathbf{S}_n$, then we call X^* and Y^* *optimal (mixed) strategies* for P_1 and P_2, respectively, and we call $E(X^*, Y^*)$ the *value* of the game (to P_1). If X^* and Y^* are optimal strategies for P_1 and P_2, respectively, then we sometimes call the ordered pair $\| X^* \quad Y^* \|$ a *solution* of the game, or, sometimes, a *strategic saddle-point.*

The intuitive adequacy of the above definitions lies in the following fact: If X^* and Y^* are mixed strategies which satisfy condition (3), then, by making use of X^*, P_1 can make sure that he will receive at least $E(X^*, Y^*)$, regardless of what P_2 does; and, similarly, by making use of Y^*, P_2 can keep P_1 from getting more than $E(X^*, Y^*)$. Thus $E(X^*, Y^*)$ is the amount which P_1 can reasonably expect to get—he can get it by playing X^*—and P_2 can hold him down to it by playing Y^*.

If it happens that the two quantities

$$v_1 = \max_{X \in \mathbf{S}_m} \min_{Y \in \mathbf{S}_n} E(X, Y)$$

and

$$v_2 = \min_{Y \in \mathbf{S}_n} \max_{X \in \mathbf{S}_m} E(X, Y)$$

both exist and are equal, then we see by Theorem 1.5 that there exist mixed strategies which satisfy condition (3)—so that the game has a value and there are optimal strategies for the two players. Thus the question of when v_1 and v_2 exist and are equal is very important for our subject; we shall show in Sec. 3 that they always exist and are equal—and thus that the definitions of this section suffice to specify the value of an arbitrary rectangular game and optimal ways of playing it. Before turning to the proof of this point, however, it is convenient to introduce some geometrical notions and theorems which will be used throughout the remainder of the book.

2. Geometrical Background. By *Euclidean n-space* (in symbols, $\mathbf{E}_n$) we shall mean the set of all n-tuples $\| x_1 \quad \cdots \quad x_n \|$, where $x_1, \cdots, x_n$ are real numbers. If $X^{(1)} = \| x_1^{(1)} \quad \cdots \quad x_n^{(1)} \|$ and $X^{(2)} = \| x_1^{(2)} \quad \cdots \quad x_n^{(2)} \|$ are two points of $\mathbf{E}_n$, then we define the *distance*, $d(X^{(1)}, X^{(2)})$, between $X^{(1)}$ and $X^{(2)}$ by the formula

$$d(X^{(1)}, X^{(2)}) = \sqrt{(x_1^{(1)} - x_1^{(2)})^2 + \cdots + (x_n^{(1)} - x_n^{(2)})^2}.$$

A subset $\mathbf{X}$ of $\mathbf{E}_n$ is called *bounded* if there exists a number M such that, for all points $X^{(1)}$ and $X^{(2)}$ in $\mathbf{X}$,

$$d(X^{(1)}, X^{(2)}) \leq M.$$

It is easily shown that a necessary and sufficient condition that a set be bounded is that it lie in some hypersphere; i.e., that there exist a point a of $\mathbf{E}_n$ and a number R such that, for every x in $\mathbf{X}$,

$$d(x, a) \leq R.$$

We call a point x of $\mathbf{E}_n$ a *limit-point* of a subset $\mathbf{A}$ of $\mathbf{E}_n$ if, for every positive ε, there exists a point y of $\mathbf{A}$, which is different from x, such that $d(x, y) < \varepsilon$. Thus if $\mathbf{A}$ is the set of points $\| x \quad y \|$ of $\mathbf{E}_2$ such that

$$x^2 + y^2 < 1,$$

then any point $\| x_0 \quad y_0 \|$ such that

$$x_0^2 + y_0^2 = 1$$

is a limit-point of $\mathbf{A}$. It should be noticed that a finite set of points has no limit-points. On the other hand, every bounded infinite set has at least one limit-point.

The *closure* of a set is the set obtained by adding to it all of its limit-points. Thus the closure of the set $\mathbf{A}$ of points mentioned above is the set of points $\| x \quad y \|$ of $\mathbf{E}_2$ such that

$$x^2 + y^2 \leq 1.$$

A finite set coincides with its own closure.

A set is called *closed* if it is equal to its own closure. Thus every finite set is closed, as is also, for example, the set consisting of all points $\| x \quad y \|$ of $\mathbf{E}_2$ such that

$$x^2 + y^2 \geq 1.$$

A set is called *open* if its complement is closed. Thus the set $\mathbf{A}$, mentioned above, is open. Some sets are neither open nor closed. The set of points x of $\mathbf{E}_1$ such that

$$0 < x \leq 1$$

is neither open nor closed.

The *interior* of a set is the complement of the closure of its complement. Thus the interior of the set of points x of $\mathbf{E}_1$ such that

$$0 < x \leq 1$$

is the set of points x of $\mathbf{E}_1$ such that

$$0 < x < 1.$$

The interior of any finite set is the empty set.

It is easily seen that the closure of any set is closed and that the interior of any set is open. The complement of any closed set is open, and the complement of any open set is closed. The set $\mathbf{E}_n$, for any n, is both open and closed, and the same is true of the empty set; these are the only sets which are both open and closed.

By the *boundary* of a set, we mean the intersection of its closure with the closure of its complement. Thus the boundary of the set $\mathbf{A}$, mentioned above, is the set of all points $\| x \quad y \|$ of $\mathbf{E}_2$ such that

$$x^2 + y^2 = 1.$$

If $\mathbf{B}$ is the set of all points $\| x \quad y \|$ of $\mathbf{E}_2$ such that both x and y are rational numbers, then the boundary of $\mathbf{B}$ is all of $\mathbf{E}_2$.

A set is called *connected* if it cannot be expressed as the union of two mutually exclusive sets, neither of which contains a limit-point of the other. Thus the set $\mathbf{A}$, mentioned above, is connected; the same is true, for instance, of the set of all points $\| x \quad y \quad z \|$ of $\mathbf{E}_3$ such that

$$3x + 2y + 5z = 7.$$

If C is the set of all points $\| x \quad y \|$ of E_2 such that

$$x \neq 0,$$

then C is not connected; for, by letting C_1 be the set of all points $\| x \quad y \|$ such that

$$x > 0,$$

and if C_2 is the set of all points $\| x \quad y \|$ such that

$$x < 0,$$

then C is the union of C_1 and C_2, and neither of the sets C_1 and C_2 contains a limit-point of the other.

Besides the above general notions, we shall also make use of some slightly more specialized concepts in regard to Euclidean space. These concepts relate to the theory of convex sets, which is itself a highly developed special branch of mathematics; but we shall here present only the rudiments of this subject, and we shall confine our attention to results that will be useful in connection with the theory of games.

Let

$$\begin{aligned} x^{(1)} &= \| x_1^{(1)} \quad \cdots \quad x_n^{(1)} \|, \\ x^{(2)} &= \| x_1^{(2)} \quad \cdots \quad x_n^{(2)} \|, \\ &\vdots \\ x^{(r)} &= \| x_1^{(r)} \quad \cdots \quad x_n^{(r)} \|, \end{aligned}$$

and

$$x = \| x_1 \quad \cdots \quad x_n \|$$

be points of E_n, let $\| a_1 \quad \cdots \quad a_r \|$ be a member of S_r (so that $a_i \geq 0$, for $i = 1, \cdots, r$, and $a_1 + \cdots + a_r = 1$) and suppose that

$$x_j = a_1 x_j^{(1)} + a_2 x_j^{(2)} + \cdots + a_r x_j^{(r)} \qquad \text{for } j = 1, \cdots, n;$$

then we say that x is a *convex linear combination* of $x^{(1)}, \cdots, x^{(r)}$ *with weights* $a_1, \cdots, a_r$, and we write

$$x = a_1 x^{(1)} + \cdots + a_r x^{(r)}.$$

Thus the point $\| 0 \quad 15 \|$ of E_2 is a convex linear combination (with weights $\frac{1}{6}$, $\frac{1}{3}$, and $\frac{1}{2}$) of the points $\| 6 \quad 12 \|$, $\| -9 \quad 15 \|$, and $\| 4 \quad 16 \|$. Similarly, the point $\| -1 \quad 2 \quad 11 \|$ of E_3 is a convex linear combination (with weights $\frac{1}{3}$ and $\frac{2}{3}$) of the points $\| 3 \quad 6 \quad 9 \|$ and $\| -3 \quad 0 \quad 12 \|$.

We call a subset X of E_n *convex* if every convex linear combination of points of X is a point of X. Thus the set A consisting of all points $\| x \quad y \|$

such that

$$x^2 + y^2 < 1$$

and the set **B** consisting of all points $\| x \quad y \|$ such that

$$x^2 + y^2 \leq 1$$

are both convex. But the set **C** consisting of all points $\| x \quad y \|$ such that

$$x^2 + y^2 = 1$$

and the set **D** consisting of all points $\| x \quad y \|$ such that

$$x^2 + y^2 \geq 1$$

are neither of them convex; for the point $\| 0 \quad 0 \|$ is a convex linear combination (with weights ½ and ½) of the points $\| 1 \quad 0 \|$ and $\| -1 \quad 0 \|$, both of which belong both to **C** and to **D**; but $\| 0 \quad 0 \|$ itself belongs neither to **C** nor to **D**.

We have defined a convex set as a set **X** such that whenever $y^{(1)}, \cdots, y^{(p)}$ are in **X** and $\| a_1 \quad \cdots \quad a_p \|$ is in $\mathbf{S}_p$, then

$$y = a_1 y^{(1)} + \cdots + a_p y^{(p)}$$

is in **X**. It is easily shown, however, that a necessary and sufficient condition that **X** be convex is that the above condition be satisfied with $p = 2$; thus **X** is convex if, whenever $y^{(1)}$ and $y^{(2)}$ are in **X** and $\| a_1 \quad a_2 \|$ is in $\mathbf{S}_2$,

$$y = a_1 y^{(1)} + a_2 y^{(2)}$$

is in **X**.

By the *convex hull* of a set **X**, we mean the intersection of all convex sets of which **X** is a subset. Thus the convex hull of the set **D** of all points $\| x \quad y \|$ such that

$$x^2 + y^2 = 1$$

is the set of all points $\| x \quad y \|$ such that

$$x^2 + y^2 \leq 1 .$$

The convex hull of the set of all points $\| x \quad y \|$ such that

$$x^2 + y^2 > 1$$

is the whole plane, $\mathbf{E}_2$.

Since $\mathbf{E}_n$ is convex, it is apparent that every set is contained in at least one convex set and hence that every set has a convex hull. It can be shown without difficulty that the convex hull of a set $\mathbf{X}$ consists of just those points which are convex linear combinations of points of $\mathbf{X}$. The following theorem, which is due to Fenchel, is somewhat more difficult to prove, but we shall omit the proof (indeed, we shall omit all proofs in this section, since the theory of convex sets is only a mathematical preliminary to the theory of games proper). The theorem will not be used in the present chapter (it will be used later, however, in order to prove Theorem 11.5).

> THEOREM 2.1. If $\mathbf{X}$ is any subset of $\mathbf{E}_n$, then any point of the convex hull of $\mathbf{X}$ can be represented as a convex linear combination of $n + 1$ points of $\mathbf{X}$. If $\mathbf{X}$ is connected, moreover, then any point of the convex hull of $\mathbf{X}$ can be represented as a convex linear combination of n points of $\mathbf{X}$.

REMARK 2.2. To illustrate the theorem, suppose first that $\mathbf{X}$ consists of the four points $\|0 \quad 0\|$, $\|0 \quad 1\|$, $\|1 \quad 1\|$, and $\|1 \quad 0\|$ of $\mathbf{E}_2$, i.e., that $\mathbf{X}$ consists of the four vertices of a certain square, as indicated in Fig. 1. Then the convex hull of $\mathbf{X}$ is clearly the entire square, including the boundary. Every point of the triangle abd can be represented as a convex linear combination of the three points a, b, and d; and similarly, every point of the triangle bcd can be represented as a convex linear combination of the three points b, c, and d. We notice that the point $\|\frac{1}{2} \quad \frac{1}{4}\|$, which is not on the boundary of the square and is not on either of the diagonals ac and bd, cannot be represented as a convex linear combination of just two points of $\mathbf{X}$.

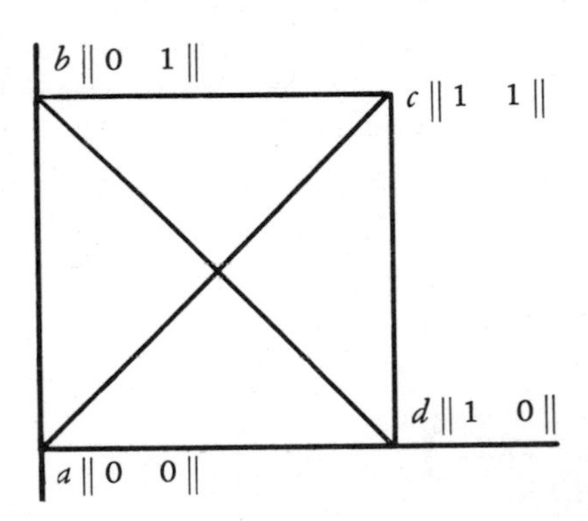

Fig. 1

On the other hand, if $\mathbf{X}$ is the *connected* set which consists of all points on the boundary of the square, then the convex hull is the same as before; but now, every point of the convex hull can be represented as a convex linear combination of *two* points of $\mathbf{X}$.

Let $a_1, \cdots, a_n$ be n real numbers, not all of which are 0, and let b be any real number; then the set of points $\|x_1 \quad \cdots \quad x_n\|$ of $\mathbf{E}_n$ such that

$$a_1x_1 + \cdots + a_nx_n = b$$

is called a *hyperplane* (of E_n). Thus the set of points $\| x_1 \quad x_2 \quad x_3 \quad x_4 \|$ such that

$$2x_1 + 3x_2 - 7x_3 + x_4 = 7$$

is a hyperplane of E_4. A hyperplane of E_3 is an ordinary plane, a hyperplane of E_2 is a line, and a hyperplane of E_1 is a point.

If

$$a_1x_1 + \cdots + a_nx_n = b$$

is the equation of a hyperplane, then, by the two *half-spaces corresponding* to the hyperplane, we mean the set of all points $\| x_1 \quad \cdots \quad x_n \|$ such that

$$a_1x_1 + \cdots + a_nx_n > b$$

and the set of all points $\| x_1 \quad \cdots \quad x_n \|$ such that

$$a_1x_1 + \cdots + a_nx_n < b.$$

Clearly, the plane and the two half-spaces are mutually exclusive, and their union is E_n. In terms of these notions, we now state a theorem which will be used later in this chapter.

> THEOREM 2.3. Let X be any closed convex subset of E_n, and let x be a point of E_n which does not belong to X; then there exists a hyperplane, P, which contains x, such that X is a subset of one of the half-spaces determined by P.

In closing this section, we introduce still another notion and formulate a theorem regarding it, which will be used in Chap. 3. If X is a subset of E_n, then by the *extremal set* of X, which we shall indicate by

$$K(\mathsf{X}),$$

we mean the set of those points x of X which cannot be represented in the form

$$x = \frac{1}{2}x_1 + \frac{1}{2}x_2,$$

where x_1 and x_2 are distinct points of X. Thus the extremal set of a closed circle (i.e., of a circle together with its boundary) is the boundary of the circle. The extremal set of a closed triangle is the set consisting of the three vertices of the triangle.

It is clear that a nonempty set can have an empty extremal set. Thus the extremal set of the whole space, E_n, is empty; indeed, every open set has an empty extremal set. But the following theorem gives a sufficient condition in order that the extremal set not be empty.

THEOREM 2.4. Let X be a nonempty, bounded, closed, and convex subset of E_n. Then $K(\mathsf{X})$ is not empty, and X is the convex hull of $K(\mathsf{X})$.

3. Proof of the Fundamental Theorem for Arbitrary Rectangular Games. In this section we shall prove the fundamental theorem (sometimes called the "min-max theorem") for rectangular games; i.e., we shall show that if the function E is defined as in Sec. 1, then, for any rectangular game, the quantities

$$\max_{X \in \mathsf{S}_m} \min_{Y \in \mathsf{S}_n} E(X, Y)$$

and

$$\min_{Y \in \mathsf{S}_n} \max_{X \in \mathsf{S}_m} E(X, Y)$$

exist and are equal. It is convenient, however, first to prove a lemma about matrices.

LEMMA 2.5. Let

$$A = \begin{Vmatrix} a_{11} & \cdots & a_{1n} \\ \vdots & & \vdots \\ a_{m1} & \cdots & a_{mn} \end{Vmatrix}$$

be any matrix. Then either (i) there exists an element $\| x_1 \quad \cdots \quad x_m \|$ of S_m such that

$$a_{1j}x_1 + a_{2j}x_2 + \cdots + a_{mj}x_m \geq 0 \qquad \text{for } j = 1, \cdots, n,$$

or (ii) there exists an element $\| y_1 \quad \cdots \quad y_n \|$ of S_n such that

$$a_{i1}y_1 + a_{i2}y_2 + \cdots + a_{in}y_n \leq 0 \qquad \text{for } i = 1, \cdots, m.$$

PROOF. In this proof we shall make use of the delta symbols of Kronecker, which are defined as follows:

$$\delta_{ij} = 0 \qquad \text{if } i \neq j,$$
$$\delta_{ii} = 1.$$

We set

$$\delta^{(1)} = \| \delta_{11} \quad \delta_{21} \quad \cdots \quad \delta_{m1} \|,$$
$$\delta^{(2)} = \| \delta_{12} \quad \delta_{22} \quad \cdots \quad \delta_{m2} \|,$$
$$\vdots$$
$$\delta^{(m)} = \| \delta_{1m} \quad \delta_{2m} \quad \cdots \quad \delta_{mm} \|.$$

Thus $\delta^{(j)}$ (for $j = 1, \cdots, m$) is the point of $\mathbf{E}_m$ whose jth coordinate is 1, and all of whose other coordinates are 0.

We also set

$$\begin{aligned} a^{(1)} &= \| a_{11} \quad \cdots \quad a_{m1} \|, \\ a^{(2)} &= \| a_{12} \quad \cdots \quad a_{m2} \|, \\ &\vdots \\ a^{(n)} &= \| a_{1n} \quad \cdots \quad a_{mn} \|. \end{aligned}$$

Thus $a^{(j)}$ (for $j = 1, \cdots, n$) is the point whose coordinates are the components of the jth column of A.

Let $\mathbf{C}$ be the convex hull of the set of $m + n$ points

$$\delta^{(1)}, \cdots, \delta^{(m)}, \; a^{(1)}, \cdots, a^{(n)}.$$

Let $z = \| 0 \quad 0 \quad \cdots \quad 0 \|$ be the origin of $\mathbf{E}_m$. We distinguish two cases, according as $z \in \mathbf{C}$ or $z \notin \mathbf{C}$.

If $z \in \mathbf{C}$, then z is a convex linear combination of the points $\delta^{(1)}, \cdots, \delta^{(m)}$, $a^{(1)}, \cdots, a^{(n)}$. Hence there is an element $\| u_1 \quad \cdots \quad u_m \quad v_1 \quad \cdots \quad v_n \|$ of $\mathbf{S}_{m+n}$ such that

$$u_1\delta^{(1)} + \cdots + u_m\delta^{(m)} + v_1a^{(1)} + \cdots + v_ma^{(n)} = z. \tag{4}$$

Equation (4) means that we have

$$u_1\delta_{i1} + \cdots + u_m\delta_{im} + v_1a_{i1} + \cdots + v_na_{in} = 0 \qquad \text{for } i = 1, \cdots, m,$$

and hence, from the definition of the delta symbols, that

$$u_i + v_1a_{i1} + \cdots + v_na_{in} = 0 \qquad \text{for } i = 1, \cdots, m. \tag{5}$$

Since $\| u_1 \quad \cdots \quad u_m \quad v_1 \quad \cdots \quad v_n \| \in \mathbf{S}_{m+n}$, we see that u_i is non-negative and hence, from (5), that

$$v_1a_{i1} + \cdots + v_na_{in} \leq 0 \qquad \text{for } i = 1, \cdots, m. \tag{6}$$

Now we notice that

$$v_1 + \cdots + v_n > 0; \tag{7}$$

for otherwise, since $v_i \geq 0$ (for $i = 1, \cdots, n$) we should have

$$v_1 = v_2 = \cdots = v_n = 0$$

and hence, by (5), also

$$u_i = 0 \qquad \text{for } i = 1, \cdots, m,$$

which would contradict the fact that $\| u_1 \quad \cdots \quad u_m \quad v_1 \quad \cdots \quad v_n \| \in \mathbf{S}_{m+n}$. Hence we can set

$$\left.\begin{aligned} y_1 &= \frac{v_1}{v_1 + \cdots + v_n}, \\ y_2 &= \frac{v_2}{v_1 + \cdots + v_n}, \\ &\vdots \\ y_n &= \frac{v_n}{v_1 + \cdots + v_n}. \end{aligned}\right\} \tag{8}$$

From (6) and (7) we then conclude that

$$y_1 a_{i1} + \cdots + y_n a_{in} \leq 0 \qquad \text{for } i = 1, \cdots, m;$$

since it is clear from (8) that $\| y_1 \quad \cdots \quad y_n \| \in \mathbf{S}_n$, we therefore see that condition (ii) of Lemma 2.5 is satisfied, so that our lemma is true in this case, i.e., in the case where $z \in \mathbf{C}$.

Now suppose, on the other hand, that $z \notin \mathbf{C}$. Then, by Theorem 2.3, there exists a hyperplane which contains z and such that all of $\mathbf{C}$ is contained in one of the corresponding half-spaces. Let the equation of this hyperplane be:

$$b_1 t_1 + \cdots + b_m t_m = b_{m+1}.$$

Since z lies on the hyperplane, we have

$$b_1 0 + \cdots + b_m 0 = b_{m+1}$$

and hence

$$b_{m+1} = 0.$$

Thus the equation of the hyperplane is

$$b_1 t_1 + \cdots + b_m t_m = 0, \tag{9}$$

and we can suppose that every point $\| t_1 \quad \cdots \quad t_m \|$ of $\mathbf{C}$ satisfies the inequality

$$b_1 t_1 + \cdots + b_m t_m > 0; \tag{10}$$

for if the points of $\mathbf{C}$ satisfy the inequality

$$c_1 t_1 + \cdots + c_m t_m < 0,$$

then we can multiply through by -1, obtaining

$$(-c_1)t_1 + \cdots + (-c_m)t_m > 0,$$

and we can then obtain (10) by replacing $-c_i$ by b_i. The inequality (10) must hold, in particular, for the points $\delta^{(1)}, \cdots, \delta^{(m)}$ of C; thus

$$b_1\delta_{1i} + \cdots + b_m\delta_{mi} > 0 \qquad \text{for } i = 1, \cdots, m,$$

which means (from the definition of the delta symbols) that

$$b_i > 0 \qquad \text{for } i = 1, \cdots, m. \tag{11}$$

Moreover, (10) must hold for the points $a^{(1)}, \cdots, a^{(n)}$; thus

$$b_1a_{1i} + \cdots + b_ma_{mi} > 0 \qquad \text{for } i = 1, \cdots, n. \tag{12}$$

From (11) it is clear that $b_1 + \cdots + b_m > 0$, and hence that we can write

$$\left.\begin{aligned} x_1 &= \frac{b_1}{b_1 + \cdots + b_m}, \\ x_2 &= \frac{b_2}{b_1 + \cdots + b_m}, \\ &\vdots \\ x_m &= \frac{b_m}{b_1 + \cdots + b_m}. \end{aligned}\right\} \tag{13}$$

From (12) and (13), we conclude that

$$x_1a_{1i} + \cdots + x_ma_{mi} > 0 \qquad \text{for } i = 1, \cdots, n,$$

and hence, a fortiori, that

$$x_1a_{1i} + \cdots + x_ma_{mi} \geq 0 \qquad \text{for } i = 1, \cdots, n. \tag{14}$$

Since it follows from (11) and (13) that $\| x_1 \quad \cdots \quad x_m \| \in \mathsf{S}_m$, the inequality (14) means that condition (i) of our lemma is satisfied, which completes the proof.

The following is the fundamental theorem of the theory of rectangular games.

THEOREM 2.6. Let

$$A = \begin{Vmatrix} a_{11} & \cdots & a_{1n} \\ \vdots & & \vdots \\ a_{m1} & \cdots & a_{mn} \end{Vmatrix}$$

be any matrix, and let the expectation function $E(X, Y)$, for any $X = \| x_1 \quad \cdots \quad x_m \|$ and any $Y = \| y_1 \quad \cdots \quad y_n \|$ that are members of S_m and S_n, respectively, be defined as follows:

$$E(X, Y) = \sum_{i=1}^{m} \sum_{j=1}^{n} a_{ij} x_i y_j .$$

Then the quantities

$$\max_{X \in \mathsf{S}_m} \min_{Y \in \mathsf{S}_n} E(X, Y)$$

and

$$\min_{Y \in \mathsf{S}_n} \max_{X \in \mathsf{S}_m} E(X, Y)$$

exist and are equal.

PROOF. For each $Y = \| y_1 \quad \cdots \quad y_n \|$, the function $E(X, Y)$ is a continuous (indeed, a linear) function of $X = \| x_1 \quad \cdots \quad x_m \|$, which is defined over the closed subset S_m of E_m; hence, we see that

$$\max_{X \in \mathsf{S}_m} E(X, Y)$$

exists for each Y in S_n. Moreover, it is easily verified that

$$\max_{X \in \mathsf{S}_m} E(X, Y)$$

is a piecewise linear function of $\| y_1 \quad \cdots \quad y_n \|$ and is continuous. Since S_n is a closed subset of E_n, we therefore conclude that

$$\min_{Y \in \mathsf{S}_n} \max_{X \in \mathsf{S}_m} E(X, Y)$$

exists. Similarly, we can show that

$$\max_{X \in \mathsf{S}_m} \min_{Y \in \mathsf{S}_n} E(X, Y)$$

exists.

If condition (i) of Lemma 2.5 holds, then there is an element $\| x_1 \quad \cdots \quad x_m \|$ of S_m such that

$$a_{1j}x_1 + a_{2j}x_2 + \cdots + a_{mj}x_m \geq 0 \qquad \text{for } j = 1, \cdots, n,$$

and hence such that, for every Y in S_n,

$$E(X, Y) = \sum_{j=1}^{n} (a_{1j}x_1 + \cdots + a_{mj}x_m) y_j \geq 0. \tag{15}$$

Since (15) holds for every Y in $\mathbf{S}_n$, we see that

$$\min_{Y \in \mathbf{S}_n} E(X, Y) \geq 0$$

and hence that

$$\max_{X \in \mathbf{S}_m} \min_{Y \in \mathbf{S}_n} E(X, Y) \geq 0. \tag{16}$$

In a similar fashion we conclude that if condition (ii) of Lemma 2.5 holds, then

$$\min_{Y \in \mathbf{S}_n} \max_{X \in \mathbf{S}_m} E(X, Y) \leq 0. \tag{17}$$

Since either condition (i) or condition (ii) of Lemma 2.5 holds, however, we conclude that at least one of the two inequalities, (15) and (16), must hold, and hence that the following condition can *not* hold:

$$\max_{X \in \mathbf{S}_m} \min_{Y \in \mathbf{S}_n} E(X, Y) < 0 < \min_{Y \in \mathbf{S}_n} \max_{X \in \mathbf{S}_m} E(X, Y). \tag{18}$$

Now let A_k be the matrix which arises from A by the subtraction of k from each element of A, that is to say,

$$\mathrm{A}_k = \left\| \begin{matrix} a_{11} - k & \cdots & a_{1n} - k \\ \vdots & & \vdots \\ a_{m1} - k & \cdots & a_{mn} - k \end{matrix} \right\|,$$

and let E_k be the expectation function for A_k, so that, for any X and any Y that are members of $\mathbf{S}_m$ and $\mathbf{S}_n$, respectively, we have

$$E_k(X, Y) = \sum_{i=1}^{m} \sum_{j=1}^{n} (a_{ij} - k) x_i y_j. \tag{19}$$

Then, just as we showed that (18) does not hold for A, we can show that the following condition does not hold for A_k:

$$\max_{X \in \mathbf{S}_m} \min_{Y \in \mathbf{S}_n} E_k(X, Y) < 0 < \min_{Y \in \mathbf{S}_n} \max_{X \in \mathbf{S}_m} E_k(X, Y). \tag{20}$$

From (19), however, it is easily seen that

$$E_k(X, Y) = E(X, Y) - k; \tag{21}$$

and from (20) and (21), we conclude that the following condition does not hold:

$$\max_{X \in \mathbf{S}_m} \min_{Y \in \mathbf{S}_n} E(X, Y) - k < 0 < \min_{Y \in \mathbf{S}_n} \max_{X \in \mathbf{S}_m} E(X, Y) - k. \tag{22}$$

Hence the following condition does not hold:

$$\max_{X \in \mathbf{S}_m} \min_{Y \in \mathbf{S}_n} E(X, Y) < k < \min_{Y \in \mathbf{S}_n} \max_{X \in \mathbf{S}_m} E(X, Y). \tag{23}$$

Since (23) is false for every k, we conclude that the following is false:

$$\max_{X \in \mathbf{S}_m} \min_{Y \in \mathbf{S}_n} E(X, Y) < \min_{Y \in \mathbf{S}_n} \max_{X \in \mathbf{S}_m} E(X, Y),$$

and hence we conclude that the following is true:

$$\max_{X \in \mathbf{S}_m} \min_{Y \in \mathbf{S}_n} E(X, Y) \geq \min_{Y \in \mathbf{S}_n} \max_{X \in \mathbf{S}_m} E(X, Y). \tag{24}$$

From Theorem 1.1, on the other hand, we have

$$\max_{X \in \mathbf{S}_m} \min_{Y \in \mathbf{S}_n} E(X, Y) \leq \min_{Y \in \mathbf{S}_n} \max_{X \in \mathbf{S}_m} E(X, Y); \tag{25}$$

and from (24) and (25), it follows that

$$\max_{X \in \mathbf{S}_m} \min_{Y \in \mathbf{S}_n} E(X, Y) = \min_{Y \in \mathbf{S}_n} \max_{X \in \mathbf{S}_m} E(X, Y),$$

which completes the proof of our theorem.

Expressed in game-theoretic terminology, Theorem 2.6 assumes the following form:

THEOREM 2.7. Every rectangular game has a value; a player of a rectangular game always has an optimal strategy.

PROOF. By Theorems 2.6 and 1.5.

4. Properties of Optimal Strategies. It is sometimes possible to determine the value of a game by an intuitive argument, or by direct inspection. In such cases it is often convenient to make use of the following theorem in order to find optimal strategies for the two players.

THEOREM 2.8. Let E be the expectation function of an $m \times n$ rectangular game whose value is v. Then a necessary and sufficient condition that a member X^* of $\mathbf{S}_m$ be an optimal strategy for P_1 is that, for every member Y of $\mathbf{S}_n$, we have

$$v \leq E(X^*, Y).$$

Similarly, a necessary and sufficient condition that a member Y^* of $\mathbf{S}_n$ be an optimal strategy for P_2 is that, for every member X of $\mathbf{S}_m$, we have

$$E(X, Y^*) \leq v.$$

PROOF. If X^* is an optimal strategy for P_1, then there is a member Y^* of $\mathbf{S}_n$ such that $\| X^* \quad Y^* \|$ is a saddle-point of E and hence such that, for every Y in $\mathbf{S}_n$,

$$v = E(X^*, Y^*) \leq E(X^*, Y),$$

as was to be shown.

Now suppose, on the other hand, that X^* is a member of $\mathbf{S}_m$ such that, for every Y in $\mathbf{S}_n$,

$$v \leq E(X^*, Y). \tag{26}$$

By Theorem 2.7, there exists a point $\| X' \quad Y' \|$ such that, for all X in $\mathbf{S}_m$ and all Y in $\mathbf{S}_n$,

$$E(X, Y') \leq E(X', Y') \leq E(X', Y); \tag{27}$$

and since, by hypothesis, v is the value of the game, we have

$$E(X', Y') = v. \tag{28}$$

From (26) and (28), we conclude that

$$E(X', Y') \leq E(X^*, Y). \tag{29}$$

Replacing Y by Y' in (29) and replacing X by X^* in the first part of (27), we obtain

$$E(X^*, Y') \leq E(X', Y') \leq E(X^*, Y');$$

thus

$$E(X', Y') = E(X^*, Y'). \tag{30}$$

From (27), (29), and (30), we now conclude that

$$E(X, Y') \leq E(X^*, Y') \leq E(X^*, Y),$$

so that $\| X^* \quad Y' \|$ is a saddle-point of E, and hence X^* is an optimal strategy for P_1, as was to be shown.

The proof of the second part of the theorem is similar.

The following theorem provides us with a quick way of checking a proposed solution of a game and also, as we shall see shortly, enables us to reduce the problem of finding solutions to a problem of elementary algebra. To simplify the statement of this theorem, we write

$$E(i, Y)$$

for

$$E(X_i, Y),$$

where X_i is the member of S_m whose ith component is 1, and hence all of whose other components are 0; and, similarly, we write

$$E(X, j)$$

for

$$E(X, Y_j),$$

where Y_j is the member of S_n whose jth component is 1. If $X = \| x_1 \quad \cdots \quad x_m \|$ and $Y = \| y_1 \quad \cdots \quad y_n \|$, then

$$E(i, Y) = \sum_{j=1}^{n} a_{ij} y_j$$

and

$$E(X, j) = \sum_{i=1}^{m} a_{ij} x_i .$$

We also note that

$$E(X, Y) = \sum_{i=1}^{m} E(i, Y) x_i = \sum_{j=1}^{n} E(X, j) y_j .$$

THEOREM 2.9. Let E be the expectation function of an $m \times n$ rectangular game, let v be a real number, and let X^* and Y^* be members of S_m and S_n, respectively. Then a necessary and sufficient condition that v be the value of the game and that X^* and Y^* be optimal strategies for P_1 and P_2, respectively, is that, for $1 \le i \le m$ and $1 \le j \le n$,

$$E(i, Y^*) \le v \le E(X^*, j) .$$

PROOF. The necessity of the condition follows directly from the definition of a saddle-point—by replacing X by X_i and Y by Y_j.

On the other hand, if the condition is satisfied, then we obtain, for any $X = \| x_1 \quad \cdots \quad x_m \|$ that is a member of S_m,

$$\sum_{i=1}^{m} E(i, Y^*) x_i \le \sum_{i=1}^{m} v x_i = v,$$

and thus

$$E(X, Y^*) \le v. \tag{31}$$

Similarly, we obtain, for any $Y = \| y_1 \quad \cdots \quad y_n \|$ that is a member of $\mathbf{S}_n$,

$$v \leq E(X^*, Y). \tag{32}$$

From (31) and (32), respectively, replacing X by X^* and replacing Y by Y^*, we obtain

$$E(X^*, Y^*) \leq v$$

and

$$v \leq E(X^*, Y^*);$$

and hence

$$v = E(X^*, Y^*). \tag{33}$$

Then from (31), (32), and (33), we have

$$E(X, Y^*) \leq E(X^*, Y^*) \leq E(X^*, Y),$$

so that $\| X^* \quad Y^* \|$ is indeed a saddle-point of E and v is the value of the game.

The following theorem is an easy consequence of Theorem 2.8; its proof will be left as an exercise.

THEOREM 2.10. Let E be the expectation function of an $m \times n$ rectangular game whose value is v. Then a necessary and sufficient condition that a member X^* of $\mathbf{S}_m$ be an optimal strategy for P_1 is that, for $1 \leq j \leq n$, we have

$$v \leq E(X^*, j).$$

Similarly, a necessary and sufficient condition that a member Y^* of $\mathbf{S}_n$ be an optimal strategy for P_2 is that, for $1 \leq i \leq m$, we have

$$E(i, Y^*) \leq v.$$

THEOREM 2.11. Let E be the expectation function of an $m \times n$ rectangular game, and let $\| X^* \quad Y^* \|$ be a solution of the game. Then

$$\max_{1 \leq i \leq m} E(i, Y^*) = \min_{1 \leq j \leq n} E(X^*, j).$$

PROOF. By Theorem 2.10 we have, for $1 \leq j \leq n$,

$$v \leq E(X^*, j),$$

where v is the value of the game; hence

$$v \leq \min_{1 \leq j \leq n} E(X^*, j).$$

Now, if we had

$$v < \min_{1 \leq j \leq n} E(X^*, j),$$

then we would have, for $1 \leq j \leq n$,

$$v < E(X^*, j),$$

and hence

$$\sum_{j=1}^{n} v y_j^* < \sum_{j=1}^{n} E(X^*, j) y_j^*$$

or

$$v < E(X^*, Y^*),$$

contrary to the hypothesis that v is the value of the game. Therefore, we conclude that

$$v = \min_{1 \leq j \leq n} E(X^*, j).$$

The proof that

$$v = \max_{1 \leq i \leq m} E(i, Y^*)$$

is similar.

The following more special theorem is frequently very useful when the solution of a given game is desired.

THEOREM 2.12. Let E be the expectation function of an $m \times n$ rectangular game whose value is v, and let $X^* = \| x_1^* \quad \cdots \quad x_m^* \|$ and $Y^* = \| y_1^* \quad \cdots \quad y_n^* \|$ be any optimal strategies for P_1 and P_2, respectively. Then, for any i such that

$$E(i, Y^*) < v,$$

we have

$$x_i^* = 0.$$

And, for any j such that

$$v < E(X^*, j),$$

we have

$$y_j^* = 0.$$

PROOF. Suppose, if possible, that, for some h,

$$E(h, Y^*) < v$$

and

$$x_h^* \neq 0;$$

then we conclude that

$$E(h, Y^*)x_h^* < vx_h^*.$$

Since, for $k = 1, \cdots, h-1, h+1, \cdots, m$, we have

$$E(k, Y^*) \leq v,$$

and hence

$$E(k, Y^*)x_k^* \leq vx_k^*,$$

we conclude that

$$\sum_{i=1}^{m} E(i, Y^*)x_i^* < \sum_{i=1}^{m} vx_i^*$$

and hence that

$$E(X^*, Y^*) < v,$$

contrary to the hypothesis that v is the value of the game.

The proof of the second part of the theorem is similar.

The proof of the following theorem will be left as an exercise.

THEOREM 2.13. Let E be the expectation function of an $m \times n$ rectangular game, and let X^* and Y^* be members of S_m and S_n, respectively. Then the following conditions are all equivalent:

(i) X^* is an optimal strategy for P_1, and Y^* is an optimal strategy for P_2.

(ii) If X is any member of S_m and Y is any member of S_n, then

$$E(X, Y^*) \leq E(X^*, Y^*) \leq E(X^*, Y).$$

(iii) If i and j are any integers such that $1 \leq i \leq m$ and $1 \leq j \leq n$, then

$$E(i, Y^*) \leq E(X^*, Y^*) \leq E(X^*, j).$$

We shall now show by some examples how the above theorems can be used to compute the values and the solutions of given games. The methods in their present form are extremely laborious and time-consuming; later, we shall establish methods which will shorten the process of computation.

EXAMPLE 2.14. We wish to find the value, and optimal strategies for the two players, of the rectangular game whose matrix is

$$\begin{Vmatrix} 1 & -1 & -1 \\ -1 & -1 & 3 \\ -1 & 2 & -1 \end{Vmatrix}.$$

By Theorem 2.9, it suffices to find numbers x_1, x_2, x_3, y_1, y_2, y_3, and v which satisfy the following conditions:

$$\begin{aligned} x_1 + x_2 + x_3 &= 1, & y_1 + y_2 + y_3 &= 1, \\ 0 \leq x_1 &\leq 1, & 0 \leq y_1 &\leq 1, \\ 0 \leq x_2 &\leq 1, & 0 \leq y_2 &\leq 1, \\ 0 \leq x_3 &\leq 1, & 0 \leq y_3 &\leq 1, \end{aligned}$$

$$\begin{gathered} (1)x_1 + (-1)x_2 + (-1)x_3 \geq v, \\ (-1)x_1 + (-1)x_2 + (2)x_3 \geq v, \\ (-1)x_1 + (3)x_2 + (-1)x_3 \geq v, \\ (1)y_1 + (-1)y_2 + (-1)y_3 \leq v, \\ (-1)y_1 + (-1)y_2 + (3)y_3 \leq v, \\ (-1)y_1 + (2)y_2 + (-1)y_3 \leq v. \end{gathered}$$

The more familiar methods of elementary algebra do not suffice to enable us to solve systems containing inequalities as well as equalities. However, we know from Theorem 2.6 that there exists a solution to the system; and we can consider separately the 2^6 possible cases that arise by replacing the signs "$\leq$" and "$\geq$" in the last six inequalities by the signs "$<$" and "$=$" and the signs "$>$" and "$=$", respectively. Suppose that we start by replacing all of these six inequalities by equalities. By doing this, we obtain:

$$x_1 - x_2 - x_3 = v, \qquad y_1 - y_2 - y_3 = v,$$
$$-x_1 - x_2 + 2x_3 = v, \qquad -y_1 - y_2 + 3y_3 = v,$$
$$-x_1 + 3x_2 - x_3 = v, \qquad -y_1 + 2y_2 - y_3 = v,$$

and, of course, also

$$x_1 + x_2 + x_3 = 1, \qquad y_1 + y_2 + y_3 = 1.$$

Using any of the familiar methods of elementary algebra, we find that this set of equations has the solution:

$$x_1 = \frac{6}{13}, \qquad x_2 = \frac{3}{13}, \qquad x_3 = \frac{4}{13},$$
$$y_1 = \frac{6}{13}, \qquad y_2 = \frac{4}{13}, \qquad y_3 = \frac{3}{13},$$
$$v = -\frac{1}{13}.$$

Since the x_1, x_2, x_3, y_1, y_2, and y_3, found in this way, also turn out to be non-negative, we have found a solution of our original set of equations and inequalities. Thus an optimal way for P_1 to play this game is to choose the numbers 1, 2, and 3 with respective probabilities $6/13$, $3/13$, and $4/13$; and an optimal way for P_2 to play is to choose the numbers 1, 2, and 3 with respective probabilities $6/13$, $4/13$, and $3/13$. The value of the game is $-1/13$; i.e., P_1 can play in such a way as to make sure of not losing more than $1/13$, and P_2 can play in such a way as to make certain that P_1 will lose at least $1/13$.

The next example shows how to take care of the difficulties which arise when the equations corresponding to the six above turn out to be inconsistent, or to have no solutions in the interval $[0, 1]$.

EXAMPLE 2.15. We wish to find the value, and optimal strategies for the two players, of the rectangular game whose matrix is

$$\begin{Vmatrix} 3 & -2 & 4 \\ -1 & 4 & 2 \\ 2 & 2 & 6 \end{Vmatrix}.$$

It suffices, again by Theorem 2.9, to find numbers x_1, x_2, x_3, y_1, y_2, y_3, and v which satisfy the following conditions:

$$x_1 + x_2 + x_3 = 1, \qquad y_1 + y_2 + y_3 = 1,$$
$$0 \leq x_1 \leq 1, \qquad 0 \leq y_1 \leq 1,$$
$$0 \leq x_2 \leq 1, \qquad 0 \leq y_2 \leq 1,$$
$$0 \leq x_3 \leq 1, \qquad 0 \leq y_3 \leq 1,$$

$$\begin{aligned} 3x_1 - x_2 + 2x_3 &\geq v, & 3y_1 - 2y_2 + 4y_3 &\leq v, \\ -2x_1 + 4x_2 + 2x_3 &\geq v, & -y_1 + 4y_2 + 2y_3 &\leq v, \\ 4x_1 + 2x_2 + 6x_3 &\geq v, & 2y_1 + 2y_2 + 6y_3 &\leq v. \end{aligned}$$

Considering first the case where all the six inequalities are replaced by equalities, we obtain

$$\begin{aligned} 3x_1 - x_2 + 2x_3 &= v, & 3y_1 - 2y_2 + 4y_3 &= v, \\ -2x_1 + 4x_2 + 2x_3 &= v, & -y_1 + 4y_2 + 2y_3 &= v, \\ 4x_1 + 2x_2 + 6x_3 &= v, & 2y_1 + 2y_2 + 6y_3 &= v, \\ x_1 + x_2 + x_3 &= 1, & y_1 + y_2 + y_3 &= 1. \end{aligned}$$

It is easily verified, however, that these equations have no solution which makes x_1, x_2, x_3, y_1, y_2, and y_3 all non-negative.

This means that we cannot obtain a solution to our problem by replacing all the inequalities by equalities. Hence we consider another case: We replace the first "$\geq$" by a "$>$" and replace the other inequalities by equalities, obtaining

$$\begin{aligned} 3x_1 - x_2 + 2x_3 &> v, & 3y_1 - 2y_2 + 4y_3 &= v, \\ -2x_1 + 4x_2 + 2x_3 &= v, & -y_1 + 4y_2 + 2y_3 &= v, \\ 4x_1 + 2x_2 + 6x_3 &= v, & 2y_1 + 2y_2 + 6y_3 &= v, \\ x_1 + x_2 + x_3 &= 1, & y_1 + y_2 + y_3 &= 1. \end{aligned}$$

Since $3x_1 - x_2 + 2x_3 > v$, we conclude, by means of Theorem 2.12, that $y_1 = 0$. Replacing y_1 by 0, we obtain a set of equations to be solved for x_1, x_2, x_3, y_2, and y_3. These equations are easily seen to be inconsistent, however, which means that we must go on to another case. (Theorem 2.6 assures us that we shall eventually hit upon a case which has a solution.) Continuing in this way, we consider finally the following case:

$$\begin{aligned} 3x_1 - x_2 + 2x_3 &= v, & 3y_1 - 2y_2 + 4y_3 &< v, \\ -2x_1 + 4x_2 + 2x_3 &= v, & -y_1 + 4y_2 + 2y_3 &= v, \\ 4x_1 + 2x_2 + 6x_3 &> v, & 2y_1 + 2y_2 + 6y_3 &= v, \\ x_1 + x_2 + x_3 &= 1, & y_1 + y_2 + y_3 &= 1. \end{aligned}$$

The strict inequality $4x_1 + 2x_2 + 6x_3 > v$ implies, by Theorem 2.12, that $y_3 = 0$, and the strict inequality $3y_1 - 2y_2 + 4y_3 < v$ implies that $x_1 = 0$.

Thus we are to solve the equations

$$-x_2 + 2x_3 = v, \qquad -y_1 + 4y_2 = v,$$
$$4x_2 + 2x_3 = v, \qquad 2y_1 + 2y_2 = v,$$
$$x_2 + x_3 = 1, \qquad y_1 + y_2 = 1.$$

This set of equations is found to have the solution

$$x_2 = 0, \qquad x_3 = 1, \qquad y_1 = \frac{2}{5}, \qquad y_2 = \frac{3}{5}, \qquad v = 2.$$

We now substitute the values

$$x_1 = 0, \qquad x_2 = 0, \qquad x_3 = 1,$$
$$y_1 = \frac{2}{5}, \qquad y_2 = \frac{3}{5}, \qquad y_3 = 0,$$
$$v = 2$$

into our original set of inequalities, and we find that they are all satisfied. Thus the value of the game is 2, the vector $\| 0 \quad 0 \quad 1 \|$ is an optimal strategy for the first player, and the vector $\| 2/5 \quad 3/5 \quad 0 \|$ is an optimal strategy for the second player.

5. Relations of Dominance. As we have seen in Sec. 3 of Chap. 1, it is sometimes possible to tell by direct inspection of the matrix of a rectangular game that certain pure strategies will never enter into optimal mixed strategies except with probability zero.

Thus, for example, if

$$\begin{Vmatrix} 1 & 7 & 2 \\ 6 & 2 & 7 \\ 5 & 1 & 6 \end{Vmatrix} \tag{34}$$

is the matrix of a rectangular game, then it is clear, on intuitive grounds, that no optimal strategy for P_1 should assign a positive probability to the third row; for, no matter what P_2 does, P_1 can do better by choosing the second row rather than the third row. Thus it appears that in order to solve this game, we could solve the game whose matrix is

$$\begin{Vmatrix} 1 & 7 & 2 \\ 6 & 2 & 7 \end{Vmatrix}. \tag{35}$$

The values of the two games should be the same: P_2 should have the same optimal strategies in both; and if $\| x_1 \quad x_2 \|$ is an optimal strategy for P_1 in the second game, then $\| x_1 \quad x_2 \quad 0 \|$ should be an optimal strategy for P_1 in the first game.

In a similar way, since every element of the first column of (35) is less than the corresponding element of the third column, and since P_2 wants to minimize the payoff, it appears that we can cross out the third column of (35), obtaining

$$\begin{Vmatrix} 1 & 7 \\ 6 & 2 \end{Vmatrix}, \tag{36}$$

and we can find solutions of the original game by solving the game whose matrix is (36). Thus, since, for the game whose matrix is (36), the value is 4, and since $\| \frac{2}{5} \quad \frac{3}{5} \|$ and $\| \frac{1}{2} \quad \frac{1}{2} \|$ are optimal strategies for P_1 and P_2, respectively, it appears reasonable to suppose that the value of the original game is also 4, that $\| \frac{2}{5} \quad \frac{3}{5} \quad 0 \|$ is an optimal strategy for P_1 in playing it, and that $\| \frac{1}{2} \quad \frac{1}{2} \quad 0 \|$ is an optimal strategy for P_2—and it is easily verified that such is indeed the case.

These considerations can be somewhat generalized to the case in which the elements of one row, although they are not all smaller than the corresponding elements of another row, are all smaller than certain convex linear combinations of the corresponding elements of other rows. Thus, for example, consider the game whose payoff matrix is

$$\begin{Vmatrix} 24 & 0 \\ 0 & 8 \\ 4 & 5 \end{Vmatrix}. \tag{37}$$

Here we notice that

$$4 < \frac{1}{4} \cdot 24 + \frac{3}{4} \cdot 0,$$

$$5 < \frac{1}{4} \cdot 0 + \frac{3}{4} \cdot 8,$$

and hence it seems that P_1 would never be wise to play the third row, for he could always do better by dividing between the first two rows (in the ratio of 1 to 3) any probability he might consider assigning to the third row. Here again it appears that we can solve the game simply by solving the game whose matrix is

$$\begin{Vmatrix} 24 & 0 \\ 0 & 8 \end{Vmatrix}.$$

Since the value of the latter game is 6, and since $\| \frac{1}{4} \quad \frac{3}{4} \|$ is an optimal strategy for each player of it, we conclude that the game whose matrix is (37) has a value of 6 also, that $\| \frac{1}{4} \quad \frac{3}{4} \quad 0 \|$ is an optimal strategy for P_1, and that $\| 1/4 \quad 3/4 \|$ is an optimal strategy for P_2.

Similarly, if each element of a certain column is *greater* than a certain convex linear combination of the corresponding elements of certain other columns, then the column in question can be eliminated. Thus we can solve the game whose matrix is

$$\begin{Vmatrix} 10 & 0 & 6 \\ 0 & 10 & 7 \end{Vmatrix}$$

by solving the game whose matrix is

$$\begin{Vmatrix} 10 & 0 \\ 0 & 10 \end{Vmatrix}.$$

In order to establish the validity of this method in a formal way, it is convenient to introduce some definitions. If $a = \| a_1 \quad \cdots \quad a_n \|$ and $b = \| b_1 \quad \cdots \quad b_n \|$ are vectors (or rows or columns of a matrix), and if $a_i \geq b_i$ (for $i = 1, \cdots, n$), then we say that *a dominates b*; if $a_i > b_i$ (for $i = 1, \cdots, n$), we say that *a strictly dominates b*. Both of these relations are transitive: if a dominates b and b dominates c, then a dominates c, and similarly for strict dominance. We notice also that the relation of dominance is reflexive, so that everything dominates itself; and the relation of strict dominance is irreflexive, so that nothing strictly dominates itself. In using these notions, however, it is important always to bear in mind the fact that since two vectors are distinct, it by no means follows that one of them dominates the other; consider, for instance, the vectors $\| 1 \quad 2 \|$ and $\| 2 \quad 1 \|$.

We introduce also the notion of an i-place extension of a mixed strategy. If $x = \| x_1 \quad \cdots \quad x_n \|$ is a member of $\mathbf{S}_n$, and $1 \leq i \leq n + 1$, then by the *i-place extension* of x we mean the vector $\| x_1 \quad \cdots \quad x_{i-1} \quad 0 \quad x_i \quad \cdots \quad x_n \|$. Thus the 2-place extension of $\| 1/10 \quad 2/10 \quad 7/10 \|$ is $\| 1/10 \quad 0 \quad 2/10 \quad 7/10 \|$; the 1-place extension is $\| 0 \quad 1/10 \quad 2/10 \quad 7/10 \|$; and the 4-place extension is $\| 1/10 \quad 2/10 \quad 7/10 \quad 0 \|$. It is clear that if x is any member of $\mathbf{S}_n$, then each i-place extension of x is a member of $\mathbf{S}_{n+1}$.

> THEOREM 2.16. Let Γ be a rectangular game whose matrix is A; suppose that, for some i, the ith row of A is dominated by some convex linear combination of the other rows of A; let A′ be the matrix obtained from A by omitting the ith row; and let Γ' be the rectangular game whose matrix is A′. Then the value of Γ' is the same as the value of Γ; every optimal strategy for P_2 in Γ' is also an optimal strategy for P_2 in Γ; and if w is any optimal strategy for P_1 in Γ' and x is the i-place extension of w, then x

is an optimal strategy for P_1 in Γ. Moreover, if the ith row of A is strictly dominated by the convex linear combination of the other rows of A, then every solution of Γ can be obtained in this way from a solution of Γ'.

PROOF. Let

$$A = \begin{Vmatrix} a_{11} & \cdots & a_{1n} \\ \vdots & & \vdots \\ a_{m1} & \cdots & a_{mn} \end{Vmatrix}.$$

We can suppose, without loss of generality, that the last row of A is dominated by a convex linear combination of the other rows. Thus there is a member $z = \| z_1 \quad \cdots \quad z_{m-1} \|$ of $\mathbf{S}_{m-1}$ such that

$$a_{mj} \leq \sum_{i=1}^{m-1} a_{ij} z_i \qquad \text{for } j = 1, \cdots, n. \tag{38}$$

Let v be the value of Γ', let $w = \| w_1 \quad \cdots \quad w_{m-1} \|$ be an optimal strategy for P_1 in Γ', and let $y = \| y_1 \quad \cdots \quad y_n \|$ be an optimal strategy for P_2 in Γ'. Then we see from Theorem 2.9 that

$$\sum_{j=1}^{n} a_{ij} y_j \leq v \qquad \text{for } i = 1, \cdots, m-1, \tag{39}$$

$$v \leq \sum_{i=1}^{m-1} a_{ij} w_i \qquad \text{for } j = 1, \cdots, n. \tag{40}$$

To prove the first part of our theorem, it will now suffice to show that v is also the value of Γ, that y is an optimal strategy for P_2 in Γ, and that $\| w_1 \quad \cdots \quad w_{m-1} \quad 0 \|$ is an optimal strategy for P_1 in Γ. By Theorem 2.9 this amounts to showing that

$$\sum_{j=1}^{n} a_{ij} y_j \leq v \qquad \text{for } i = 1, \cdots, m, \tag{39a}$$

and that

$$v \leq \sum_{i=1}^{m-1} a_{ij} w_i + a_{mj} 0 \qquad \text{for } j = 1, \cdots, n. \tag{40a}$$

Since (40*a*) is an obvious consequence of (40), we conclude that it suffices to prove (39*a*); and hence, by (39), it suffices to show that

$$\sum_{j=1}^{n} a_{mj}y_j \leq v. \tag{41}$$

But, by using (38) and (39), we have immediately

$$\sum_{j=1}^{n} a_{mj}y_j \leq \sum_{j=1}^{n} \sum_{i=1}^{m-1} a_{ij}z_i y_j$$

$$= \sum_{i=1}^{m-1} \sum_{j=1}^{n} a_{ij}y_j z_i \leq \sum_{i=1}^{m-1} v z_i = v,$$

as was to be shown.

To prove the second part of the theorem, we need only to notice that if the equality holds in none of the relations of (38), then we have

$$\sum_{j=1}^{n} a_{mj}y_j < \sum_{j=1}^{n} \sum_{i=1}^{m-1} a_{ij}z_i y_j \leq v,$$

and hence, by Theorem 2.12, every optimal strategy for P_1 in Γ will have its mth component equal to zero.

The proof of the next theorem is similar, and it will therefore be omitted.

THEOREM 2.17. Let Γ be a rectangular game whose matrix is A; suppose that, for some j, the jth column of A dominates some convex linear combination of the other columns of A; let A′ be the matrix obtained from A by omitting the jth column; and let Γ' be the rectangular game whose matrix is A′. Then the value of Γ' is the same as the value of Γ; every optimal strategy for P_1 in Γ' is also an optimal strategy for P_1 in Γ; and if z is any optimal strategy for P_2 in Γ', and y is the j-place extension of z, then y is an optimal strategy for P_2 in Γ. Furthermore, if the jth column of A strictly dominates the convex linear combination of other columns of A, then every solution of Γ can be obtained in this way from a solution of Γ'.

As a somewhat more complicated example of the application of these theorems, we have:

EXAMPLE 2.18. The payoff matrix of a rectangular game is

$$\begin{Vmatrix} 3 & 2 & 4 & 0 \\ 3 & 4 & 2 & 4 \\ 4 & 2 & 4 & 0 \\ 0 & 4 & 0 & 8 \end{Vmatrix}.$$

Here we notice that the first row is dominated by the third row; crossing out this row we obtain

$$\begin{Vmatrix} 3 & 4 & 2 & 4 \\ 4 & 2 & 4 & 0 \\ 0 & 4 & 0 & 8 \end{Vmatrix}.$$

In the new matrix, the first column dominates the third; crossing out the first column we obtain

$$\begin{Vmatrix} 4 & 2 & 4 \\ 2 & 4 & 0 \\ 4 & 0 & 8 \end{Vmatrix}.$$

In this matrix, no row (or column) dominates another row (or column); but we notice that the first column dominates a certain convex linear combination of the second and third columns, for we have

$$4 > \frac{1}{2} \cdot 2 + \frac{1}{2} \cdot 4,$$

$$2 = \frac{1}{2} \cdot 4 + \frac{1}{2} \cdot 0,$$

$$4 = \frac{1}{2} \cdot 0 + \frac{1}{2} \cdot 8.$$

Thus we can omit the first column, obtaining

$$\begin{Vmatrix} 2 & 4 \\ 4 & 0 \\ 0 & 8 \end{Vmatrix}.$$

In this matrix, in turn, the first row is dominated by a convex linear combination of the other two rows, for

$$2 = \frac{1}{2} \cdot 4 + \frac{1}{2} \cdot 0,$$

$$4 = \frac{1}{2} \cdot 0 + \frac{1}{2} \cdot 8.$$

Hence our matrix reduces to

$$\begin{Vmatrix} 4 & 0 \\ 0 & 8 \end{Vmatrix}.$$

The value of the game with this 2×2 matrix is $8/3$, and an optimal strategy for the first player (and for the second player) is $\| 2/3 \quad 1/3 \|$. The value of the original game is therefore also 8/3; since we obtained the 2×2 matrix from the original 4×4 matrix by crossing out the first two rows and the first two columns, we conclude that an optimal strategy for the first player (and for the second player) in the original games is $\| 0 \quad 0 \quad 2/3 \quad 1/3 \|$.

REMARK 2.19. From Theorems 2.16 and 2.17 it will be noticed that whenever we cross out a row which is strictly dominated (or a column which strictly dominates), we obtain a matrix which leads to exactly the same set of solutions as could be obtained by solving the original game. This is not the case, however, when there is a relation of dominance which is not strict; in such a case we can "lose" some of the solutions of the original matrix. Thus, for example, consider the game whose payoff matrix is

$$\begin{Vmatrix} 0 & 0 & 0 \\ 0 & 1 & -1 \\ 0 & -1 & 1 \end{Vmatrix}.$$

Here it is easily verified that any member of $\mathbf{S}_3$ of the form $\| a \quad b \quad b \|$ will be an optimal strategy for P_1. On the other hand, if we cross out the first row (which is dominated, but not strictly dominated, by a convex linear combination of the other two rows), and if we then cross out the first column, we obtain the matrix

$$\begin{Vmatrix} 1 & -1 \\ -1 & 1 \end{Vmatrix},$$

for which P_1 has the unique optimal strategy $\| 1/2 \quad 1/2 \|$; thus by this method we would obtain for the original game only the one optimal strategy: $\| 0 \quad 1/2 \quad 1/2 \|$.

6. A Graphical Method of Solution. In this section we shall explain a graphical method of finding the solutions of a rectangular game. This method is very easily applied to games having $2 \times n$ or $m \times 2$ matrices. It can also be applied (if one is skillful at drawing 3-dimensional diagrams) to games having $3 \times n$ and $m \times 3$ matrices; but it becomes impracticable in the case of $m \times n$ matrices where both m and n are greater than 3. We shall illustrate the method by some examples of the solution of $2 \times n$ games.

EXAMPLE 2.20. Consider the game whose payoff matrix is

$$\begin{matrix} & \boxed{1} & \boxed{2} & \boxed{3} \\ \textcircled{1} & 2 & 3 & 11 \\ \textcircled{2} & 7 & 5 & 2 \end{matrix}.$$

Here we have given the names "①," "②," "[1]," etc., to the various strategies of the two players. If P_1 uses the mixed strategy $\| x \quad 1 - x \|$ and if P_2 uses the pure strategy [1], then the expected payoff to P_1 will be

$$2x + 7(1 - x) = 7 - 5x.$$

Similarly, if P_2 uses pure strategy [2], then the expected payoff to P_1 is

$$3x + 5(1 - x) = 5 - 2x;$$

and if P_2 uses pure strategy [3], then the expected payoff to P_1 is

$$11x + 2(1 - x) = 2 + 9x.$$

We now plot, over the interval [0, 1], the three lines

$$y = 7 - 5x, \qquad y = 5 - 2x, \qquad y = 2 + 9x,$$

indicating them, respectively, by [1], [2], and [3]; in this way we obtain Fig. 2. For each choice of x by P_1, he can be certain of getting at least

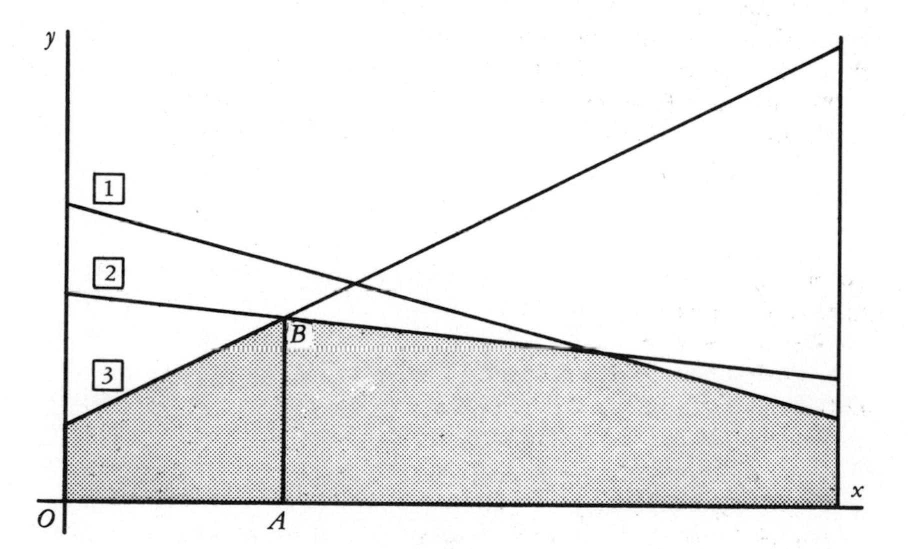

Fig. 2

the minimum of the ordinates of the three lines at x. Thus for P_1 to choose an optimal x means for him to choose an x which will make the minimum of the three ordinates as large as possible; hence, from the figure, it is apparent that the optimal x will be the segment OA and that the value of the game is AB. We can therefore find an optimal strategy for P_1 (in this game, moreover, we see from the figure that there is only one optimal strategy for P_1) and the value of the game by solving simultaneously the equations

$$y = 5 - 2x, \qquad y = 2 + 9x.$$

By doing this, we find that the optimal strategy for P_1 is $\| 3/11 \quad 8/11 \|$ and that the value of the game is $49/11$. From the figure, moreover, it is clear that no optimal mixed strategy for P_2 will contain his strategy $\boxed{1}$; hence we can find an optimal strategy for P_2 by means of the matrix

$$\begin{Vmatrix} 3 & 11 \\ 5 & 2 \end{Vmatrix},$$

so that an optimal strategy for P_2 is $\| 0 \quad 9/11 \quad 2/11 \|$.

REMARK 2.21. We notice that the value of the game in the above example is found as follows: we take the maximum ordinate of the convex set which is bounded above by the various lines. The same method is used for any $2 \times n$ game. In the case of an $m \times 2$ game, the method of graphing is of course similar, but now the value of the game is the minimum ordinate of the convex set which is bounded below by the various lines.

We turn now to an example where P_1 has many optimal strategies.

EXAMPLE 2.22. Consider the game whose payoff matrix is

$$\begin{Vmatrix} 2 & 4 & 11 \\ 7 & 4 & 2 \end{Vmatrix}.$$

Numbering the strategies as in Example 2.20 and plotting the appropriate lines as in that example, we obtain Fig. 3.

Here it is seen immediately that the value of the game is 4 and that any x will be optimal for P_1, so long as it satisfies $OA_1 \leq x \leq OA_2$. We find OA_1 to be $2/9$ by solving simultaneously the equations of $\boxed{2}$ and $\boxed{3}$; and we find OA_2 to be $3/5$ by solving simultaneously the equations of $\boxed{1}$ and $\boxed{2}$. Thus an optimal strategy for P_1 is any couple $\| x \quad 1 - x \|$ where $2/9 \leq x \leq 3/5$. An optimal strategy for P_2 in this game is $\| 0 \quad 1 \quad 0 \|$.

REMARK 2.23. In Example 2.22 we found that the set of optimal mixed strategies for P_1 consisted of the points of a straight line-segment; it is clear that this will always be the case for a $2 \times n$ game (though, of course, the segment may collapse to a single point, as in Example 2.20). It seems reasonable to suppose, moreover, that for the case of an $m \times n$ game, the set of optimal strategies for each player will still be the natural generalization of a line-segment to higher dimensional space, namely, the convex hull of a finite set of points (a "hyperpolyhedron"). A detailed proof that this is the case will be found in the next chapter.

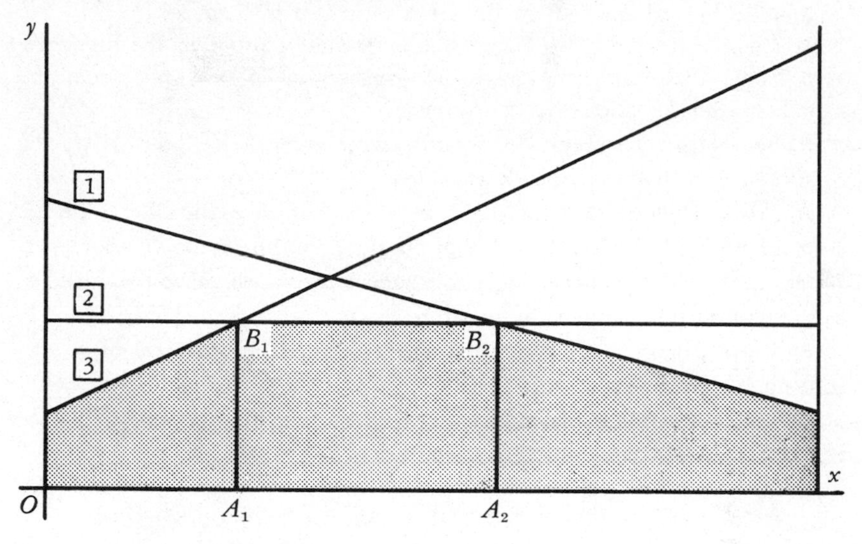

Fig. 3

HISTORICAL AND BIBLIOGRAPHICAL REMARKS

The earliest proof of the fundamental theorem (our Theorem 2.6) was given in von Neumann [1]; this proof depended on the use of the Brouwer Fixed-Point Theorem of topology. Later, Ville [1] gave an elementary proof. And in von Neumann and Morgenstern [1] a simple proof was given, depending on the theory of convex sets; the proof given above is essentially that of von Neumann and Morgenstern. Other proofs of this theorem, or generalizations of it, are to be found in Bohnenblust and Karlin [1], Brown and von Neumann [1], and Weyl [2].

For additional information regarding the theory of convex sets, the student is referred to Glezerman and Pontryagin [1] or Bonnesen and Fenchel [1].

EXERCISES

1. Let **A** be the set of all points of $\mathbf{E}_3$ whose coordinates $\| x \quad y \quad z \|$ satisfy the conditions

$$x^2 + y^2 + z^2 < 4, \qquad x - y + z \geq 0.$$

Find the closure of **A**, the interior of **A**, and the boundary of **A**.

2. Show that every convex set is connected. Give an example of a connected set which is not convex.

3. Find the extremal set of the set **A** described in Exercise 1.

4. Construct a proof of Theorem 2.1, or find a proof in the literature.

5. Show that Theorem 2.3 is no longer true if we omit from the hypothesis the condition that **X** is convex.

6. Show that Theorem 2.3 is no longer true if we omit from the hypothesis the condition that **X** is closed.

7. Prove Theorem 2.3 for the case where $n = 2$. (Hint: Let y be the point of **X** which is closest to x, and consider the line through x and perpendicular to the line connecting x to y.) Now try to generalize this proof so as to establish the theorem for an arbitrary n.

8. Prove Theorem 2.4 for the special case that the boundary of **X** contains no straight line-segments.

9. Find the values of the rectangular games whose matrices are as follows, and find optimal strategies for the two players:

$$\text{(a)} \qquad \begin{Vmatrix} 0 & 1 & 2 \\ 2 & 0 & 1 \\ 1 & 2 & 0 \end{Vmatrix},$$

$$\text{(b)} \qquad \begin{Vmatrix} 0 & 5 & -4 \\ 3 & 9 & -6 \\ 3 & -1 & 2 \end{Vmatrix},$$

$$\text{(c)} \qquad \begin{Vmatrix} 1 & -1 & 2 \\ 2 & 3 & 1 \end{Vmatrix}.$$

10. Given the $m \times m$ rectangular game whose matrix is

$$\mathsf{A} = \begin{Vmatrix} a_{11} & \cdots & a_{1m} \\ \vdots & & \vdots \\ a_{m1} & \cdots & a_{mm} \end{Vmatrix},$$

where

$$a_{ij} = 1 \qquad \text{if } i \neq j,$$

$$a_{ii} = -1,$$

show by means of Theorem 2.9 that an optimal way for each player to play is to choose each of the numbers from 1 to m with equal probability, i.e., to play the mixed strategy $\| 1/m \quad 1/m \quad \cdots \quad 1/m \|$, and show that

$$v = \frac{m-2}{m}.$$

11. A square matrix

$$A = \begin{Vmatrix} a_{11} & \cdots & a_{1m} \\ \vdots & & \vdots \\ a_{m1} & \cdots & a_{mm} \end{Vmatrix}$$

is called *skew symmetric* if $a_{ij} = -a_{ji}$ (for $i = 1, \cdots, m$ and $j = 1, \cdots, m$), so that, in particular, $a_{ii} = -a_{ii}$ (for $i = 1, \cdots, m$), and hence $a_{ii} = 0$. Show that the value of a rectangular game with a skew symmetric matrix is 0, and that if $\| X^* \quad Y^* \|$ is a strategic saddle-point of such a game, then so is $\| Y^* \quad X^* \|$.

12. Prove Theorem 2.10.

13. In Example 2.15 we found that an optimal strategy for P_1 is $\| 0 \quad 0 \quad 1 \|$; i.e., the best thing he can do is always to choose the number 3. Since the first two elements of the last row of the matrix are equal, it might seem that it would be immaterial what mixed strategy $\| y_1 \quad y_2 \quad y_3 \|$ was employed by P_2, so long as he took $y_3 = 0$. Why is this not the case?

14. Give an example to show that

$$E(X^*, Y^*) = \max_{X \in S_m} \min_{Y \in S_n} E(X, Y) = \min_{Y \in S_n} \max_{X \in S_m} E(X, Y)$$

is not a sufficient condition in order that $\| X^* \quad Y^* \|$ be a solution of a rectangular game.

15. The matrix of a certain rectangular game is

$$A = \begin{Vmatrix} a_{11} & \cdots & a_{1n} \\ \vdots & & \vdots \\ a_{m1} & \cdots & a_{mm} \end{Vmatrix},$$

and the matrix of a second game is

$$A = \begin{Vmatrix} ka_{11} & \cdots & ka_{1n} \\ \vdots & & \vdots \\ ka_{m1} & \cdots & ka_{mn} \end{Vmatrix},$$

where k is a positive constant. Show that the two games have the same optimal strategies, and that if v_1 is the value of the first game and v_2 is the value of the second game, then

$$v_2 = kv_1.$$

16. Show that the statement made in Exercise 15 is no longer true if we omit the condition that k is positive.

17. An $m \times m$ matrix is called a *latin square* if each row and each column contains each of the integers from 1 to m, e. g.,

$$\begin{Vmatrix} 1 & 3 & 2 & 4 \\ 2 & 4 & 3 & 1 \\ 3 & 1 & 4 & 2 \\ 4 & 2 & 1 & 3 \end{Vmatrix}.$$

Show that an $m \times m$ game whose matrix is a latin square has the value $(m + 1)/2$.

18. Show that the value of a game is unique.

19. Show that $\|0 \quad 0 \quad 5/12 \quad 0 \quad 4/12 \quad 0 \quad 3/12 \quad 0 \quad 0\|$ is an optimal strategy for each player in three-finger Morra (as described in Exercise 17 of Chap. 1).

20. Find, by making use of the notion of dominance, a solution of the rectangular game whose matrix is as follows:

$$\begin{Vmatrix} 0 & 0 & 0 & 0 & 0 & 0 \\ 4 & 2 & 0 & 2 & 1 & 1 \\ 4 & 3 & 1 & 3 & 2 & 2 \\ 4 & 3 & 7 & -5 & 1 & 2 \\ 4 & 3 & 4 & -1 & 2 & 2 \\ 4 & 3 & 3 & -2 & 2 & 2 \end{Vmatrix}.$$

21. Make use of the graphical method explained in Sec. 5 in order to find a solution of the rectangular game whose matrix is as follows:

$$\begin{Vmatrix} 19 & 15 & 17 & 16 \\ 0 & 20 & 15 & 5 \end{Vmatrix}.$$

22. Solve graphically the game whose payoff matrix is

$$\begin{Vmatrix} 2 & 7 \\ 3 & 5 \\ 11 & 2 \end{Vmatrix}.$$

How is the graph for this game related to the graph of Example 2.20? (In this connection, see Remark 2.21.)

CHAPTER 3

THE SOLUTIONS OF A RECTANGULAR GAME

1. The Set of Solutions. In this chapter we shall study the set of all solutions of an arbitrary rectangular game Γ; since a couple $\| X \quad Y \|$ is a solution of Γ if, and only if, X and Y are optimal strategies for the first and second player, respectively, it will suffice to study the set $\mathsf{T}_1(\Gamma)$ of optimal strategies for P_1 and the set $\mathsf{T}_2(\Gamma)$ of optimal strategies for P_2. We are going to see that if Γ is an $m \times n$ game, then $\mathsf{T}_1(\Gamma)$ is simply the convex hull of a certain finite set of points of m-space and, similarly, that $\mathsf{T}_2(\Gamma)$ is the convex hull of a certain finite set of points of n-space. Thus $\mathsf{T}_1(\Gamma)$ and $\mathsf{T}_2(\Gamma)$ have a simple geometrical characterization—they are hyperpolyhedra.

The proof of the above will lead us also to a general method for finding all the solutions of a given rectangular game—a method which is shorter, and better adapted to machine computation, than the method indicated at the end of Chap. 2.

REMARK 3.1. Some of the proofs of this chapter will probably be found to be more difficult than those in the rest of the book, especially for students who are not very familiar with matrix algebra. It may be convenient to omit, at least on the first reading, the proofs of Lemma 3.5 and Theorem 3.6. It is also possible simply to leave out Secs. 2 and 3 altogether, since nothing in the remainder of the text depends on them.

LEMMA 3.2. If Γ is an $m \times n$ rectangular game, then $\mathsf{T}_1(\Gamma)$ and $\mathsf{T}_2(\Gamma)$ are nonempty, bounded, convex, and closed subsets of m-space and n-space, respectively.

PROOF. We prove Lemma 3.2 only for $\mathsf{T}_1(\Gamma)$; the proof for $\mathsf{T}_2(\Gamma)$ is similar.

By Theorem 2.3, $\mathsf{T}_1(\Gamma)$ is nonempty.

Since $\mathsf{T}_1(\Gamma)$ is a subset of S_m, in order to show that $\mathsf{T}_1(\Gamma)$ is bounded, it suffices to show that S_m is bounded. But this is obvious, since every member of S_m is a vector $\| x_1 \quad \cdots \quad x_m \|$ whose components satisfy

$$x_1 + \cdots + x_m = 1,$$

with

$$x_i \geq 0 \qquad \text{for } i = 1, \cdots, m,$$

and hence

$$x_1^2 + \cdots + x_m^2 \leq 1,$$

so that S_m lies in the hypersphere of radius 1 whose center is at the origin.

Let the payoff matrix of Γ be

$$\mathrm{A} = \begin{Vmatrix} a_{11} & \cdots & a_{1n} \\ \vdots & & \vdots \\ a_{m1} & \cdots & a_{mn} \end{Vmatrix},$$

and let the expectation function of Γ be E, so that if $X = \| x_1 \quad \cdots \quad x_m \| \in \mathsf{S}_m$ and $Y = \| y_1 \quad \cdots \quad y_n \| \in \mathsf{S}_n$; then

$$E(X, Y) = \sum_{j=1}^{n} \sum_{i=1}^{m} a_{ij} x_i y_j. \tag{1}$$

Let v be the value of the game.

To see that $\mathsf{T}_1(\Gamma)$ is convex, let $X^{(1)}, \cdots, X^{(r)}$ be any members of $\mathsf{T}_1(\Gamma)$, let $\alpha = \| \alpha_1 \quad \cdots \quad \alpha_r \|$ be any member of S_r, and let

$$X^* = \alpha_1 X^{(1)} + \cdots + \alpha_r X^{(r)}; \tag{2}$$

we are to show that $X^* \in \mathsf{T}_1(\Gamma)$. Since $X^{(1)}, \cdots, X^{(r)}$ are all in $\mathsf{T}_1(\Gamma)$, we see by Theorem 2.8 that, for every member Y of S_n,

$$v \leq E(X^{(i)}, Y) \qquad \text{for } i = 1, \cdots, r. \tag{3}$$

Since $\alpha_i \geq 0$, we conclude from (3) that

$$\alpha_i v \leq \alpha_i E(X^{(i)}, Y) \qquad \text{for } i = 1, \cdots, r,$$

and hence that

$$\alpha_1 v + \cdots + \alpha_r v \leq \alpha_1 E(X^{(1)}, Y) + \cdots + \alpha_r E(X^{(r)}, Y),$$

or

$$v \leq \alpha_1 E(X^{(1)}, Y) + \cdots + \alpha_r E(X^{(r)}, Y). \tag{4}$$

From (1) and (2), however, we see that

$$\begin{aligned} \alpha_1 E(X^{(1)}, Y) + \cdots + \alpha_r E(X^{(r)}, Y) &= E(\alpha_1 X^{(1)} + \cdots + \alpha_r X^{(r)}, Y) \\ &= E(X^*, Y); \end{aligned} \tag{5}$$

and, from (4) and (5), it follows that

$$v \leq E(X^*, Y). \tag{6}$$

Since (6) holds for every member Y of S_n, it follows from Theorem 2.8 that X^* belongs to $\mathsf{T}_1(\Gamma)$, as was to be shown.

To prove that $\mathsf{T}_1(\Gamma)$ is closed, let $X^{(1)}, X^{(2)}, \cdots$ be a sequence of members of $\mathsf{T}_1(\Gamma)$, which converges to the vector X^*. Since S_m is closed, it is clear that $X^* \in \mathsf{S}_m$; we are to show that $X^* \in \mathsf{T}_1(\Gamma)$. Since $X^{(i)} \in \mathsf{T}_1(\Gamma)$, we see by Theorem 2.8 that, for every member Y of S_n,

$$v \leq E(X^{(i)}, Y). \tag{7}$$

However, since $E(X, Y)$ is a linear (and hence a continuous) function of the components of X, we conclude from (7) that

$$v \leq E(X^*, Y),$$

so that, again by Theorem 2.8, $X^* \in \mathsf{T}_1(\Gamma)$, as was to be shown.

REMARK 3.3. From the convexity of $\mathsf{T}_1(\Gamma)$, it follows that if $\mathsf{T}_1(\Gamma)$ has more than one member, then it has infinitely many members; and similarly for $\mathsf{T}_2(\Gamma)$. Thus a rectangular game has either just one solution, or infinitely many solutions.

From Lemma 3.2, we see that the sets $\mathsf{T}_1(\Gamma)$ and $\mathsf{T}_2(\Gamma)$ satisfy the hypothesis of Theorem 2.4. Thus the set $\mathsf{K}[\mathsf{T}_i(\Gamma)]$ (for $i = 1, 2$) is non-empty, and every member of $\mathsf{T}_i(\Gamma)$ is a convex linear combination of members of $\mathsf{K}[\mathsf{T}_i(\Gamma)]$. Hence, in order to find all members of $\mathsf{T}_i(\Gamma)$, it suffices to find all members of $\mathsf{K}[\mathsf{T}_i(\Gamma)]$. We are going to show later in the chapter that $\mathsf{K}[\mathsf{T}_i(\Gamma)]$ is a finite set.

2. Some Properties of Matrices. In the statements and proofs of the next two lemmas we shall make extensive use of the theory of matrices. By I_n, we shall mean the $n \times n$ matrix, each element of whose main diagonal is 1, and all of whose other elements are 0; thus

$$\mathrm{I}_1 = \begin{Vmatrix} 1 \end{Vmatrix},$$

$$\mathrm{I}_2 = \begin{Vmatrix} 1 & 0 \\ 0 & 1 \end{Vmatrix},$$

$$\mathrm{I}_3 = \begin{Vmatrix} 1 & 0 & 0 \\ 0 & 1 & 0 \\ 0 & 0 & 1 \end{Vmatrix}.$$

By J_n, we shall mean the vector (i.e., the row matrix) which has n elements, each of which is 1; thus

$$J_1 = \| 1 \|,$$
$$J_2 = \| 1 \quad 1 \|,$$
$$J_3 = \| 1 \quad 1 \quad 1 \|.$$

By O_n, we shall mean the vector which has n elements, each of which is 0; thus

$$O_1 = \| 0 \|,$$
$$O_2 = \| 0 \quad 0 \|,$$
$$O_3 = \| 0 \quad 0 \quad 0 \|.$$

By A^T, we mean the transpose of the matrix A; thus if

$$A = \begin{Vmatrix} 1 & 2 & 3 \\ 4 & 5 & 6 \end{Vmatrix},$$

then

$$A^T = \begin{Vmatrix} 1 & 4 \\ 2 & 5 \\ 3 & 6 \end{Vmatrix}.$$

For each n, we have

$$I_n^T = I_n;$$

but J_n^T is a column matrix:

$$J_1^T = \| 1 \|,$$

$$J_2^T = \begin{Vmatrix} 1 \\ 1 \end{Vmatrix},$$

$$J_3^T = \begin{Vmatrix} 1 \\ 1 \\ 1 \end{Vmatrix}.$$

If A is a matrix and k is a number, we denote the (scalar) product of A and k by Ak, or kA; thus

$$3 \begin{Vmatrix} 1 & 2 \\ 3 & -2 \\ 7 & 0 \end{Vmatrix} = \begin{Vmatrix} 1 & 2 \\ 3 & -2 \\ 7 & 0 \end{Vmatrix} 3 = \begin{Vmatrix} 3 & 6 \\ 9 & -6 \\ 21 & 0 \end{Vmatrix}.$$

We denote the product of the matrices A and B by $A \cdot B$, or simply AB;

thus if

$$A = \begin{Vmatrix} 1 & 2 & 3 \\ 1 & 3 & 4 \end{Vmatrix},$$

$$B = \begin{Vmatrix} 1 & 2 \\ 0 & 1 \\ 5 & 6 \end{Vmatrix},$$

then

$$A \cdot B = AB = \begin{Vmatrix} 1 \cdot 1 + 2 \cdot 0 + 3 \cdot 5 & 1 \cdot 2 + 2 \cdot 1 + 3 \cdot 6 \\ 1 \cdot 1 + 3 \cdot 0 + 4 \cdot 5 & 1 \cdot 2 + 3 \cdot 1 + 4 \cdot 6 \end{Vmatrix}$$

$$= \begin{Vmatrix} 16 & 22 \\ 21 & 29 \end{Vmatrix}.$$

If A is a square matrix of order n, then by A^{-1} we shall mean the matrix (if it exists) such that

$$A \cdot A^{-1} = A^{-1} \cdot A = I_n;$$

A^{-1} is called the *inverse* of A. Thus the inverse of

$$\begin{Vmatrix} 1 & 2 \\ 3 & 7 \end{Vmatrix}$$

is

$$\begin{Vmatrix} 7 & -2 \\ -3 & 1 \end{Vmatrix}.$$

A matrix A has an inverse if, and only if, it is nonsingular, i.e., if, and only if, the determinant of A, which we indicate by $|A|$, is different from 0. If A is a matrix having only one element a, and if $a \neq 0$, then A^{-1} is the matrix whose only element is $1/a$. Thus

$$\| 10 \|^{-1} = \left\| \frac{1}{10} \right\|.$$

The matrix $\| 0 \|$, of course, does not possess an inverse.

If $\| a \|$ is a matrix having only one element, we sometimes omit the double vertical bars indicative of a matrix and write simply

$$a$$

instead of

$$\| a \|.$$

If A is a square matrix, we indicate the *adjoint* of A by adj A. The adjoint is obtained by replacing the element in the ith row and the jth column of A by the cofactor of the element in the jth row and the ith column. Thus

$$\operatorname{adj}\begin{Vmatrix} 1 & 2 & 3 \\ 2 & 4 & 6 \\ 1 & 1 & 1 \end{Vmatrix} = \begin{Vmatrix} -2 & 1 & 0 \\ 4 & -2 & 0 \\ -2 & 1 & 0 \end{Vmatrix}.$$

If A is a matrix having only one element a, then (regardless of what the value of a is) we set adj A $= \| 1 \|$; thus

$$\operatorname{adj} \| 1 \| = \operatorname{adj} \| -3 \| = \operatorname{adj} \| 0 \| = \| 1 \| = 1.$$

It will be noticed that the adjoint of every square matrix A exists; if A is nonsingular (so that A^{-1} also exists), we have

$$A^{-1} | A | = \operatorname{adj} A,$$

where $| A |$ is the determinant of A.

If A and B are two matrices having the same number of rows and the same number of columns, we indicate the sum of A and B by A + B, and the difference between A and B by A − B. Thus

$$\begin{Vmatrix} 1 & 2 & 3 \\ 4 & 5 & 6 \end{Vmatrix} + \begin{Vmatrix} 0 & -2 & 3 \\ 4 & -6 & 0 \end{Vmatrix} = \begin{Vmatrix} 1+0 & 2-2 & 3+3 \\ 4+4 & 5-6 & 6+0 \end{Vmatrix}$$

$$= \begin{Vmatrix} 1 & 0 & 6 \\ 8 & -1 & 6 \end{Vmatrix},$$

and

$$\begin{Vmatrix} 1 & 2 & 3 \\ 4 & 5 & 6 \end{Vmatrix} - \begin{Vmatrix} 0 & -2 & 3 \\ 4 & -6 & 0 \end{Vmatrix} = \begin{Vmatrix} 1-0 & 2-(-2) & 3-3 \\ 4-4 & 5-(-6) & 6-0 \end{Vmatrix}$$

$$= \begin{Vmatrix} 1 & 4 & 0 \\ 0 & 11 & 6 \end{Vmatrix}.$$

Multiplication of matrices is distributive with respect to addition and subtraction, so that, for all A, B, and C,

$$A \cdot (B + C) = A \cdot B + A \cdot C,$$

and

$$A \cdot (B - C) = A \cdot B - A \cdot C.$$

By a *submatrix* of a matrix A, we mean a matrix B which is identical with A, or can be obtained from A by crossing out certain rows and columns. Thus the matrix

$$\begin{Vmatrix} 1 & 2 & 3 \\ 4 & 5 & 6 \end{Vmatrix}$$

has the following submatrices:

$$\begin{Vmatrix} 1 & 2 & 3 \\ 4 & 5 & 6 \end{Vmatrix},$$

$$\begin{Vmatrix} 1 & 2 \\ 4 & 5 \end{Vmatrix}, \quad \begin{Vmatrix} 1 & 3 \\ 4 & 6 \end{Vmatrix}, \quad \begin{Vmatrix} 2 & 3 \\ 5 & 6 \end{Vmatrix},$$

$$\begin{Vmatrix} 1 & 2 & 3 \end{Vmatrix}, \quad \begin{Vmatrix} 4 & 5 & 6 \end{Vmatrix},$$

$$\begin{Vmatrix} 1 \\ 4 \end{Vmatrix}, \quad \begin{Vmatrix} 2 \\ 5 \end{Vmatrix}, \quad \begin{Vmatrix} 3 \\ 6 \end{Vmatrix},$$

$$\begin{Vmatrix} 1 & 2 \end{Vmatrix}, \quad \begin{Vmatrix} 1 & 3 \end{Vmatrix}, \quad \begin{Vmatrix} 2 & 3 \end{Vmatrix}, \quad \begin{Vmatrix} 4 & 5 \end{Vmatrix}, \quad \begin{Vmatrix} 4 & 6 \end{Vmatrix}, \quad \begin{Vmatrix} 5 & 6 \end{Vmatrix},$$

$$\begin{Vmatrix} 1 \end{Vmatrix}, \quad \begin{Vmatrix} 2 \end{Vmatrix}, \quad \begin{Vmatrix} 3 \end{Vmatrix}, \quad \begin{Vmatrix} 4 \end{Vmatrix}, \quad \begin{Vmatrix} 5 \end{Vmatrix}, \quad \begin{Vmatrix} 6 \end{Vmatrix}.$$

In order to simplify one of our later proofs, it is convenient here to establish a special result regarding matrices.

LEMMA 3.4. Let

$$A = \begin{Vmatrix} a_{11} & \cdots & a_{1r} \\ \vdots & & \vdots \\ a_{r1} & \cdots & a_{rr} \end{Vmatrix}$$

be a square matrix of order r, and let

$$A_x = \begin{Vmatrix} a_{11} + x & \cdots & a_{1r} + x \\ \vdots & & \vdots \\ a_{r1} + x & \cdots & a_{rr} + x \end{Vmatrix},$$

where x is any real number, be the matrix obtained from A by adding x to each of its elements. Then

$$J_r \operatorname{adj} A_x = J_r \operatorname{adj} A; \tag{8}$$

$$|A_x| = |A| + xJ_r \operatorname{adj} A\, J_r^T. \tag{9}$$

PROOF. The equation

$$\begin{vmatrix} a_{11}+b_1 & a_{12} & \cdots & a_{1r} \\ a_{21}+b_2 & a_{22} & \cdots & a_{2r} \\ \vdots & \vdots & & \vdots \\ a_{r1}+b_r & a_{r2} & \cdots & a_{rr} \end{vmatrix} =$$

$$\begin{vmatrix} a_{11} & a_{12} & \cdots & a_{1r} \\ a_{21} & a_{22} & \cdots & a_{2r} \\ \vdots & \vdots & & \vdots \\ a_{r1} & a_{r2} & \cdots & a_{rr} \end{vmatrix} + \begin{vmatrix} b_1 & a_{12} & \cdots & a_{1r} \\ b_2 & a_{22} & \cdots & a_{2r} \\ \vdots & \vdots & & \vdots \\ b_r & a_{r2} & \cdots & a_{rr} \end{vmatrix} \tag{10}$$

is easily seen to be an identity; and the analogous identity, of course, holds with respect to an arbitrary row (or column).

The following equation is seen to be true by multiplying the first row of the left member by x, and subtracting from each of the other rows:

$$\begin{vmatrix} 1 & 1 & \cdots & 1 \\ a_{21}+x & a_{22}+x & \cdots & a_{2r}+x \\ \vdots & \vdots & & \vdots \\ a_{r1}+x & a_{r2}+x & \cdots & a_{rr}+x \end{vmatrix} = \begin{vmatrix} 1 & 1 & \cdots & 1 \\ a_{21} & a_{22} & \cdots & a_{2r} \\ \vdots & \vdots & & \vdots \\ a_{r1} & a_{r2} & \cdots & a_{rr} \end{vmatrix}. \tag{11}$$

The analogous equation, of course, also holds for an arbitrary row (or column).

Since it is easily verified, however, that

$$J_r \operatorname{adj} A = \left\| \begin{vmatrix} 1 & 1 & \cdots & 1 \\ a_{21} & a_{22} & \cdots & a_{2r} \\ \vdots & \vdots & & \vdots \\ a_{r1} & a_{r2} & \cdots & a_{rr} \end{vmatrix} \cdots \begin{vmatrix} a_{11} & \cdots & a_{1r} \\ \vdots & & \vdots \\ a_{r-1,1} & \cdots & a_{r-1,r} \\ 1 & \cdots & 1 \end{vmatrix} \right\|, \tag{12}$$

we see that (8) of our lemma is an immediate consequence of (11).

To establish (9), we notice that (10) implies that

$$|A_x| = |A| + \begin{vmatrix} x & x & \cdots & x \\ a_{21} & a_{22} & \cdots & a_{2r} \\ \vdots & \vdots & & \vdots \\ a_{r1} & a_{r2} & \cdots & a_{rr} \end{vmatrix} + \cdots + \begin{vmatrix} a_{11} & \cdots & a_{1r} \\ \vdots & & \vdots \\ a_{r-1,1} & \cdots & a_{r-1,r} \\ x & \cdots & x \end{vmatrix}$$

$$= |A| + x\left[\begin{vmatrix} 1 & 1 & \cdots & 1 \\ a_{21} & a_{22} & \cdots & a_{2r} \\ \vdots & \vdots & & \vdots \\ a_{r1} & a_{r2} & \cdots & a_{rr} \end{vmatrix} + \cdots + \begin{vmatrix} a_{11} & \cdots & a_{1r} \\ \vdots & & \vdots \\ a_{r-1,1} & \cdots & a_{r-1,r} \\ 1 & \cdots & 1 \end{vmatrix}\right], \tag{13}$$

and from (12) we see that the expression in square brackets in (13) is equal to $J_r \operatorname{adj} A \, J_r^T$.

3. The Determination of All Solutions.

LEMMA 3.5. Let Γ be a rectangular game whose payoff matrix is A, suppose that $v(\Gamma)$, the value of the game, is different from 0, and let $X \in \mathsf{T}_1(\Gamma)$ and $Y \in \mathsf{T}_2(\Gamma)$. Then a necessary and sufficient condition that $X \in \mathsf{K}[\mathsf{T}_1(\Gamma)]$ and $Y \in \mathsf{K}[\mathsf{T}_2(\Gamma)]$ is that there exist a nonsingular submatrix B of A such that

$$v = \frac{1}{J_r B^{-1} J_r^T}, \tag{14}$$

$$\dot{X} = \frac{J_r B^{-1}}{J_r B^{-1} J_r^T}, \tag{15}$$

$$\dot{Y} = \frac{J_r (B^{-1})^T}{J_r B^{-1} J_r^T}, \tag{16}$$

where r is the order of B, $\dot{X}$ is the vector obtained from X by deleting the elements corresponding to the rows deleted to obtain B from A, and $\dot{Y}$ is the vector obtained from Y by deleting the elements corresponding to the columns deleted to obtain B from A.

PROOF. Let

$$A = \begin{Vmatrix} a_{11} & \cdots & a_{1n} \\ \vdots & & \vdots \\ a_{m1} & \cdots & a_{mn} \end{Vmatrix}.$$

To see that the condition is sufficient, suppose, if possible, that there is a nonsingular matrix B satisfying conditions (14), (15), and (16) and that $X \in \mathsf{K}[\mathsf{T}_1(\Gamma)]$ and $Y \in \mathsf{K}[\mathsf{T}_2(\Gamma)]$ are not both true, so that either $X \notin \mathsf{K}[\mathsf{T}_1(\Gamma)]$ or $Y \notin \mathsf{K}[\mathsf{T}_2(\Gamma)]$. We shall show that the supposition that $X \notin \mathsf{K}[\mathsf{T}_1(\Gamma)]$ leads to a contradiction; the proof is similar if we suppose that $Y \notin \mathsf{K}[\mathsf{T}_2(\Gamma)]$.

Since the problem of solving a game is not essentially affected by interchanging rows or interchanging columns, we can suppose without loss of generality that B is situated in the upper left-hand corner of A, i.e., that

$$\mathrm{B} = \begin{Vmatrix} a_{11} & \cdots & a_{1r} \\ \vdots & & \vdots \\ a_{r1} & \cdots & a_{rr} \end{Vmatrix}.$$

Thus, letting $X = \| x_1 \quad \cdots \quad x_m \|$, we see that $\dot{X} = \| x_1 \quad \cdots \quad x_r \|$. Now, since $X \in \mathsf{S}_m$, we have $x_i \geq 0$ (for $i = 1, \cdots, m$, and hence, a fortiori, for $i = 1, \cdots, r$). Moreover, from (15) we conclude that

$$\sum_{i=1}^{r} x_i = \dot{X} \cdot J_r^T = \frac{J_r \mathrm{B}^{-1} J_r^T}{J_r \mathrm{B}^{-1} J_r^T} = 1.$$

Hence

$$\dot{X} \in \mathsf{S}_r \tag{17}$$

and

$$x_i = 0 \qquad \text{for } r < i \leq m. \tag{18}$$

Since we are supposing that $X \notin \mathsf{K}[\mathsf{T}_1(\Gamma)]$, there are distinct elements $U = \| u_1 \quad \cdots \quad u_m \|$ and $W = \| w_1 \quad \cdots \quad w_m \|$ of $\mathsf{T}_1(\Gamma)$ such that

$$X = \frac{1}{2}(U + W),$$

that is to say, such that

$$x_i = \frac{1}{2}(u_i + w_i) \qquad \text{for } i = 1, \cdots, m. \tag{19}$$

From (17), (18), and (19), we see that

$$u_i = w_i = 0 \qquad \text{for } r < i \leq m. \tag{20}$$

Hence we obtain

$$u_1 a_{1j} + \cdots + u_r a_{rj} = u_1 a_{1j} + \cdots + u_m a_{mj} \qquad \text{for } j = 1, \cdots, m. \tag{21}$$

Since $\| u_1 \quad \cdots \quad u_m \| \in \mathsf{T}_1(\Gamma)$, however,

$$u_1 a_{1j} + \cdots + u_m a_{mj} \geq v,$$

and hence, from (21),

$$u_1 a_{1j} + \cdots + u_r a_{rj} \geq v. \tag{22}$$

Similarly, it can be concluded that

$$w_1 a_{1j} + \cdots + w_r a_{rj} \geq v. \tag{23}$$

From (14) and (15), we have

$$\dot{X}B = \frac{J_r B^{-1} B}{J_r B^{-1} J_r^T} = \frac{J_r}{J_r B^{-1} J_r} = vJ_r,$$

and hence

$$x_1 a_{1j} + \cdots + x_r a_{rj} = v \qquad \text{for } j = 1, \cdots, r,$$

so that, making use of (19),

$$\frac{1}{2}(u_1 + w_1)a_{1j} + \cdots + \frac{1}{2}(u_r + w_r)a_{rj} = v,$$

or

$$[u_1 a_{1j} + \cdots + u_r a_{rj}] + [w_1 a_{1j} + \cdots + w_r a_{rj}] = 2v. \tag{24}$$

From (22), (23), and (24), we then have

$$[u_1 a_{1j} + \cdots + u_r a_{rj}] = [w_1 a_{1j} + \cdots + w_r a_{rj}] = v \qquad \text{for } j = 1, \cdots, r, \tag{25}$$

and hence

$$U \cdot B = W \cdot B,$$

which implies that

$$U \cdot B - W \cdot B = O_r,$$

and hence that

$$(U - W) \cdot B = O_r. \tag{26}$$

Since we have supposed, however, that U and W are distinct vectors, we see that

$$U - W \neq O_r; \tag{27}$$

and, from (26) and (27), it follows that B is a singular matrix, contrary to hypothesis.

It remains to be shown that the condition is also necessary. Suppose, then, that $X \in K[T_1(\Gamma)]$ and $Y \in K[T_2(\Gamma)]$; we shall construct a nonsingular matrix B which will satisfy (14), (15), and (16).

Without loss of generality, we can suppose that the rows of A are arranged in such a way that

$$x_i \neq 0 \qquad \text{for } i \leq m', \tag{28}$$

and

$$x_i = 0 \qquad \text{for } i > m'; \tag{29}$$

and, similarly, that the columns of A are arranged in such a way that

$$y_j \neq 0 \qquad \text{for } j \leq n', \tag{30}$$

and

$$y_j = 0 \qquad \text{for } j > n'. \tag{31}$$

By Theorem 2.9, we then conclude that

$$a_{1j}x_1 + \cdots + a_{mj}x_m = v \qquad \text{for } j = 1, \cdots, n', \tag{32}$$

and

$$a_{i1}y_1 + \cdots + a_{in}y_n = v \qquad \text{for } i = 1, \cdots, m'. \tag{33}$$

Without loss of generality, moreover, we can also suppose that the rows, and the columns, of A are arranged in such a way that, for some $n'' \geq n'$ and some $m'' \geq m'$,

$$a_{1j}x_1 + \cdots + a_{mj}x_m = v \qquad \text{for } j = 1, \cdots, n'', \tag{34}$$

$$a_{1j}x_i + \cdots + a_{mj}x_m > v \qquad \text{for } j > n'', \tag{35}$$

$$a_{i1}y_1 + \cdots + a_{in}y_n = v \qquad \text{for } i = 1, \cdots, m'', \tag{36}$$

$$a_{i1}y_1 + \cdots + a_{in}y_n < v \qquad \text{for } i > m''. \tag{37}$$

If j is any integer in the set $\{1, \cdots, n\}$, let D_j be the vector $\| a_{1j} \quad \cdots \quad a_{m'j} \|$. We are going to define recursively some sets A_0, A_1, $\cdots$, $\mathsf{A}_{n''-n'}$ of integers. We first set

$$\mathsf{A}_0 = \{1, \cdots, n'\}.$$

Now, supposing that A_k has been defined and that

$$\mathsf{A}_k = \{k_1, \cdots, k_u\},$$

we distinguish two cases.

CASE 1. There exists an integer j in $\{1, \cdots, n''\} - \mathsf{A}_k$ such that the vector D_j is linearly independent of the vectors $D_{k_1}, \cdots, D_{k_u}$, i.e., such that there do not exist constants $c_1, \cdots, c_{k_u}$ satisfying

$$D_j = c_1 D_{k_1} + \cdots + c_{k_u} D_{k_u}.$$

CASE 2. There exists no such integer j.

In Case 1, letting j_0 be the smallest integer j in $\{1, \cdots, n''\} - \mathsf{A}_k$ such that D_j is linearly independent of $D_{k_1}, \cdots, D_{k_u}$, we set

$$\mathsf{A}_{k+1} = \mathsf{A}_k \cup \{j_0\}.$$

In Case 2, we set

$$\mathsf{A}_{k+1} = \mathsf{A}_k.$$

We now put

$$\mathsf{A} = \mathsf{A}_{n''-n'}.$$

Without loss of generality, we can suppose that the columns of A are arranged in such an order that, for some t,

$$\mathsf{A} = \{1, \cdots, n', n' + 1, \cdots, t\}.$$

Then it is easily seen that, for $n' < j \leq t$, the vector D_j is linearly independent of the vectors $D_1, \cdots, D_{j-1}, D_{j+1}, \cdots, D_t$; and, moreover, if $t < j \leq n''$, then D_j is linearly dependent on $D_1, \cdots, D_t$.

Similarly, if we set

$$C_i = \| a_{i1} \quad \cdots \quad a_{in'} \|,$$

where i is any member of $\{1, \cdots, m\}$, then, by analogous considerations, it is seen that we can suppose without loss of generality that there exists an integer s satisfying $m' \leq s \leq m''$ such that, for $m' < i \leq s$, the vector C_i is linearly independent of the vectors $C_1, \cdots, C_{i-1}, C_{i+1}, \cdots, C_s$; while if $s < i \leq m''$, then C_i is linearly dependent on $C_1, \cdots, C_s$.

We now set

$$B = \begin{Vmatrix} a_{11} & \cdots & a_{1t} \\ \vdots & & \vdots \\ a_{s1} & \cdots & a_{st} \end{Vmatrix}.$$

Now suppose, if possible, that B is singular, so that either the rows of B are linearly dependent, or the columns of B are linearly dependent. We shall treat only of the case in which the rows are linearly dependent; the proof is similar for the other case.

If i is any integer in $\{1, \cdots, m\}$, we set

$$B_i = \left\| a_{i1} \quad \cdots \quad a_{it} \right\|.$$

Since the rows of B are dependent, there are constants $c_1, \cdots, c_s$, not all of which are 0, such that

$$c_1 B_1 + \cdots + c_s B_s = O_t. \tag{38}$$

Moreover, for $i > m'$, we must have $c_i = 0$; for, otherwise, we could divide (38) through by c_i, and we could then conclude that B_i was linearly dependent on $B_1, \cdots, B_{i-1}, B_{i+1}, \cdots, B_s$ and hence (referring back to the definition of C_i and noticing that C_i is a subvector of B_i) that C_i was linearly dependent on $C_1, \cdots, C_{i-1}, C_{i+1}, \cdots, C_s$, contrary to construction. By (28) and (29), we therefore conclude that

$$c_i = 0 \qquad \text{whenever } x_i = 0 \qquad \text{for } i = 1, \cdots, t. \tag{39}$$

By (36), we have

$$\left\| y_1 \quad \cdots \quad y_n \right\| \cdot \left\| \begin{matrix} a_{i1} \\ \vdots \\ a_{in} \end{matrix} \right\| = v \qquad \text{for } i = 1, \cdots, m',$$

and hence, by (31), and the fact that $n' \leq t$,

$$\left\| y_1 \quad \cdots \quad y_t \right\| \cdot \left\| \begin{matrix} a_{i1} \\ \vdots \\ a_{it} \end{matrix} \right\| = v;$$

thus, setting $Y^* = \left\| y_1 \quad \cdots \quad y_t \right\|$,

$$Y^* \cdot (B_i)^T = v \qquad \text{for } i = 1, \cdots, m'. \tag{40}$$

Using (40) and (38), we now see that

$$\begin{aligned} v \sum_{i=1}^{s} c_i &= \sum_{i=1}^{s} v c_i = \sum_{i=1}^{s} Y^*(B_i)^T c_i \\ &= Y^* \sum_{i=1}^{s} (B_i)^T c_i = Y^* \cdot (O_t)^T = 0, \end{aligned}$$

and hence, since by hypothesis $v \neq 0$,

$$\sum_{i=1}^{s} c_i = 0. \tag{41}$$

We now set $c_i = 0$, for $s < i \leq m$, and

$$C = \| c_1 \quad \cdots \quad c_m \|. \tag{42}$$

From (39), and the definition of c_i, for $i > s$,

$$c_i = 0 \qquad \text{whenever } x_i = 0 \qquad \text{for } i = 1, \cdots, m. \tag{43}$$

For every real number α, we define a vector X_α by setting

$$X_\alpha = X + \alpha \cdot C. \tag{44}$$

Since not every component of C is 0, it is clear that $X_\alpha = X$ if, and only if, $\alpha = 0$; more generally, $X_\alpha = X_\beta$ if, and only if, $\alpha = \beta$.

From (41), it is seen that, if $X_\alpha = \| z_1 \quad \cdots \quad z_m \|$, then

$$\begin{aligned} \sum_{i=1}^{m} z_i &= \sum_{i=1}^{m} (x_i + \alpha c_i) \\ &= \sum_{i=1}^{m} x_i + \alpha \sum_{i=1}^{m} c_i = 1 + \alpha \cdot 0 = 1. \end{aligned} \tag{45}$$

Moreover, from (43), it is seen that by choosing α sufficiently small we can ensure that $z_i = x_i + \alpha c_i \geq 0$ (for $i = 1, \cdots, m$). Hence there is an ε_1 such that

$$X_\alpha \in \mathbf{S}_m \qquad \text{for } |\alpha| < \varepsilon_1. \tag{46}$$

Letting $\mathrm{A}^{(j)}$ be the jth column of A, we see by (43) and (38) that, for $j = 1, \cdots, t$,

$$\begin{aligned} C \cdot \mathrm{A}^{(j)} &= \| c_1 \quad \cdots \quad c_m \| \cdot \left\| \begin{matrix} a_{1j} \\ \vdots \\ a_{mj} \end{matrix} \right\| \\ &= c_1 a_{1j} + \cdots + c_m a_{mj} = c_1 a_{1j} + \cdots + c_s a_{sj} = 0. \end{aligned} \tag{47}$$

If $t < j \leq n'$, then, as we saw above, D_j is linearly dependent on $D_1, \cdots, D_t$. Thus there are constants $d_1, \cdots, d_t$ such that

$$D_j = d_1 D_1 + \cdots + d_t D_t.$$

Hence, using (47),

$$
\begin{aligned}
C \cdot \mathrm{A}^{(j)} &= \| c_1 \quad \cdots \quad c_m \| \left\| \begin{matrix} a_{1j} \\ \vdots \\ a_{mj} \end{matrix} \right\| \\
&= \| c_1 \quad \cdots \quad c_{m'} \| \left\| \begin{matrix} a_{1j} \\ \vdots \\ a_{m'j} \end{matrix} \right\| \\
&= \| c_1 \quad \cdots \quad c_{m'} \| \cdot D_j^T \\
&= \| c_1 \quad \cdots \quad c_{m'} \| \cdot [d_1 D_1^T + \cdots + d_t D_t^T] \\
&= \| c_1 \quad \cdots \quad c_m \| \, [d_1 \mathrm{A}^{(1)} + \cdots + d_t \mathrm{A}^{(t)}] \\
&= C[d_1 \mathrm{A}^{(1)} + \cdots + d_t \mathrm{A}^{(t)}] \\
&= d_1 C\mathrm{A}^{(1)} + \cdots + d_t C\mathrm{A}^{(t)} = 0. \qquad (48)
\end{aligned}
$$

Thus, from (47) and (48),

$$C \cdot \mathrm{A}^{(j)} = 0 \qquad \text{for } j = 1, \cdots, n'';$$

and hence we have

$$
\begin{aligned}
X_\alpha \cdot \mathrm{A}^{(j)} &= (X + \alpha \cdot C)\mathrm{A}^{(j)} \\
&= X \cdot \mathrm{A}^{(j)} + \alpha \cdot C \cdot \mathrm{A}^{(j)} = X \cdot \mathrm{A}^{(j)} = v \\
&\qquad \text{for } j = 1, \cdots, n''. \qquad (49)
\end{aligned}
$$

(Equation (49) holds for all α.)

For all $j > n''$, we have

$$X \cdot \mathrm{A}^{(j)} > v;$$

hence we can find an ε_2 such that

$$
\begin{aligned}
X_\alpha \cdot \mathrm{A}^{(j)} &= (X + \alpha C)\mathrm{A}^{(j)} \\
&= X \cdot \mathrm{A}^{(j)} + \alpha C \cdot \mathrm{A}^{(j)} > v \qquad \text{for } |\alpha| < \varepsilon_2. \qquad (50)
\end{aligned}
$$

Taking ε to be the minimum of ε_1 and ε_2, we now conclude from (46), (49), and (50) that, for $|\alpha| < \varepsilon$, we have $X_\alpha \in \mathbf{S}_m$ and $X_\alpha \cdot \mathrm{A}^{(j)} \geq v$ (for $j = 1, \cdots, n$), so that $X_\alpha \in \mathbf{T}_1(\Gamma)$. In particular, we have, setting $\beta = \frac{1}{2}\varepsilon$,

$$X_\beta \in \mathbf{T}_1(\Gamma) \qquad \text{and} \qquad X_{-\beta} \in \mathbf{T}_1(\Gamma).$$

Since

$$X = \frac{1}{2}(X_\beta + X_{-\beta}),$$

however, we then conclude that $X \notin \mathsf{K}[\mathsf{T}_1(\Gamma)]$, contrary to hypothesis. Since we have thus arrived at a contradiction, we conclude that B is actually nonsingular, as was to be shown.

To complete the proof, it remains only to show that B satisfies (14), (15), and (16). Since B is nonsingular, $s = t$. Letting $r = s = t$, then

$$\mathrm{B} = \begin{Vmatrix} a_{11} & \cdots & a_{1r} \\ \vdots & & \vdots \\ a_{r1} & \cdots & a_{rr} \end{Vmatrix}.$$

Let $\dot{X} = \| x_1 \quad \cdots \quad x_r \|$ and $\dot{Y} = \| y_1 \quad \cdots \quad y_r \|$. Then $\dot{X}$ is the vector obtained from X by deleting the elements corresponding to the rows deleted in A to obtain B; the vector $\dot{Y}$ is obtained similarly. By (34), we now see, a fortiori, that

$$a_{1j}x_1 + \cdots + a_{mj}x_m = v \qquad \text{for } j = 1, \cdots, r,$$

and hence, since $x_i = 0$, for $i > r$, that

$$a_{1j}x_1 + \cdots + a_{rj}x_r = v \qquad \text{for } j = 1, \cdots, r. \tag{51}$$

From (51), it follows that

$$\dot{X} \cdot \mathrm{B} = v \cdot J_r,$$

and hence, since B is nonsingular, that

$$\dot{X} = \dot{X} \cdot \mathrm{B} \cdot \mathrm{B}^{-1} = v \cdot J_r \cdot \mathrm{B}^{-1}. \tag{52}$$

Similarly, using (36), we conclude that

$$a_{i1}y_1 + \cdots + a_{ir}y_r = v \qquad \text{for } i = 1, \cdots, r,$$

and hence that

$$\dot{Y} \cdot \mathrm{B}^T = v \cdot J_r,$$

and hence that

$$\dot{Y} = \dot{Y}\mathrm{B}^T \cdot (\mathrm{B}^T)^{-1} = v \cdot J_r \cdot (\mathrm{B}^T)^{-1} = vJ_r(\mathrm{B}^{-1})^T. \tag{53}$$

From (52), we obtain

$$v[J_r \mathrm{B}^{-1} J_r^T] = (vJ_r) \cdot (\mathrm{B}^{-1} J_r^T) = \dot{X} \cdot \mathrm{B} \cdot \mathrm{B}^{-1} J_r^T = \dot{X} J_r^T = 1 \tag{54}$$

and hence

$$v = \frac{1}{J_r \mathrm{B}^{-1} J_r^T},$$

which is (14) of our lemma. Equations (15) and (16) follow immediately from (52) and (53) by means of (54), which completes the proof.

THEOREM 3.6. Let Γ be a rectangular game whose payoff matrix is A, and suppose that $X \in \mathsf{T}_1(\Gamma)$ and $Y \in \mathsf{T}_2(\Gamma)$. Then a necessary and sufficient condition that $X \in \mathsf{K}[\mathsf{T}_1(\Gamma)]$ and $Y \in \mathsf{K}[\mathsf{T}_2(\Gamma)]$ is that there exist a square submatrix B of A, of order r, such that $J_r(\text{adj B})J_r^T \neq 0$, and

$$v = \frac{|\mathrm{B}|}{J_r(\text{adj B})J_r^T}, \tag{55}$$

$$\dot{X} = \frac{J_r \text{ adj B}}{J_r(\text{adj B})J_r^T}, \tag{56}$$

$$\dot{Y} = \frac{J_r(\text{adj B})^T}{J_r(\text{adj B})J_r^T}, \tag{57}$$

where $\dot{X}$ is the vector obtained from X by deleting the elements corresponding to the rows deleted to obtain B from A, and $\dot{Y}$ is the vector obtained from Y by deleting the elements corresponding to the columns deleted to obtain B from A.

PROOF. In case $v(\Gamma) \neq 0$, the theorem follows easily from Lemma 3.5, if we remember that, for a nonsingular matrix B,

$$\mathrm{B}^{-1} = \frac{\text{adj B}}{|\mathrm{B}|}.$$

In case $v(\Gamma) = 0$, we define a new game Γ' with a matrix A′, by adding a number $b \neq 0$ to each element of A; thus, if

$$\mathrm{A} = \begin{Vmatrix} a_{11} & \cdots & a_{1n} \\ \vdots & & \vdots \\ a_{m1} & \cdots & a_{mn} \end{Vmatrix},$$

then we set

$$a'_{ij} = a_{ij} + b \qquad \text{for } 1 \leq i \leq m \text{ and } 1 \leq j \leq n, \tag{58}$$

and

$$A' = \begin{Vmatrix} a'_{11} & \cdots & a'_{1n} \\ \vdots & & \vdots \\ a'_{m1} & \cdots & a'_{mn} \end{Vmatrix} = \begin{Vmatrix} a_{11} + b & \cdots & a_{1n} + b \\ \vdots & & \vdots \\ a_{m1} + b & \cdots & a_{mn} + b \end{Vmatrix}.$$

Then it is easily verified, by means of Theorem 2.6, that

$$v(\Gamma') = v(\Gamma) + b = b \neq 0; \tag{59}$$

and, moreover, that

$$\mathsf{T}_1(\Gamma') = \mathsf{T}_1(\Gamma) \tag{60}$$

and

$$\mathsf{T}_2(\Gamma') = \mathsf{T}_2(\Gamma), \tag{61}$$

so that we of course also have

$$\mathsf{K}[\mathsf{T}_1(\Gamma')] = \mathsf{K}[\mathsf{T}_1(\Gamma)] \tag{62}$$

and

$$\mathsf{K}[\mathsf{T}_2(\Gamma')] = \mathsf{K}[\mathsf{T}_2(\Gamma)]. \tag{63}$$

Since, by (59), $v(\Gamma') \neq 0$, our theorem is true for Γ' (as was shown in the first paragraph). Moreover, by (60) and (61) we see that $X \in \mathsf{T}_1(\Gamma')$ and $Y \in \mathsf{T}_2(\Gamma')$. Hence a necessary and sufficient condition that $X \in \mathsf{K}[\mathsf{T}_1(\Gamma')]$ and $Y \in \mathsf{K}[\mathsf{T}_2(\Gamma')]$ is that there exist a square submatrix B′ of A′, of order r, such that

$$\left.\begin{aligned} J_r(\text{adj } B')J_r^T &\neq 0, \\ v(\Gamma') &= \frac{|B'|}{J_r(\text{adj } B')J_r^T}, \\ \dot{X} &= \frac{J_r \text{ adj } B'}{J_r(\text{adj } B')J_r^T}, \\ \dot{Y} &= \frac{J_r(\text{adj } B')^T}{J_r(\text{adj } B')J_r^T}, \end{aligned}\right\} \tag{64}$$

where $\dot{X}$ and $\dot{Y}$ are the vectors obtained from X and Y by deleting the elements corresponding to the rows and columns deleted to obtain B′ from A′.

By (62) and (63), we see that the existence of a matrix satisfying (64) is also a necessary and sufficient condition that $X \in \mathsf{K}[\mathsf{T}_1(\Gamma)]$ and $Y \in \mathsf{K}[\mathsf{T}_2(\Gamma)]$.

Now let B be the matrix obtained from B′ by subtracting b from each element of B′; it is clear that B is a square submatrix of A, which can be obtained from A by deleting the same rows and columns which are deleted from A′ in order to obtain B′. Then, by (59) and Lemma 3.4, it is seen that the conditions (64) are equivalent to

$$J_r\,(\text{adj B})J_r^T \neq 0,$$
$$v(\Gamma) = \frac{|\text{B}|}{J_r\,(\text{adj B})J_r^T},$$
$$\dot{X} = \frac{J_r \text{ adj B}}{J_r\,(\text{adj B})J_r^T},$$
$$\dot{Y} = \frac{J_r(\text{adj B})^T}{J_r\,(\text{adj B})J_r^T},$$

and this completes the proof of our theorem.

THEOREM 3.7. If Γ is any rectangular game, then the sets $\mathsf{K}[\mathsf{T}_1(\Gamma)]$ and $\mathsf{K}[\mathsf{T}_2(\Gamma)]$ are finite, and $\mathsf{T}_1(\Gamma)$ and $\mathsf{T}_2(\Gamma)$ are the convex hulls of $\mathsf{K}[\mathsf{T}_1(\Gamma)]$ and $\mathsf{K}[\mathsf{T}_2(\Gamma)]$, respectively; thus $\mathsf{T}_1(\Gamma)$ and $\mathsf{T}_2(\Gamma)$ are hyperpolyhedra having the points of $\mathsf{K}[\mathsf{T}_1(\Gamma)]$ and $\mathsf{K}[\mathsf{T}_2(\Gamma)]$, respectively, as vertices.

PROOF. By Theorem 3.6, every member of $\mathsf{K}[\mathsf{T}_1(\Gamma)]$ can be obtained from a square submatrix of the matrix A of Γ, and a given square submatrix cannot lead to more than one member of $\mathsf{K}[\mathsf{T}_1(\Gamma)]$. Since a matrix A has only a finite number of submatrices, it therefore follows that $\mathsf{K}[\mathsf{T}_1(\Gamma)]$ is a finite set. The proof of the finiteness of $\mathsf{K}[\mathsf{T}_2(\Gamma)]$ is similar. The rest of the theorem follows from Lemma 3.2.

REMARK 3.8. Theorem 3.6 provides us with a convenient, systematic way of finding all solutions of a rectangular game. The method is as follows: Given a rectangular game with a payoff matrix A, we consider, in turn, each square submatrix B of A. For a submatrix B of order r, we determine v, $\dot{X}$, and $\dot{Y}$ by the formulas given in Theorem 3.6 and decide first whether $\dot{X}$ and $\dot{Y}$ belong to S_r. If not, we reject B and go on to another square submatrix. If $\dot{X}$ and $\dot{Y}$ both belong to S_r, then we form X and Y, by adding to $\dot{X}$ and $\dot{Y}$ the zero components which correspond to the rows and columns deleted from A to obtain B, and determine by Theorem 2.9 whether $X \in \mathsf{T}_1(\Gamma)$ and $Y \in \mathsf{T}_2(\Gamma)$. If not, then again we reject B and go on to another square submatrix of A. If so, then $\|X \quad Y\|$ is a solution and, moreover, by Theorem 3.6, $X \in \mathsf{K}[\mathsf{T}_1(\Gamma)]$ and $Y \in \mathsf{K}[\mathsf{T}_2(\Gamma)]$. In this way we can determine all the members of $\mathsf{K}[\mathsf{T}_1(\Gamma)]$ and $\mathsf{K}[\mathsf{T}_2(\Gamma)]$, and hence find all members of $\mathsf{T}_1(\Gamma)$ and $\mathsf{T}_2(\Gamma)$ by forming convex linear combinations of members of $\mathsf{K}[\mathsf{T}_1(\Gamma)]$ and $\mathsf{K}[\mathsf{T}_2(\Gamma)]$, respectively.

In this connection, it should be noticed that we need never consider submatrices of order 1, since such submatrices determine solutions by the formulas of Theorem 3.6 if, and only if, the corresponding points are saddle-points of the original matrix.

We shall now illustrate this procedure by some examples.

EXAMPLE 3.9. Consider the game Γ whose payoff matrix is

$$A = \begin{Vmatrix} 2 & 4 & 0 \\ 1 & 0 & 4 \end{Vmatrix}.$$

Since the matrix has no saddle-point, we start by considering the three submatrices of order 2:

$$B = \begin{Vmatrix} 2 & 4 \\ 1 & 0 \end{Vmatrix},$$

$$C = \begin{Vmatrix} 2 & 0 \\ 1 & 4 \end{Vmatrix},$$

$$D = \begin{Vmatrix} 4 & 0 \\ 0 & 4 \end{Vmatrix}.$$

It is readily verified that

$$|B| = -4$$

and that

$$\text{adj } B = \begin{Vmatrix} 0 & -4 \\ -1 & 2 \end{Vmatrix};$$

hence

$$(\text{adj } B)^T = \begin{Vmatrix} 0 & -1 \\ -4 & 2 \end{Vmatrix}.$$

Thus

$$J_2(\text{adj } B) = \begin{Vmatrix} -1 & -2 \end{Vmatrix},$$

and

$$J_2(\text{adj } B)^T = \begin{Vmatrix} -4 & 1 \end{Vmatrix},$$

and

$$J_2(\text{adj } B)J_2^T = -3.$$

Substituting in the formulas of Theorem 3.6, we then find that

$$v = \frac{-4}{-3} = \frac{4}{3},$$

$$\dot{X} = \frac{\| -1 \quad -2 \|}{-3} = \left\| \frac{1}{3} \quad \frac{2}{3} \right\|,$$

$$\dot{Y} = \frac{\| -4 \quad 1 \|}{-3} = \left\| \frac{4}{3} \quad -\frac{1}{3} \right\|.$$

Since the second component of $\dot{Y}$ is negative, however, $\dot{Y} \notin \mathbf{S}_2$; hence we conclude that the matrix B does not yield a solution of our game. (The fact that $\dot{Y} \notin \mathbf{S}_2$ could also, it may be noted, have been concluded from the fact that the components of $J_2(\text{adj B})^T$ were not all of the same sign.)

Turning now to the matrix C, we find that the formulas of Theorem 3.6 yield

$$v = \frac{8}{5},$$

$$\dot{X} = \left\| \frac{3}{5} \quad \frac{2}{5} \right\|,$$

$$\dot{Y} = \left\| \frac{4}{5} \quad \frac{1}{5} \right\|.$$

Here $\dot{X} \in \mathbf{S}_2$ and $\dot{Y} \in \mathbf{S}_2$. Since no rows were deleted from A to obtain C,

$$X = \dot{X} = \left\| \frac{3}{5} \quad \frac{2}{5} \right\|.$$

Since the second column was deleted from A to obtain C,

$$Y = \left\| \frac{4}{5} \quad 0 \quad \frac{1}{5} \right\|.$$

Now, testing the quantities v, X, and Y by Theorem 2.9, we find

$$E(X, 1) = \frac{3}{5} \cdot 2 + \frac{2}{5} \cdot 1 = \frac{8}{5} = v,$$

$$E(X, 2) = \frac{3}{5} \cdot 4 + \frac{2}{5} \cdot 0 = \frac{12}{5} > v,$$

$$E(X, 3) = \frac{3}{5} \cdot 0 + \frac{2}{5} \cdot 4 = \frac{8}{5} = v,$$

$$E(1, Y) = \frac{4}{5} \cdot 2 + 0 \cdot 4 + \frac{1}{5} \cdot 0 = \frac{8}{5} = v,$$

$$E(2, Y) = \frac{4}{5} \cdot 1 + 0 \cdot 0 + \frac{1}{5} \cdot 4 = \frac{8}{5} = v.$$

Thus this X and this Y indeed constitute a solution.

Turning now, finally, to D, we obtain from Theorem 3.6

$$v = 2,$$

$$\dot{X} = \left\| \frac{1}{2} \quad \frac{1}{2} \right\|,$$

$$\dot{Y} = \left\| \frac{1}{2} \quad \frac{1}{2} \right\|,$$

and hence

$$X = \left\| \frac{1}{2} \quad \frac{1}{2} \right\|,$$

$$Y = \left\| \frac{1}{2} \quad 0 \quad \frac{1}{2} \right\|.$$

When we test the quantities X, Y, and v by Theorem 2.9, however, we find that

$$E(1, X) = \frac{1}{2} \cdot 2 + \frac{1}{2} \cdot 1 = \frac{3}{2} < v,$$

and hence that this X and this Y do not constitute a solution. (This could have been foreseen, by the way, from the fact that the value of v obtained here was different from that obtained from matrix C, which had already been seen to give a solution.)

Thus this game has a unique solution:

$$v = \frac{8}{5},$$

$$X = \left\| \frac{3}{5} \quad \frac{2}{5} \right\|,$$

$$Y = \left\| \frac{4}{5} \quad 0 \quad \frac{1}{5} \right\|,$$

a result which can easily be verified by the graphical method developed in Chap. 2.

EXAMPLE 3.10. Consider the game Γ whose payoff matrix is

$$A = \left\| \begin{matrix} 2 & 1 & 0 \\ 0 & 1 & 2 \end{matrix} \right\|.$$

Here we set

$$B = \begin{Vmatrix} 2 & 1 \\ 0 & 1 \end{Vmatrix},$$

$$C = \begin{Vmatrix} 2 & 0 \\ 0 & 2 \end{Vmatrix},$$

$$D = \begin{Vmatrix} 1 & 0 \\ 1 & 2 \end{Vmatrix}.$$

Applying the formulas of Theorem 3.6 to B, we find that

$$v = 1,$$

$$\dot{X} = \begin{Vmatrix} \frac{1}{2} & \frac{1}{2} \end{Vmatrix},$$

$$\dot{Y} = \begin{Vmatrix} 0 & 1 \end{Vmatrix},$$

and hence that

$$X = \begin{Vmatrix} \frac{1}{2} & \frac{1}{2} \end{Vmatrix},$$

$$Y = \begin{Vmatrix} 0 & 1 & 0 \end{Vmatrix}.$$

From Theorem 2.6, we see that these quantities are indeed a solution of the original matrix.

Similarly, using Theorem 3.6 and matrix C, we obtain

$$v = 1,$$

$$X = \begin{Vmatrix} \frac{1}{2} & \frac{1}{2} \end{Vmatrix},$$

$$Y = \begin{Vmatrix} \frac{1}{2} & 0 & \frac{1}{2} \end{Vmatrix},$$

and, using Theorem 2.6, we again verify that this is a solution.

Finally, using matrix D, we get

$$v = 1,$$

$$X = \begin{Vmatrix} \frac{1}{2} & \frac{1}{2} \end{Vmatrix},$$

$$Y = \begin{Vmatrix} 0 & 1 & 0 \end{Vmatrix},$$

which is the same solution as that obtained by using matrix B.

Thus, for this game, $\mathsf{K}[\mathsf{T}_1(\Gamma)]$ contains only one member, namely, $\| \frac{1}{2} \quad \frac{1}{2} \|$, and hence $\mathsf{T}_1(\Gamma)$ contains only one member. The set $\mathsf{K}[\mathsf{T}_2(\Gamma)]$, however, contains two members, namely, $\|0 \quad 1 \quad 0\|$ and $\|\frac{1}{2} \quad 0 \quad \frac{1}{2}\|$. Hence the general member of $\mathsf{T}_2(\Gamma)$ can be written:

$$\alpha_1 \|0 \quad 1 \quad 0\| + \alpha_2 \left\| \frac{1}{2} \quad 0 \quad \frac{1}{2} \right\|,$$

or simply

$$\left\| \frac{1}{2}\alpha_2 \quad \alpha_1 \quad \frac{1}{2}\alpha_2 \right\|,$$

where $\|\alpha_1 \quad \alpha_2\| \in \mathsf{S}_2$. This means that any strategy will be optimal for P_2, so long as it makes him play the first and third columns with equal frequencies.

EXAMPLE 3.11. Consider the game whose payoff matrix is

$$A = \left\| \begin{matrix} -1 & 3 & -3 \\ 2 & 0 & 3 \\ 2 & 1 & 0 \end{matrix} \right\|.$$

Here we have

$$\text{adj } A = \left\| \begin{matrix} -3 & -3 & 9 \\ 6 & 6 & -3 \\ 2 & 7 & -6 \end{matrix} \right\|,$$

which yields a solution,

$$X = \left\| \frac{1}{3} \quad \frac{2}{3} \quad 0 \right\|, \qquad Y = \left\| \frac{1}{5} \quad \frac{3}{5} \quad \frac{1}{5} \right\|.$$

Of the nine 2×2 submatrices, it can be seen that only one yields an additional solution, namely,

$$B = \left\| \begin{matrix} 3 & -3 \\ 0 & 3 \end{matrix} \right\|,$$

for which

$$\dot{X} = \left\| \frac{1}{3} \quad \frac{2}{3} \right\|, \qquad \dot{Y} = \left\| \frac{2}{3} \quad \frac{1}{3} \right\|;$$

hence

$$X = \left\| \frac{1}{3} \quad \frac{2}{3} \quad 0 \right\|, \qquad Y = \left\| 0 \quad \frac{2}{3} \quad \frac{1}{3} \right\|.$$

Thus P_1 has the unique optimal strategy

$$X = \left\| \frac{1}{3} \quad \frac{2}{3} \quad 0 \right\|,$$

while P_2 has the set of optimal strategies

$$Y = \alpha_1 \left\| \frac{1}{5} \quad \frac{3}{5} \quad \frac{1}{5} \right\| + \alpha_2 \left\| 0 \quad \frac{2}{3} \quad \frac{1}{3} \right\|,$$

where $\| \alpha_1 \quad \alpha_2 \|$ is an arbitrary element of $\mathbf{S}_2$. The value of the game is 1.

REMARK 3.12. It should be remarked that the method we have developed in this chapter for solving rectangular games, though highly systematic (and of course much quicker than the rough-and-ready method employed in Chap. 2), still leads to laborious calculations in the case of games with large payoff matrices; for the number of square submatrices possessed by a matrix of high order is itself a very large number and it is exceedingly laborious to compute the adjoint of a square matrix of high order. Indeed, the number of arithmetic operations which this method necessitates increases so rapidly with the order of the matrix that it appears unlikely that it would be feasible, even with the use of modern electronic computing machinery, to find solutions of games of order, say, 100×100.

REMARK 3.13. From the results of this chapter, it is clear that for many rectangular games there are infinitely many optimal strategies for one or both players. It is natural to inquire whether, in such a case, it is possible still to discriminate among the various optimal strategies, i.e., whether reasons can be advanced for considering some of them better than others. This can, indeed, be done in various ways, one of which will now be explained.

Suppose that, whenever the first player chooses a mixed strategy X and the second chooses a pure strategy j, the expectation of the first player is

$$E(X, j).$$

We say that a mixed strategy X *dominates* a mixed strategy X' if, for every pure strategy j for P_2,

$$E(X, j) \geq E(X', j)$$

and there exists at least one strategy j for P_2 such that

$$E(X, j) > E(X', j).$$

We call X a *best* strategy if it is optimal and is not dominated by any other strategy. (Dominance and best strategies for P_2 are defined analogously.)

The intuitive justification of these definitions lies in the consideration that if X dominates X', then, regardless of what P_2 decides to do, P_1 will do at least as well by using X as by using X'; furthermore, if P_2 makes certain kinds of mistakes, then P_1 will do better with X than with X'. Thus a dominating strategy takes better advantage of possible mistakes of the opponent than does a dominated strategy. Thus there would seem to be little reason ever to play a strategy which was dominated by another; so a player might do well to choose his strategy from our class of "best" strategies.

It can easily be shown that for every rectangular game there exist best strategies. The proof of this will be left as an exercise.

EXAMPLE 3.14. Consider the game whose payoff matrix is

$$\begin{Vmatrix} 2 & 3 & 4 \\ 2 & 5 & 6 \end{Vmatrix}.$$

Since each of the elements of the first column is a saddle-point, it is seen that *every* mixed strategy for the first player is optimal. We see that if P_1 uses strategy $\| x_1 \quad x_2 \|$, then his expectation, depending on which column P_2 plays, is

$$E(\| x_1 \quad x_2 \|, 1) = 2x_1 + 2x_2 = 2,$$

$$E(\| x_1 \quad x_2 \|, 2) = 3x_1 + 5x_2,$$

$$E(\| x_1 \quad x_2 \|, 3) = 4x_1 + 6x_2.$$

In particular,

$$E(\| 0 \quad 1 \|, 1) = 2,$$

$$E(\| 0 \quad 1 \|, 2) = 5,$$

$$E(\| 0 \quad 1 \|, 3) = 6.$$

Since, for all $\| x_1 \quad x_2 \|$ in $\mathbf{S}_2$,

$$2 \geq 2,$$

$$5 \geq 3x_1 + 5x_2,$$

$$6 \geq 4x_1 + 6x_2,$$

we conclude that $\| 0 \quad 1 \|$ is a best strategy for P_1 and, indeed, the only best strategy.

There is only one optimal strategy for P_2, namely, $\| 1 \quad 0 \quad 0 \|$, and hence also only one best strategy.

HISTORICAL AND BIBLIOGRAPHICAL REMARKS

The material in this chapter has been taken mostly from Shapley and Snow [1].

A problem rather closely related to the one with which we have been concerned here is the following: to find the conditions that must be satisfied by two sets X and Y in order that, for some rectangular game Γ, $\mathsf{X} = \mathsf{T}_1(\Gamma)$ and $\mathsf{Y} = \mathsf{T}_2(\Gamma)$. This problem is solved in Bohnenblust, Karlin, and Shapley [1] and in Gale and Sherman [1].

A complete discussion of matrices can be found in the following two books: Bôcher [1] and MacDuffee [1].

EXERCISES

1. Find, by the methods of this chapter, all the solutions of the games with the following matrices:

(a) $$\begin{Vmatrix} 3 & 6 \\ 5 & 5 \\ 9 & 3 \end{Vmatrix},$$

(b) $$\begin{Vmatrix} 0 & 0 & 0 \\ 0 & -1 & 1 \\ 0 & 1 & -1 \end{Vmatrix},$$

(c) $$\begin{Vmatrix} 1 & 2 & 3 \\ 6 & 1 & 2 \\ 3 & 5 & 0 \end{Vmatrix}.$$

2. Find all solutions of the game whose matrix is

$$\begin{Vmatrix} 6 & 4 & -6 & 19 \\ 1 & 0 & 3 & -5 \\ 2 & 0 & 2 & 3 \\ -5 & -2 & 16 & -35 \end{Vmatrix}$$

by making use of the notion of dominance and then applying the methods developed in the present chapter to the resulting 3×3 matrix.

3. Show that the game whose matrix is

$$\begin{Vmatrix} a & 0 & 0 \\ 0 & b & 0 \\ 0 & 0 & c \end{Vmatrix},$$

where $a > b > c > 0$, has a unique solution. Find this unique solution and the value of the game. What can be said about the solution if $a > b > c$ and $c < 0$?

4. Let Γ_c be the rectangular game whose matrix is as follows:

$$\begin{Vmatrix} c & c & c \\ c & 3 & 4 \\ c & 5 & 1 \end{Vmatrix}.$$

For what values of c is $\mathsf{T}_1(\Gamma_c)$ an infinite set? For what values of c is $\mathsf{T}_2(\Gamma_c)$ an infinite set? Show that, for all c, $v(\Gamma_c) = c$.

5. Show that the game with the matrix

$$\begin{Vmatrix} a & 0 & 1 \\ 1 & a & 0 \\ 0 & 1 & a \end{Vmatrix}$$

has a unique solution.

6. Show that if the elements of the matrix of a game are integers, then the value of the game is a rational number.

7. Let E be the expectation function of an $m \times n$ rectangular game Γ, let $X^{(1)}, \cdots, X^{(r)}$ be elements of $\mathsf{T}_1(\Gamma)$, let Y be an element of $\mathsf{T}_2(\Gamma)$, and let $\| \alpha_1 \quad \cdots \quad \alpha_r \|$ belong to S_r. Show that

$$\alpha_1 E(X^{(1)}, Y) + \cdots + \alpha_r E(X^{(r)}, Y) = E(\alpha_1 X^{(1)} + \cdots + \alpha_r X^{(r)}, Y).$$

8. Find all solutions of two-finger Morra as described in Sec. 3 of Chap. 1.

9. Let a *network game* be defined as follows: There are n points, some pairs of which are connected by oriented line-segments, as shown in Fig. 1. The players simultaneously each select a point; if they select the same point, or different points not connected by a segment, the payoff is zero; if they select a connected pair, the player who chooses the head of the arrow in Fig. 1 collects 1 from his opponent.

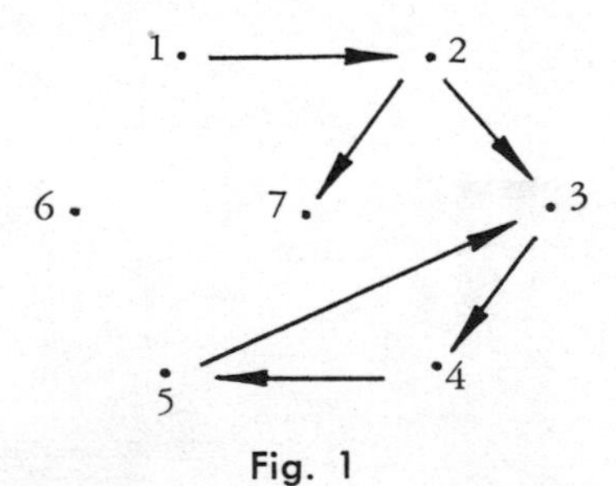

Fig. 1

Solve the network game associated with Fig. 2.

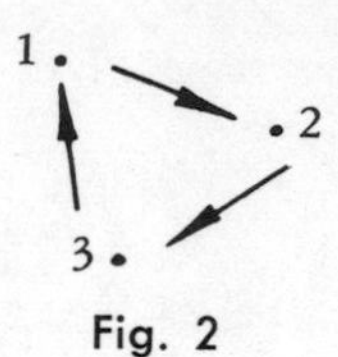

Fig. 2

10. Solve the network games (see previous exercise) associated with Figs. 3 and 4.

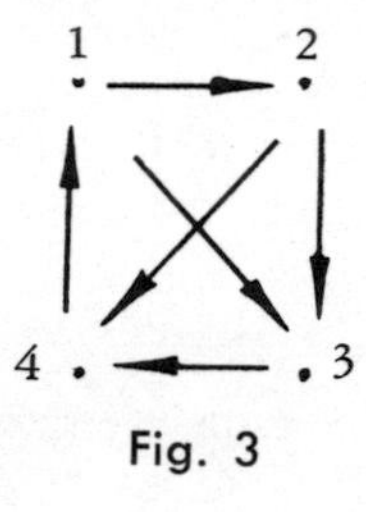

Fig. 3

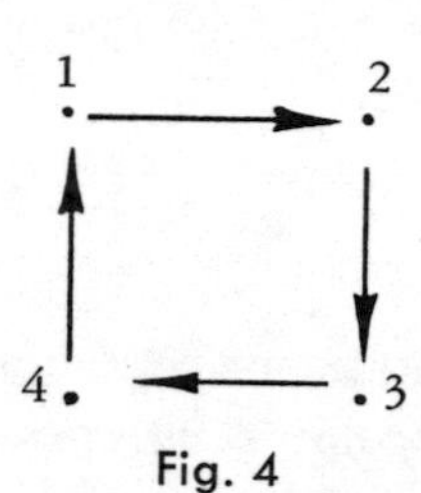

Fig. 4

11. Find a best strategy for each player of the rectangular game whose matrix is

$$\begin{Vmatrix} 1 & 2 & 4 \\ 5 & 3 & 1 \\ 14 & 7 & -1 \end{Vmatrix}.$$

12. Show that each player of every rectangular game has at least one best strategy.

13. Show that the set of all best strategies for a given player is closed.

CHAPTER 4

A METHOD OF APPROXIMATING THE VALUE OF A GAME

In this chapter we shall explain an approximate method of solving rectangular games which will enable us to find the value of such games to any desired degree of accuracy and also to approximate to optimal strategies. The labor involved in applying this method appears to increase, roughly speaking, only as a linear function of the number of rows and columns of the matrix, so that, for a game with a very large matrix, this method is much quicker than the exact method explained in Chap. 3. Since the proof that the method in question actually leads to an approximate solution is somewhat involved, and since the whole topic is a rather special one, we shall omit proofs altogether.

Our approximate method is founded on the following intuitive considerations. Suppose that two people play a long sequence of plays of a given game where neither knows an optimal strategy—because they are ignorant of game theory, perhaps, or because the matrix of the game is too large for them to be able to make the required computations. Then it might happen that each of them decides to behave, in successive plays of the game, as if he were dealing with inanimate nature instead of with a rational opponent—i.e., each of them might always play in such a way as to maximize his expectation under the assumption that "the future will resemble the past." If one knows what each player did on the first play, then this maximization principle leads to a determinate sequence of plays of the game; at each point of the sequence one can calculate upper and lower bounds for the value of the game as well as an approximation to an optimal strategy for each player.

Rather than attempt to give a detailed and precise description of this method in general, we shall explain it merely for a certain 3×3 game. Let the matrix of the game be

$$\begin{Vmatrix} 1 & 2 & 3 \\ 4 & 0 & 1 \\ 2 & 3 & 0 \end{Vmatrix}.$$

Let the three strategies available to P_1 be A, B, and C: thus to say that P_1 chooses strategy A means that he chooses the first row; to say that he chooses strategy B means that he chooses the second row; and so on. Similarly, let the three strategies available to P_2 (the three columns) be α, β, and γ. We first fill in a gap in the description of our approximation method by imposing the condition that, in the first play, P_1 choose A and P_2 choose α. Now, at the second play, P_2 knows that P_1 chose A in the first play, and, since he supposes P_1 will behave in the same way in the future, he is confronted with the following situation:

$$\begin{array}{rccc} \text{If } P_2 \text{ chooses} & \alpha & \beta & \gamma, \\ \text{he will get} & -1 & -2 & -3; \end{array}$$

since P_2 wants to maximize his expectation, he therefore decides to play α again. Similarly, since P_1, at the beginning of the second play, knows that P_2 chose α in the first play, he is confronted (under the supposition that P_2 will behave in the future as he has in the past) with the following situation:

$$\begin{array}{rccc} \text{If } P_1 \text{ chooses} & A & B & C, \\ \text{he will get} & 1 & 4 & 2. \end{array}$$

As the largest of these numbers is 4, P_1 then chooses B on the second play.

Thus, at the end of the second play, P_1 has played A once and B once. We now write Table 1, which shows that P_2 would have done better if he had chosen β on both of the first two plays; therefore, he decides to play β on the third play.

Table 1

Play	P_1 Chooses	Amount Obtained by P_2, Choosing		
		α	β	γ
1	A	-1	-2	-3
2	B	-4	0	-1
TOTALS		-5	-2	-4

Similarly, we obtain Table 2 for P_1, which shows that P_1 again decides to choose B.

Table 2

Play	P_2 Chooses	Amount Obtained by P_1, Choosing A	B	C
1	α	1	4	2
2	α	1	4	2
TOTALS		2	8	4

We now write out Table 3, which corresponds to the first three plays but sums the expected amounts of each player. It can be seen that on the fourth play P_2 will again choose β and P_1 will again choose B.

Table 3

Play	P_1 Chooses	Total Expectation of P_2			P_2 Chooses	Total Expectation of P_1		
		α	β	γ		A	B	C
1	A	−1	−2	−3	α	1	4	2
2	B	−5	−2	−4	α	2	8	4
3	B	−9	−2	−5	β	4	8	7

We now extend this table further, obtaining Table 4.

Table 4

FIRST EIGHT PLAYS OF THE GAME

Play	P_1 Chooses	P_2 Chooses	P_1 Expects			P_2 Expects		
			A	B	C	α	β	γ
1	A	α	1	4	2	−1	−2	−3
2	B	α	2	8	4	−5	−2	−4
3	B	β	4	8	7	−9	−2	−5
4	B	β	6	8	10	−13	−2	−6
5	C	β	8	8	13	−15	−5	−6
6	C	β	10	8	16	−17	−8	−6
7	C	γ	13	9	16	−19	−11	−6
8	C	γ	16	10	16	−21	−14	−6

At the end of the eighth play our rule, as heretofore formulated, is not sufficiently explicit to enable us to determine what should be the next choice for P_1. For the maximum, namely 16, of the numbers

$$16, \ 10, \ 16$$

is assumed twice, and therefore the rule does not determine whether P_1 should next choose A or C. We resolve all difficulties of this kind by agreeing that, in all such cases, the player in question shall choose the strategy whose letter comes alphabetically first. Thus, in this case, P_1 is to choose A; if his expectations had been

$$10, \ 16, \ 16,$$

then he would have chosen B. With this new convention we can extend Table 4 and write Table 5.

Table 5

PLAYS NINE THROUGH TWENTY

Play	P_1 Chooses	P_2 Chooses	P_1 Expects			P_2 Expects		
			A	B	C	α	β	γ
9	A	γ	19	11	16	−22	−16	−9
10	A	γ	22	12	16	−23	−18	−12
11	A	γ	25	13	16	−24	−20	−15
12	A	γ	28	14	16	−25	−22	−18
13	A	γ	31	15	16	−26	−24	−21
14	A	γ	34	16	16	−27	−26	−24
15	A	γ	37	17	16	−28	−28	−27
16	A	γ	40	18	16	−29	−30	−30
17	A	α	41	22	18	−30	−32	−33
18	A	α	42	26	20	−31	−34	−36
19	A	α	43	30	22	−32	−36	−39
20	A	α	44	34	24	−33	−38	−42

For each i, we now let $\bar{v}_i$ be the maximum of the numbers in the row of the table under the heading "P_1 Expects." Thus, from Tables 4 and 5, we have

$$\begin{array}{llll}
\bar{v}_1 = 4, & \bar{v}_6 = 16, & \bar{v}_{11} = 25, & \bar{v}_{16} = 40, \\
\bar{v}_2 = 8, & \bar{v}_7 = 16, & \bar{v}_{12} = 28, & \bar{v}_{17} = 41, \\
\bar{v}_3 = 8, & \bar{v}_8 = 16, & \bar{v}_{13} = 31, & \bar{v}_{18} = 42, \\
\bar{v}_4 = 10, & \bar{v}_9 = 19, & \bar{v}_{14} = 34, & \bar{v}_{19} = 43, \\
\bar{v}_5 = 13, & \bar{v}_{10} = 22, & \bar{v}_{15} = 37, & \bar{v}_{20} = 44.
\end{array}$$

And, for each i, we let $\overline{v}_i$ be the negative of the maximum of the numbers in the row of the table under the heading "P_2 Expects." Thus, from Tables 4 and 5, we have

$$\begin{array}{llll}
\underline{v}_1 = 1, & \underline{v}_6 = 6, & \underline{v}_{11} = 15, & \underline{v}_{16} = 29, \\
\underline{v}_2 = 2, & \underline{v}_7 = 6, & \underline{v}_{12} = 18, & \underline{v}_{17} = 30, \\
\underline{v}_3 = 2, & \underline{v}_8 = 6, & \underline{v}_{13} = 21, & \underline{v}_{18} = 31, \\
\underline{v}_4 = 2, & \underline{v}_9 = 9, & \underline{v}_{14} = 24, & \underline{v}_{19} = 32, \\
\underline{v}_5 = 5, & \underline{v}_{10} = 12, & \underline{v}_{15} = 27, & \underline{v}_{20} = 33.
\end{array}$$

Now it can be shown that, for every game Γ, if the value of Γ is v, then the numbers $\overline{v}_i$ and $\underline{v}_i$ calculated in this way satisfy the inequality (for all i):

$$\frac{\underline{v}_i}{i} \leq v \leq \frac{\overline{v}_i}{i}.$$

Thus we have for our game, for instance,

$$1 = \frac{1}{1} \leq v \leq \frac{4}{1} = 4,$$

$$1 = \frac{2}{2} \leq v \leq \frac{8}{2} = 4,$$

$$\frac{2}{3} \leq v \leq \frac{8}{3},$$

$$\frac{6}{4} \leq v \leq \frac{10}{4} = \frac{5}{2},$$

$$1 = \frac{5}{5} \leq v \leq \frac{13}{5},$$

$$\cdots\cdots\cdots\cdots\cdots\cdots$$

The best inequalities we obtain in this way are

$$v \leq \frac{\overline{v}_8}{8} = \frac{16}{8} = 2,$$

and

$$v \geq \frac{\underline{v}_{16}}{16} = \frac{29}{16} = 1.8125.$$

Thus we have obtained a fairly good approximation to the value of the game (which, in fact, is 1.85).

It can also be shown that v is the greatest lower bound of $\overline{v}_i/i$ and the least upper bound of $\underline{v}_i/i$. This fact ensures that, by carrying the approximation method far enough, we can find the value of v to any desired degree of accuracy.

By considering the number of times each pure strategy is played in i steps of the above approximation method, we can also find an approximation to an optimal strategy. Thus, in the first eight rows of Tables 4 and 5, P_1 plays strategy A once, strategy B three times, and strategy C four times; hence an approximation to an optimal strategy for P_1 is $\| 1/8 \quad 3/8 \quad 4/8 \|$. For each i, we let

$$X^{(i)} = \| x_1^{(i)} \quad x_2^{(i)} \quad x_3^{(i)} \|$$

be the strategy for P_1 found in this way; similarly, we let

$$Y^{(i)} = \| y_1^{(i)} \quad y_2^{(i)} \quad y_3^{(i)} \|$$

be the strategy for P_2 found in the analogous way. Thus

$$X^{(1)} = \| 1 \quad 0 \quad 0 \|, \qquad Y^{(1)} = \| 1 \quad 0 \quad 0 \|,$$

$$X^{(2)} = \left\| \frac{1}{2} \quad \frac{1}{2} \quad 0 \right\|, \qquad Y^{(2)} = \| 1 \quad 0 \quad 0 \|,$$

$$X^{(3)} = \left\| \frac{1}{3} \quad \frac{2}{3} \quad 0 \right\|, \qquad Y^{(3)} = \left\| \frac{2}{3} \quad \frac{1}{3} \quad 0 \right\|,$$

$$\vdots \qquad\qquad\qquad \vdots$$

$$X^{(20)} = \left\| \frac{13}{20} \quad \frac{3}{20} \quad \frac{4}{20} \right\|, \qquad Y^{(20)} = \left\| \frac{6}{20} \quad \frac{4}{20} \quad \frac{10}{20} \right\|.$$

(It can be verified that the exact optimal strategies are $X^* = \| 11/20 \quad 4/20 \quad 5/20 \|$ and $Y^* = \| 8/20 \quad 7/20 \quad 5/20 \|$.)

It can be shown that if there is a unique optimal strategy X for P_1, then the sequence

$$X^{(1)}, X^{(2)}, \cdots, X^{(i)}, \cdots \tag{1}$$

converges to X; similarly, if there is a unique optimal strategy for P_2, then a corresponding sequence converges. But if P_1 has more than one optimal strategy, then it may happen that (1) is not convergent. It can be shown, however, that, in any case, every convergent subsequence of (1) converges to an optimal strategy for P_1.

HISTORICAL AND BIBLIOGRAPHICAL REMARKS

The method described in this chapter of approximating solutions of rectangular games was formulated in Brown [1]. The process has been proved to be convergent in Robinson [1].

EXERCISES

1. Using the method explained in this chapter, find, to an accuracy of two places of decimals, the value of the game whose matrix is

$$\begin{Vmatrix} 0 & 2 & 0 \\ 0 & 0 & 2 \\ 2 & 0 & 0 \end{Vmatrix}.$$

2. Find the value of the game whose matrix is

$$\begin{Vmatrix} 0 & -1 & 0 & -2 & 3 \\ 1 & 1 & -3 & 1 & 0 \\ -2 & 2 & 0 & 3 & 1 \\ 0 & -1 & 1 & 0 & -1 \end{Vmatrix}.$$

3. Let

$$\begin{Vmatrix} a_{11} & \cdots & a_{1n} \\ \vdots & & \vdots \\ a_{m1} & \cdots & a_{mn} \end{Vmatrix}$$

be the matrix of a game, where $\| 1 \quad 1 \|$ is a saddle-point—i.e., where a_{11} is the minimum of the first row and the maximum of the first column. Show that the method of this chapter gives a_{11} as the value of the game.

4. Show that, for the case of a 2×2 game with a saddle-point (regardless of whether the saddle-point is in the upper left-hand corner), the method of this chapter leads to the true value of the game.

5. Show by the method of this chapter that the game whose matrix is

$$\begin{Vmatrix} 1 & -1 \\ -1 & 1 \end{Vmatrix}$$

has a value of zero.

CHAPTER 5

GAMES IN EXTENSIVE FORM

1. Normal Form and Extensive Form. Thus far we have succeeded in "solving" the problem of two-person rectangular games, i.e., we have given an intuitively acceptable definition of value and of optimal strategies for such a game, have proved that solutions always exist, and have even shown how they can be computed. But the arguments to which we have had to resort, though mostly of an elementary character, have not always been quite trivial; and the reader is perhaps appalled at the magnitude of the task of extending these results to more general games—i.e., to games where there can be more than two players, where some of the moves perhaps involve the application of chance devices, where the players can each have several moves instead of one, and where their knowledge of what has gone before can vary from move to move.

The situation is not quite so bad as it seems at first glance, however, for it turns out that the problem of solving any game where the players always make their choices from finite sets (we shall call such games themselves *finite*) is always identical with the problem of solving some rectangular game. Thus our results for rectangular games can be applied, more generally, to any zero-sum two-person game. This process of finding a rectangular game equivalent to an arbitrary game is called *normalization*, and the resulting rectangular game is said to be in *normal form*; when it is desirable to make a distinction, we shall speak of arbitrary games as being in *extensive form.*

We begin with an example of a game in extensive form.

EXAMPLE 5.1. In move I, player P_1 chooses a number x from the set $\{1, 2\}$; in move II, player P_2, having been informed what number x was chosen in move I, in turn chooses a number y from $\{1, 2\}$; in move III, player P_1, having been informed of what number y was chosen, and still remembering what x he himself chose in move I, chooses a number z from $\{1, 2\}$. After the three numbers x, y, and z have been chosen, P_2 pays P_1 the amount $M(x, y, z)$, where M is the function defined as follows:

$$\begin{aligned} M(1, 1, 1) &= -2, & M(2, 1, 1) &= 5, \\ M(1, 1, 2) &= -1, & M(2, 1, 2) &= 2, \\ M(1, 2, 1) &= 3, & M(2, 2, 1) &= 2, \\ M(1, 2, 2) &= -4, & M(2, 2, 2) &= 6. \end{aligned}$$

In order to explain how to reduce games to rectangular form, it is necessary first to introduce the general notion of a strategy, which we have previously considered only for rectangular games. By a *strategy* for a given player in a given game we mean a complete set of directions which tell him exactly how to act under all conceivable circumstances of the play, or, properly speaking, for any conceivable state of information which he could possess at any point of any play. Thus, for example, one possible strategy for P_1 to use in playing the game in question would be to choose 1 on both moves (regardless of what P_2 does in move II). Another strategy for P_1 would be to choose 1 in move I, and then, in move III, to choose the same number as was chosen by P_2 in move II. A possible strategy for P_2 would be to choose 1 in move II, regardless of what choice was made by P_1 in move I; another strategy for P_2 would be to choose 1 in move II if 2 was chosen in move I and to choose 2 if 1 was chosen in move I.

(It is worth while to notice that in common language the word "strategy" is often used to mean a clever way of proceeding; but we shall call any recipe for playing, no matter how foolish, a "strategy." The chief problem of the theory of games is thus the problem of distinguishing good strategies from bad.)

Now it is easily seen that player P_2 has precisely four strategies available in this game, i.e., he has exactly as many strategies as there are ways of mapping the set $\{1, 2\}$ into itself. If we adopt the notation f_{ij} to mean the function such that

$$f_{ij}(1) = i, \qquad f_{ij}(2) = j,$$

then the four strategies for P_2 are f_{11}, f_{12}, f_{21}, and f_{22}. To say that P_2 plays strategy f_{21}, for example, means that he has decided that he will choose 2 in move II if P_1 chose 1 in move I, and that he will choose 1 in move II if P_1 chose 2 in move I.

A strategy for P_1, on the other hand, must tell him what to do on both move I and move III. Since nothing has preceded move I, he has no past knowledge of the course of the play which would enable him to distinguish cases; hence his strategy must simply tell him either to choose 1 or to choose 2. For move III, his strategy must tell him which z to choose for each possible choice of x and y. Thus a strategy for P_1 can be represented as a system:

$$\left\| i_0 \quad \left\| i_{11} \quad i_{12} \quad i_{21} \quad i_{22} \right\| \right\|,$$

where i_0 is the number he is to choose in move I and where i_{jk} is the number he is to choose in move III, in case j was chosen in move I and k was chosen in move II. (Since each of the i's has two possible values, there are altogether thirty-two possible strategies for P_1.) To say that P_1 uses the strategy

$$\|1 \quad \|2 \quad 1 \quad 2 \quad 1\| \ \|,$$

for instance, means that P_1 first begins by choosing 1 in move I; then, if 1 was chosen in move I and 1 was chosen also by P_2 in move II, player P_1 is to choose 2 in move III; and so on for the other three possibilities so far as regards move III.

In the above game we have described strategies for P_1 subject to a certain redundancy: thus the second "2" in the example fulfills no useful function, since it means that if 1 was chosen in move II, and if 2 was chosen in move I (which it could not have been, as the given strategy directs P_1 to choose 1 in move I), then 2 is to be chosen in move III. (Hence this "2" tells P_1 what to do in a case which can never arise if he is using the given strategy!) This redundancy could be eliminated by considering a strategy for P_1 to be a system $\|i \quad \|i_1 \quad i_2\| \ \|$, where i is the number P_1 is to choose on move I, and i_1 and i_2 are the numbers he is to choose on move III according as P_2 has chosen 1 or 2 on move II. Thus in this case we could simplify the description of the strategies for P_1 by neglecting the fact that he remembers his first move.

The presence of these redundancies does no real harm, however, and the description of strategies for the general case would be greatly complicated if we were to attempt always to avoid them. For a treatment of this whole problem, see Krentel, McKinsey, and Quine [1] and Dalkey [1].

To choose a possible strategy for a game, and then to play according to that strategy, amounts to making all possible decisions before the beginning of the play. Once each player has chosen a strategy, no other choices are necessary for him; for the strategy tells him what to do at all points where, if he had not chosen a strategy, he would have had to make a decision. The choice of strategies determines the outcome of the game, so the actual play could be carried out by a computing machine.

Thus suppose that, in the game under consideration, player P_1 decides to use strategy

$$\|1 \quad \|2 \quad 1 \quad 2 \quad 2\| \ \|,$$

and player P_2 decides to use strategy f_{21}. From the strategy for P_1, we see that P_1 chooses 1 in move I; from the strategy for P_2, we see that P_2 chooses 2 in move II. Going back to the strategy for P_1, we conclude, finally, that P_1 chooses 1 in move III. Then, since $M(1, 2, 1) = 3$, we see that, if this particular pair of strategies is used by the two players, P_2 will have to pay 3 to P_1. By reasoning in a similar way for the other possible pairs of strategies, we can write down the matrix of strategies (Matrix 1) for the rectangular game to which our given game reduces.

MATRIX 1

	f_{11}	f_{12}	f_{21}	f_{22}
‖1 ‖1 1 1 1‖ ‖	−2	−2	3	3
‖1 ‖1 1 1 2‖ ‖	−2	−2	3	3
‖1 ‖1 1 2 1‖ ‖	−2	−2	3	3
‖1 ‖1 1 2 2‖ ‖	−2	−2	3	3
‖1 ‖1 2 1 1‖ ‖	−2	−2	−4	−4
‖1 ‖1 2 1 2‖ ‖	−2	−2	−4	−4
‖1 ‖1 2 2 1‖ ‖	−2	−2	−4	−4
‖1 ‖1 2 2 2‖ ‖	−2	−2	−4	−4
‖1 ‖2 1 1 1‖ ‖	−1	−1	3	3
‖1 ‖2 1 1 2‖ ‖	−1	−1	3	3
‖1 ‖2 1 2 1‖ ‖	−1	−1	3	3
‖1 ‖2 1 2 2‖ ‖	−1	−1	3	3
‖1 ‖2 2 1 1‖ ‖	−1	−1	−4	−4
‖1 ‖2 2 1 2‖ ‖	−1	−1	−4	−4
‖1 ‖2 2 2 1‖ ‖	−1	−1	−4	−4
‖1 ‖2 2 2 2‖ ‖	−1	−1	−4	−4
‖2 ‖1 1 1 1‖ ‖	5	2	5	2
‖2 ‖1 1 1 2‖ ‖	5*	6	5*	6
‖2 ‖1 1 2 1‖ ‖	2	2	2	2
‖2 ‖1 1 2 2‖ ‖	2	6	2	6
‖2 ‖1 2 1 1‖ ‖	5	2	5	2
‖2 ‖1 2 1 2‖ ‖	5*	6	5*	6
‖2 ‖1 2 2 1‖ ‖	2	2	2	2
‖2 ‖1 2 2 2‖ ‖	2	6	2	6
‖2 ‖2 1 1 1‖ ‖	5	2	5	2
‖2 ‖2 1 1 2‖ ‖	5*	6	5*	6
‖2 ‖2 1 2 1‖ ‖	2	2	2	2
‖2 ‖2 1 2 2‖ ‖	2	6	2	6
‖2 ‖2 2 1 1‖ ‖	5	2	5	2
‖2 ‖2 2 1 2‖ ‖	5*	6	5*	6
‖2 ‖2 2 2 1‖ ‖	2	2	2	2
‖2 ‖2 2 2 2‖ ‖	2	6	2	6

This matrix has saddle-points, which we have indicated by asterisks. The value of the game to P_1 is 5. Any one of the four strategies

$$\| 2 \quad \| 1 \quad 1 \quad 1 \quad 2 \| \; \|, \qquad \| 2 \quad \| 1 \quad 2 \quad 1 \quad 2 \| \; \|,$$

$$\| 2 \quad \| 2 \quad 1 \quad 1 \quad 2 \| \; \|, \qquad \| 2 \quad \| 2 \quad 2 \quad 1 \quad 2 \| \; \|$$

is optimal for P_1. Either of the strategies, f_{11} or f_{21}, is optimal for P_2. Expressed in terms of the game in its original form, this means that one optimal way for P_1 to play is as follows: in move I, to choose 2; in move III, to choose the same number that was chosen by P_2 in move II. An optimal way for P_2 to play is to choose 1 in move II, regardless of what P_1 did in move I. Another optimal way for P_2 to play is, in move II, to choose the opposite number to that chosen by P_1 in move I.

2. Graphical Representation. It is sometimes suggestive to give a graphical representation of a game. This can be done by what is called a "tree"—i.e., a plane figure consisting of a finite number of rising line-segments, where each vertex is connected to just one vertex on a lower level, and where there is but one vertex at the lowest level. (The term "tree," however, is used in a somewhat more general sense in topology.) The vertices represent the various moves, and we attach symbols to them to indicate which player makes the corresponding move. Thus the game just considered can be represented by Fig. 1.

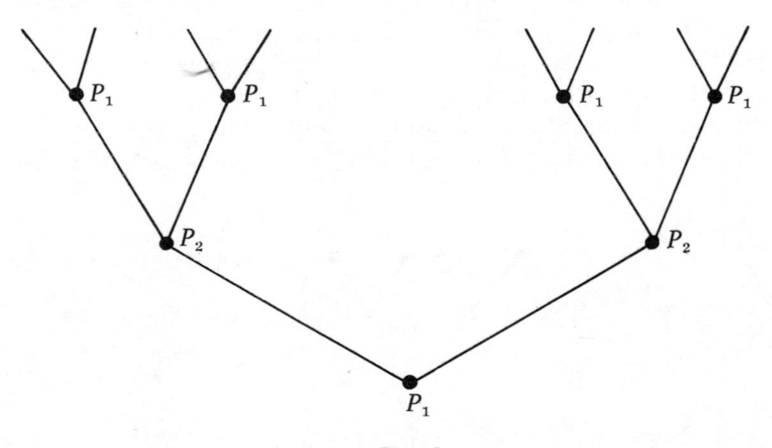

Fig. 1

The bottom vertex of this figure is marked "P_1" to indicate that player P_1 makes the first move; the two vertices on the second level represent move II, which is made by P_2; the four vertices on the third level represent move III, which is made by P_1. The two lines rising from the lowest vertex represent the two choices available to P_1 in move I, and we think of these as corresponding to 1 and 2, when going counterclockwise; thus the line with positive slope represents choice 1 by P_1, and the line with negative slope represents choice 2 by P_1. If there had been three alternatives available to P_1 in move I, the game would be represented as in Fig. 2.

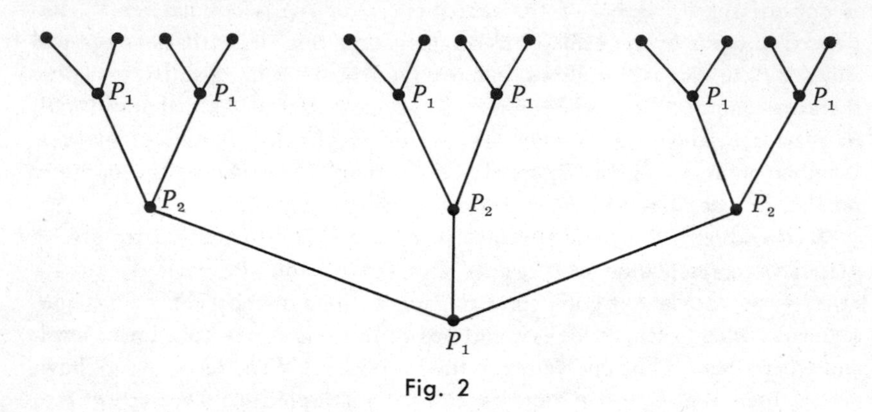

Fig. 2

In this graphical representation, a complete play of the game is represented by a path going from the bottom vertex to one of the top vertices. Since the number of such paths corresponds in a one-to-one way with the number of top vertices, we see immediately from the figures that there are eight possible plays of the game represented by Fig. 1 and that there are twelve possible plays of the game represented by Fig. 2.

In drawing such graphs it is usually convenient to write simply "1" instead of "P_1" and to write "2" instead of "P_2." Thus we obtain Fig. 3 from Fig. 2.

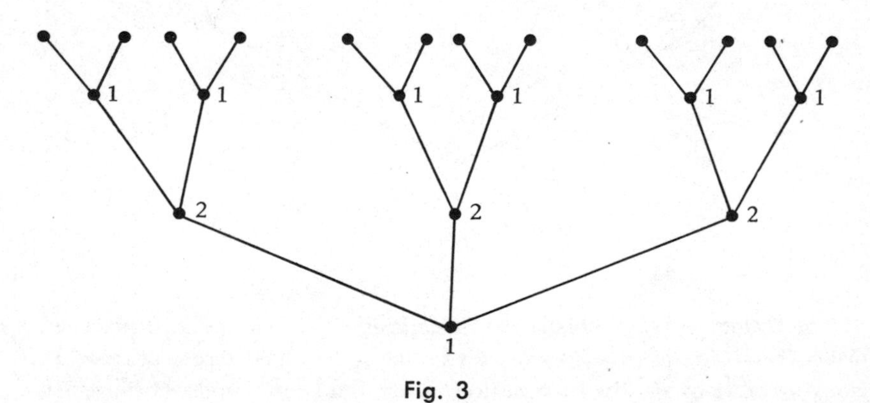

Fig. 3

The game considered in Example 5.1 is what is called a game with "perfect information," i.e., a game where each player is always informed about the whole history of the preceding play. The fact that the strategy matrix for this game turned out to have a saddle-point is no accident; we shall prove in Chap. 6 that every game with perfect information has this property.

3. Information Sets. Our next example is of a game where the information allowed is not perfect. In this game player P_1, when he makes move III, is not informed of what choice P_2 made in move II, and he even forgets what he himself did in move I. This lapse of memory on the part of P_1 could be realized in practice by arranging that P_1 be a team of two persons, the first of whom makes move I, and the second, move III.

EXAMPLE 5.2. In move I, player P_1 chooses a number x from the set $\{1, 2\}$; in move II, player P_2, having been informed what number x was chosen in move I, in turn chooses a number y from $\{1, 2\}$; in move III, player P_1, not having been informed of what number y was chosen, and having forgotten what number x was chosen, chooses a number z from $\{1, 2\}$. After the three numbers x, y, and z have been chosen, P_2 pays P_1 the amount $M(x, y, z)$, where M is the function defined in Example 5.1.

In order to obtain a graphical representation of this game, it is necessary in some way to indicate the fact that P_1 is in ignorance of the past moves when he makes move III. The important thing here is that, in making move III, P_1 does not know at exactly which vertex of Fig. 1 he is situated. This additional factor is introduced in Fig. 4, where the four vertices among which P_1 cannot distinguish in move III are enclosed in a broken line.

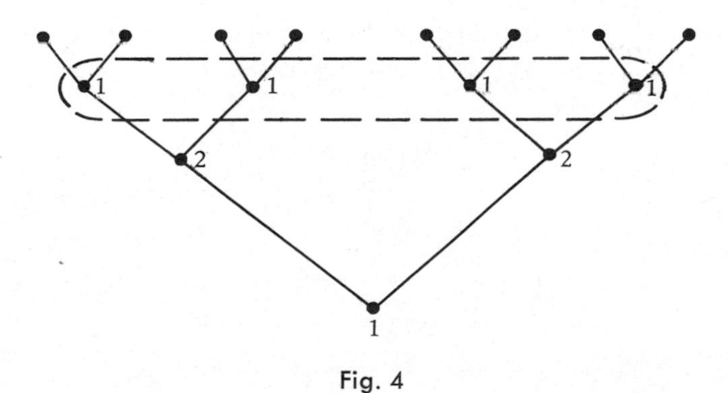

Fig. 4

We shall ordinarily modify this representation slightly, however, by indicating, by means of broken lines, some partition of the whole set of vertices (exclusive of the topmost vertices, which represent end points of the play, where no choices are to be made). In this way we obtain Fig. 5.

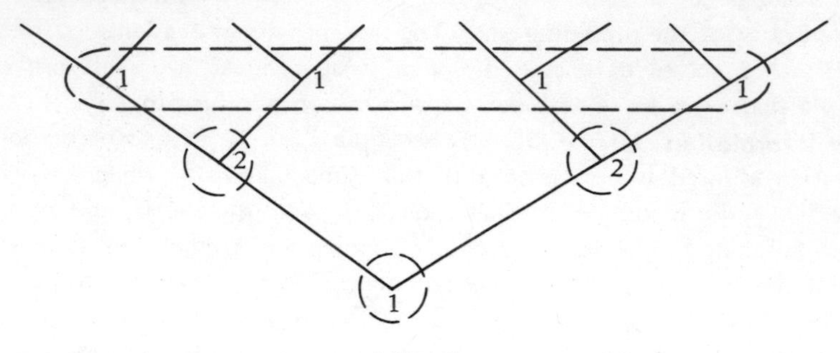

Fig. 5

Here the fact that the lowermost vertex is in a region all by itself indicates that P_1, in move I, knows exactly at what point in the tree he is situated; and a similar situation exists for the two vertices corresponding to move II. We shall refer to the sets into which the vertices of the set are partitioned in this way as the *information sets.*

We turn now to the enumeration of the strategies for this game. It is clear that the strategies available to P_2 are the same as in Example 5.1, namely, the four functions f_{11}, f_{12}, f_{21}, and f_{22}. Since P_1 is never informed about the preceding play, however, a strategy for him is now simply one of the four ordered pairs of numbers: $\|1 \quad 1\|$, $\|1 \quad 2\|$, $\|2 \quad 1\|$, $\|2 \quad 2\|$. To say that P_1 plays strategy $\|1 \quad 2\|$, for instance, means that he will choose 1 in move I and 2 in move III.

Making use of the definition of M, it is now easy to write down the matrix of strategies (Matrix 2) for our new game:

MATRIX 2

	f_{11}	f_{12}	f_{21}	f_{22}
$\|1 \quad 1\|$	-2	-2	3	3
$\|1 \quad 2\|$	-1	-1	-4	-4
$\|2 \quad 1\|$	5	2	5	2
$\|2 \quad 2\|$	2	6	2	6

It will be noticed that this game has no saddle-point. By making use of Theorem 2.9, the student can verify that the value of the game is $26/7$, that an optimal strategy for P_1 is $\|0 \quad 0 \quad 4/7 \quad 3/7\|$, and that an optimal strategy for P_2 is $\| 4/7 \quad 3/7 \quad 0 \quad 0 \|$. Thus P_1 can expect to get less from playing this game than he could expect to get from playing the game in Example 5.1; this is, of course, not surprising, since he is now in an intuitively less advantageous position.

REMARK 5.3. It will be noticed that the matrix for the normal form of the game in Example 5.2 is only 4×4, whereas that for the game in 5.1 is 32×4. It is generally true that decreasing the amount of information available will make the available strategies less numerous and thus will decrease the size of the matrix. This sometimes appears paradoxical to students, on the basis that decreasing one's information should make things more difficult rather than less. But the size of the matrix is no reliable index of the difficulty of playing the game; and it is true almost universally, besides, that the less knowledge we have, the easier we find it to make up our minds (a deaf man has less trouble deciding on a wife than has a man with normal hearing).

Other modifications of Example 5.1 can be obtained by other alterations of the information sets.

EXAMPLE 5.4. On move I, player P_1 chooses x from $\{1, 2\}$. In move II, player P_2, without knowing x, chooses y from $\{1, 2\}$. In move III, player P_1, knowing both x and y, chooses z from $\{1, 2\}$. The payoff function is the same as in Example 5.1.

For this game we obtain the graph shown in Fig. 6; here we include the two vertices corresponding to move II within the same information set, because in move II player P_2 does not know at which of these points he is

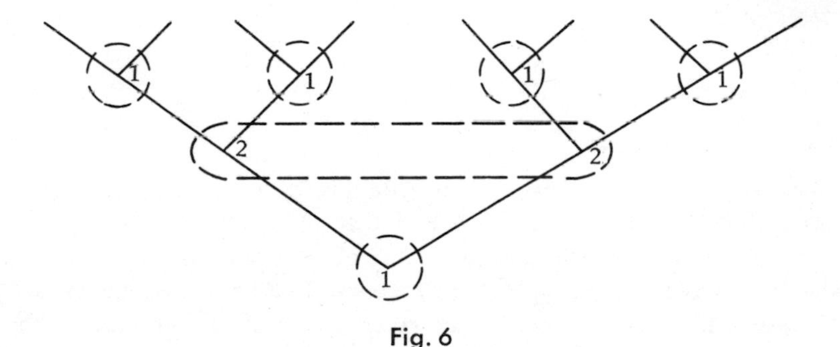

Fig. 6

situated. The enumeration of the strategies for this game, and the determination of the strategy matrix, will be left as an exercise.

EXAMPLE 5.5. In move I, player P_1 chooses x from $\{1, 2\}$. In move II, player P_2, without knowing x, chooses y from $\{1, 2\}$. In move III, player P_1, without knowing either x or y, chooses z from $\{1, 2\}$. The payoff function is the same as in Example 5.1.

For this game we obtain the graph shown in Fig. 7.

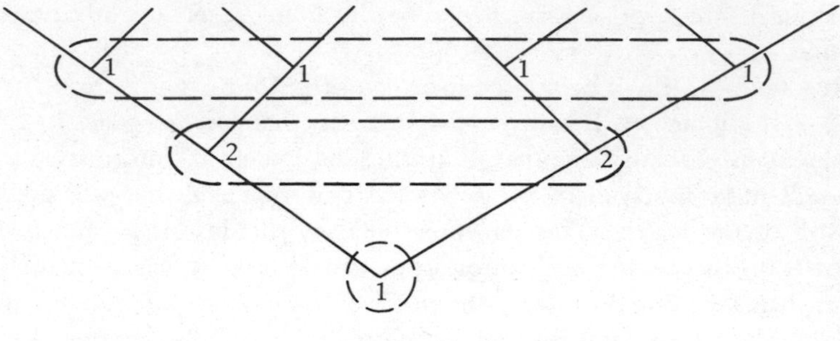

Fig. 7

In this game a strategy for P_1 consists simply of a pair of numbers $\| i \quad j \|$, where i is his choice for move I and j is his choice for move III. A strategy for P_2 is a number i, which represents his choice for move II. Thus there are four strategies for P_1 and two strategies for P_2. The strategy matrix is Matrix 3.

MATRIX 3

	1	2
$\| 1 \quad 1 \|$	-2	3
$\| 1 \quad 2 \|$	-1	-4
$\| 2 \quad 1 \|$	5	2
$\| 2 \quad 2 \|$	2	6

One can easily verify that the value of this game is $26/7$, that an optimal strategy for P_1 is $\| 0 \quad 0 \quad 4/7 \quad 3/7 \|$, and that an optimal strategy for P_2 is $\| 4/7 \quad 3/7 \|$. Thus it happens that the value of this game is the same as the value of the game in Example 5.2; so we see that the knowledge which P_2 possesses in that example does him no good. This is merely an accident, however, which arises from the particular numbers used to define the payoff function. It is clear that, in general, P_2 will be able to do better in a game of the type in 5.2 than in one of the type in 5.5.

It might be thought that Example 5.5 represents the most extreme state of ignorance in which the two players could actually find themselves; for neither, at any point, knows what have been the past choices. But this is not quite the case, since P_1, when he makes move III, knows at least that two moves have preceded this one. The following example shows that situations can actually occur in which even this knowledge is not available.

EXAMPLE 5.6. The game is a two-person game, where P_1 is a single human being, but where P_2 consists of a team of two human beings, A and B. These three individuals are isolated from each other in separate rooms and are not allowed to communicate with each other during the play. At the beginning of the play, the umpire goes to the room occupied by P_1 and asks him to pick a number x from the set $\{1, 2\}$. If P_1 picks 1, then the umpire goes to the room occupied by A and asks him to pick a number y from $\{1, 2\}$; on the other hand, if P_1 picks 2, the umpire goes to the room occupied by B and asks him to pick a number y from $\{1, 2\}$. After y is chosen, the umpire goes to the room occupied by the remaining member of P_2 and asks him to pick a number z from $\{1, 2\}$. After the three numbers are chosen, P_2 pays P_1 the amount $M(x, y, z)$, where M is defined as follows:

$$M(1, 1, 1) = 0, \qquad M(2, 1, 1) = 4,$$
$$M(1, 1, 2) = 2, \qquad M(2, 1, 2) = 0,$$
$$M(1, 2, 1) = 6, \qquad M(2, 2, 1) = 5,$$
$$M(1, 2, 2) = 8, \qquad M(2, 2, 2) = 6.$$

In connection with this game, it is important to realize that when a member of P_2 is asked to make his choice, he does not know whether he is making the second or the third move of the game, for he does not know which choice P_1 made. The game is graphed in Fig. 8.

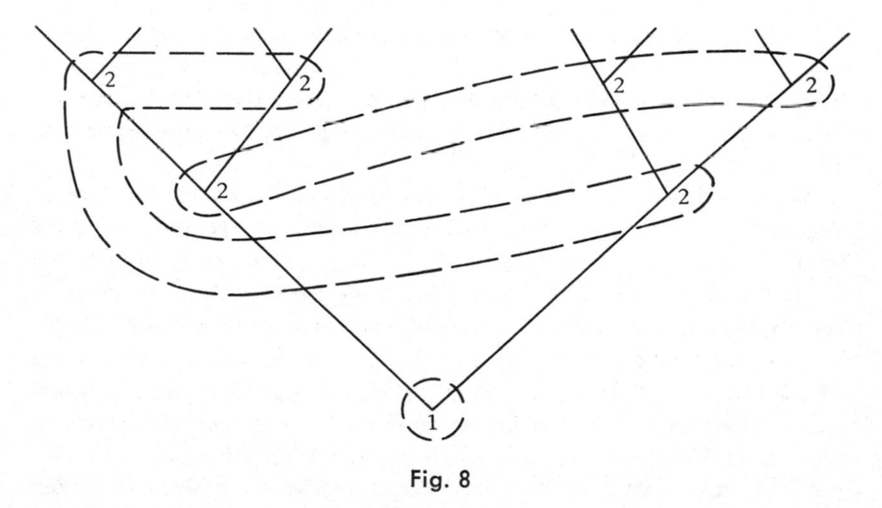

Fig. 8

There are just two strategies available to P_1, namely, he can choose either 1 or 2.

There are four strategies available to P_2: A and B can both choose 1; or A can choose 1 and B can choose 2; or A can choose 2 and B can choose 1; or both of them can choose 2. We represent these four strategies respectively by the ordered couples

$$\| 1 \quad 1 \|, \qquad \| 1 \quad 2 \|, \qquad \| 2 \quad 1 \|, \qquad \| 2 \quad 2 \|.$$

To make it clear how the payoffs for the various strategies are computed, suppose, for instance, that P_1 plays strategy 2 and that P_2 plays strategy $\| 1 \quad 2 \|$. On the first move, P_1 chooses $x = 2$ and hence the umpire goes first to the room occupied by B, who chooses $y = 2$. Finally, the umpire goes to the room occupied by A, who chooses $z = 1$. Hence the payoff is $M(2, 2, 1) = 5$.

Matrix 4 gives the payoff for all the possible combinations of strategies. The value of the game is $^{12}/_5$; an optimal strategy for P_1 is $\| ^2/_5 \quad ^3/_5 \|$ and an optimal strategy for P_2 is $\| ^3/_5 \quad 0 \quad ^2/_5 \quad 0 \|$.

MATRIX 4

	$\| 1 \quad 1 \|$	$\| 1 \quad 2 \|$	$\| 2 \quad 1 \|$	$\| 2 \quad 2 \|$
1	0	2	6	8
2	4	5	0	6

4. Chance Moves. We want also to consider games in which chance moves occur, i.e., games in which some of the choices are made by means of chance devices rather than by the personal decisions of the players. Chance moves occur in many common parlor games; e.g., in most card games the dealing of the cards is done at random.

Chance moves can intervene in games in three ways: (1) by affecting the payoff, (2) by affecting the size or nature of the sets from which the players can make their choices, and (3) by determining the order in which the players will make their moves. We shall consider three examples to illustrate these three phenomena, and in each case we shall show how to enumerate the strategies so as to reduce the game to rectangular form.

EXAMPLE 5.7. In move I, a coin is tossed; in move II, player P_1, having been informed whether the coin came up heads or tails, chooses a number x from the set $\{1, 2\}$; in move III, player P_2, not being informed of the outcome of the toss of the coin, but being informed of what number x was chosen in move II, chooses a number y from $\{1, 2\}$. We now represent heads by "1" and tails by "2." Then, if the choices in the three moves were respectively u, x, and y, player P_1 is paid $M(u, x, y)$, where M is the function defined in

Example 5.1. (Thus, since $M(1, 1, 1) = -2$, we see that, in case the coin comes up heads, and P_1 and P_2 both choose 1, P_1 pays 2 to P_2.)

The game is represented in Fig. 9, where we have written the symbol "0" beside the lowest vertex to indicate that this move is made by chance (instead of by player 1 or player 2). (Merely for the sake of generality, we enclose this vertex also in a circle, as if it were an information set—though of course chance knows nothing.)

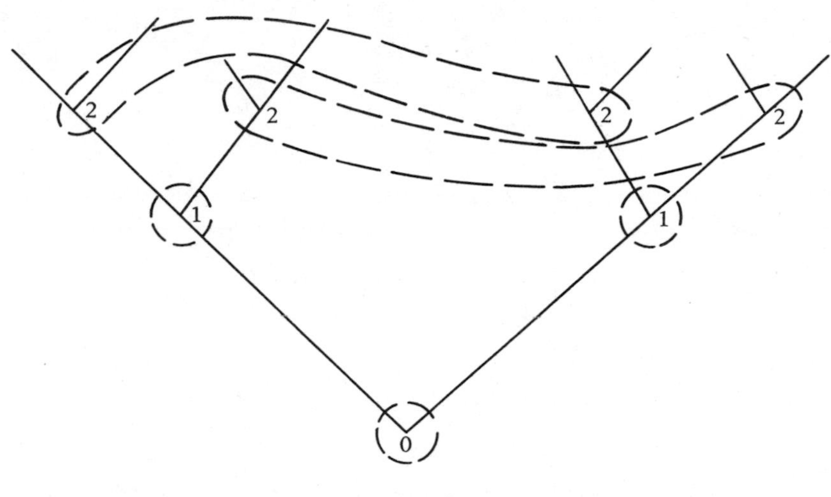

Fig. 9

A strategy for P_1 is a function which tells him whether to choose 1 or 2, according to whether the coin shows heads or tails; thus the strategy is a function f_{ij} (as defined in the discussion of Example 5.1) which maps $\{1, 2\}$ into itself.

Similarly, a strategy for P_2 is one of the functions f_{ij}, which tells him what to do according to what P_1 has done in move II.

Now suppose, for instance, that P_1 uses strategy f_{21} and that P_2 uses strategy f_{12}. Then we must distinguish two cases, according to whether the coin shows heads or tails. If it shows heads, then (remembering that "1" represents heads) the strategy f_{21} tells P_1 to choose 2 and the strategy f_{12} tells P_2 also to choose 2; thus, since $M(1, 2, 2) = -4$, the payoff to P_1 is -4. On the other hand, if the coin comes up tails, then P_1 will choose 1, and hence P_2 will choose 1; since $M(2, 1, 1) = 5$, the payoff to P_1 will therefore be 5. Since, now, we are supposing that the coin is a "true" one (i.e., one for which the probability of heads, and hence also of tails, is ½), it is seen that the mathematical expectation of P_1 is

$$(-4)\cdot\frac{1}{2}+5\cdot\frac{1}{2}=\frac{1}{2}.$$

It is natural to regard this mathematical expectation as the payoff to P_1 if these particular strategies are chosen.

In an analogous way we can calculate the payoff to P_1 for other pairs of strategies; thus we obtain Matrix 5.

MATRIX 5

	f_{11}	f_{12}	f_{21}	f_{22}
f_{11}	$\frac{3}{2}$	$\frac{3}{2}$	$\frac{1}{2}$	$\frac{1}{2}$
f_{12}	0	2	$\frac{1}{2}$	$\frac{5}{2}$
f_{21}	4	$\frac{1}{2}$	$\frac{5}{2}$	-1
f_{22}	$\frac{5}{2}$	1	$\frac{5}{2}$	1

The matrix has no saddle-point. The calculation of the value and optimal mixed strategies will be left as an exercise.

In the following example we have a simple game where the number of alternatives available to one of the players depends on chance.

EXAMPLE 5.8. In move I, player P_1 chooses a number x from $\{1, 2\}$; in move II, a number y is chosen from $\{1, 2\}$ by means of a chance device such that the probability that 1 will be chosen is ¼ (and hence, the probability that 2 will be chosen is ¾); in move III, player P_2, being informed of y, but not of x, chooses a number z from $\{1, 2\}$ in case $y = 1$ and from $\{1, 2, 3\}$ in case $y = 2$. After the three moves have been made, P_2 pays P_1 the amount $M(x, y, z)$, where M is a function defined as follows:

$$\begin{aligned} M(1,1,1) &= 2, & M(2,1,1) &= 0,\\ M(1,1,2) &= -2, & M(2,1,2) &= 5,\\ M(1,2,1) &= 1, & M(2,2,1) &= -1,\\ M(1,2,2) &= 0, & M(2,2,2) &= -3,\\ M(1,2,3) &= -2, & M(2,2,3) &= 3. \end{aligned}$$

(Here we do not give any value to $M(1, 1, 3)$ or to $M(2, 1, 3)$, since P_2 cannot choose 3 in case the chance device chose 1.)

The graph of this game is shown in Fig. 10. Here we have attached the symbols "¼" and "¾" to the two alternatives between which the chance device chooses.

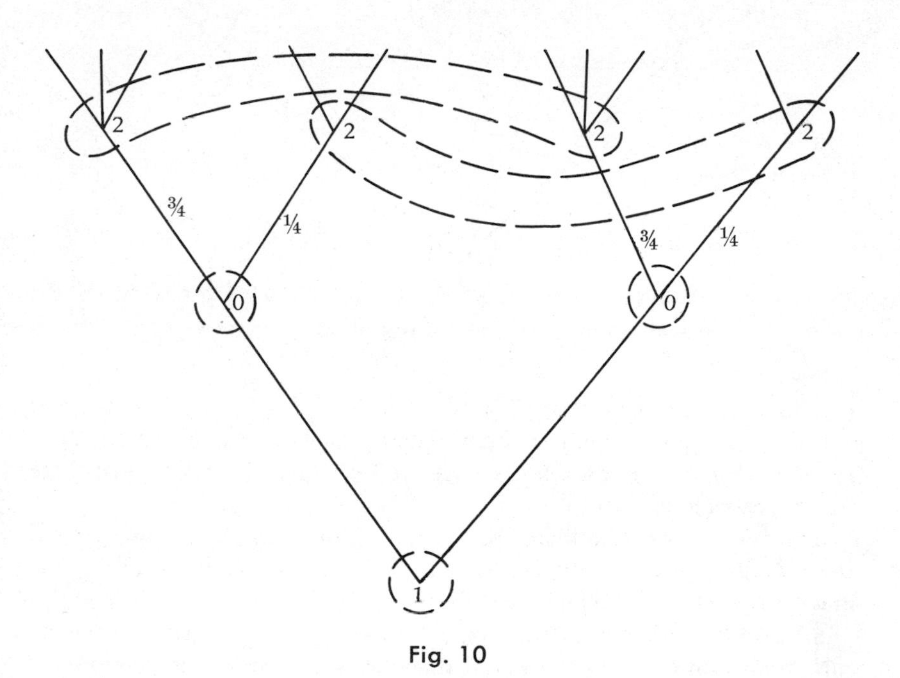

Fig. 10

The problem of enumerating the strategies and writing down the strategy matrix will be left as an exercise.

The next example is of a game where a chance move determines which player will make the next move.

Example 5.9. In move I, player P_1 chooses a number x from $\{1, 2\}$; in move II, a number y is chosen from the set $\{1, 2\}$ by means of a chance device such that the probability that 1 will be chosen is ⅕ (and hence the probability that 2 will be chosen is ⅘); in case 1 is chosen in move II, then in the last move P_2, having been informed of x and y, chooses a number z from $\{1, 2\}$; in case 2 was chosen in move II, however, the last move is made by P_1, who, having been informed of x and y, chooses a number z from $\{1, 2\}$. After the three moves, P_1 is paid the amount $M(x, y, z)$, where M is the function defined in Example 5.1.

The graph for this game is given in Fig. 11.

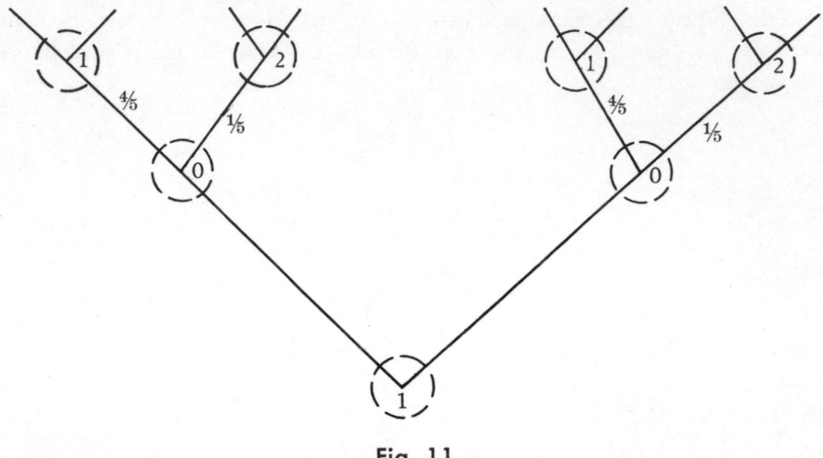

Fig. 11

Since there are three possible information sets for P_1 in this game, and since he has a choice from just two alternatives at each information set, there are altogether $2^3 = 8$ possible strategies for P_1. Similarly, we see that there are four possible strategies for P_2.

The problem of calculating the strategy matrix is left as an exercise. Since the game is a game with perfect information, it will be found that the strategy matrix has a saddle-point.

5. Games with More Than Two Players. So far we have considered only two-person games; but it is clear that our remarks apply with appropriate modifications to n-person games, where $n > 2$. In particular, a graph can be drawn for such a game, entirely analogous to the graphs already described for two-person games. And, by the introduction of the notion of a strategy, it is possible to reduce an n-person game to normal form (though, of course, we have not, up to this point, described any way of "solving" these games in normal form). These remarks will be illustrated by our next example.

EXAMPLE 5.10. In move I, a chance mechanism, which assigns probabilities ⅓ and ⅔, respectively, to the two numbers 1 and 2, chooses a number x from the set $\{1, 2\}$. If $x = 1$, then in move II, player P_1, knowing x, chooses a number y from $\{1, 2, 3\}$. If $x = 2$, then in move II, player P_2, knowing x, chooses a number y from $\{1, 2, 3\}$. If $y = 1$, then in move III, player P_3, knowing y, but not x, chooses a number z from $\{1, 2\}$. If $y \neq 1$, then in move III, player P_4, knowing x and knowing whether $y = 1$ or $y \neq 1$, chooses a number z from $\{1, 2\}$. After x, y, and z have been chosen, the payoffs to players P_1, P_2, P_3, and P_4 are respectively $M_1(x, y, z)$,

$M_2(x, y, z)$, $M_3(x, y, z)$, and $M_4(x, y, z)$, where M_1, M_2, M_3, and M_4 are certain real-valued functions.

The graph of this game is given in Fig. 12.

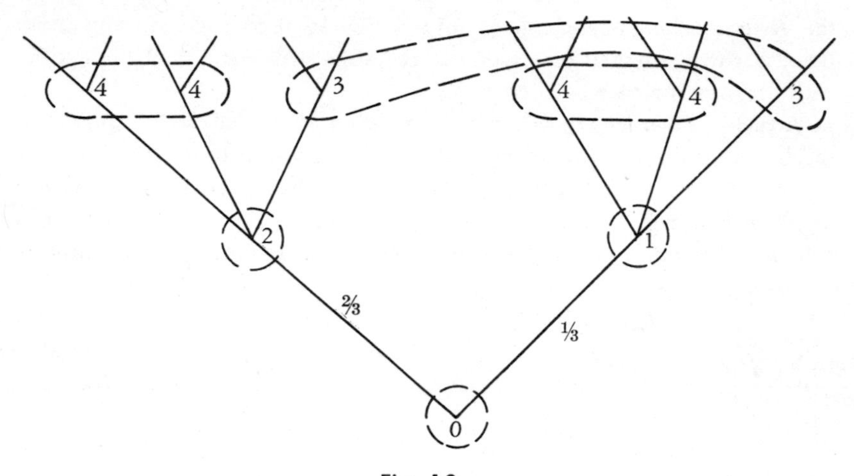

Fig. 12

A strategy for P_1 is simply one of the three numbers 1, 2, and 3, and similarly for P_2. A strategy for P_3 is simply one of the numbers 1 and 2. A strategy for P_4 must tell him whether to choose $z = 1$ or $z = 2$ in case $x = 1$ or $x = 2$. Thus we can represent the four possible strategies for P_4 by f_{11}, f_{12}, f_{21}, and f_{22}. To say that P_4 uses strategy f_{21}, for instance, means that he will choose $z = 2$ if $x = 1$, and $z = 1$ if $x = 2$.

In order to see how to calculate the elements of the strategy matrix, let us suppose, for instance, that P_1 uses his strategy 1, that P_2 uses his strategy 3, that P_3 uses strategy 2, and that P_4 uses strategy f_{21}. As in the discussion of Example 5.7, it is necessary to consider separately the cases in which $x = 1$ and $x = 2$.

If $x = 1$, then move II is made by P_1, and hence $y = 1$. Then move III is made by P_3, and hence $z = 2$. Consequently, in this case the payoff to player P_i (for $i = 1, 2, 3, 4$) is $M_i(1, 1, 2)$.

If $x = 2$, then move II is made by P_2, and hence $y = 3$. Then move III is made by P_4, and thus $z = f_{21}(x) = f_{21}(2) = 1$. Hence in this case the payoff to player P_i (for $i = 1, 2, 3, 4$) is $M_i(2, 3, 1)$.

Since the probability that $x = 1$ is ⅓, we conclude that, if the players use the strategies in question, the expected payoff to player P_i is

$$\frac{1}{3} M_i(1, 1, 2) + \frac{2}{3} M_i(2, 3, 1).$$

6. Restrictions on Information Sets. From Example 5.6 we see that it is not the case that every vertex belonging to a given information set must be at the same level (i.e., must have the same number of vertices preceding it). Thus it might be thought that it is unnecessary to put any restrictions on the vertices belonging to a given information set, except the obvious conditions that they must all correspond to the same player and must present him with the same number of alternatives.

Actually, however, an additional condition is needed for a game to be capable of realization: It must not happen that any play intersect the same information set more than once. This means that a game cannot have a graph such as Fig. 13, for instance, where the three vertices A, B, and C constitute a play which has two vertices, A and C, in common with an information set. Similarly, the condition excludes the possibility that a one-person game will have the simple graph given in Fig. 14.

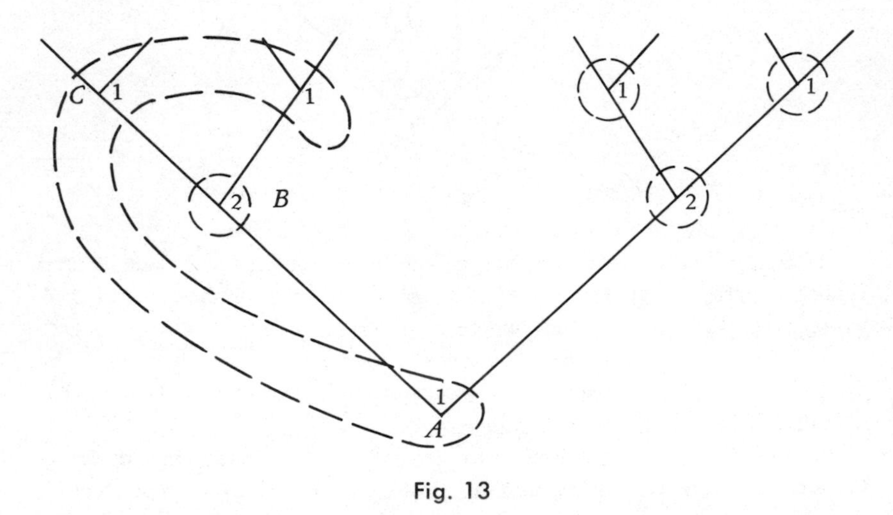

Fig. 13

To see why this condition must be imposed, suppose, for instance, that we were to try to arrange for an actual play of a game having Fig. 14 for its graph. It is clear that we would need to have two human beings to take the part of player P_1; for if there were only one, the second time he was asked to choose he would remember that he had already been asked, and hence he would know where he was situated on the graph, which is contrary to the fact that the information set requires that he not have such knowledge.

Hence, suppose that there are two human beings, A and B, who, together, constitute P_1, and suppose that they are put into separate rooms and are not allowed to communicate with each other. After the beginning of the play, the umpire is to go into one room and ask the person in that room to choose a number from $\{1, 2\}$; he then goes into the other room and asks the other person to choose a number from $\{1, 2\}$. The amount that P_1 will then be paid is $M(x, y)$, where x is the first number chosen, y is the second number chosen, and M is a certain real-valued function defined over the Cartesian product of $\{1, 2\}$ by itself.

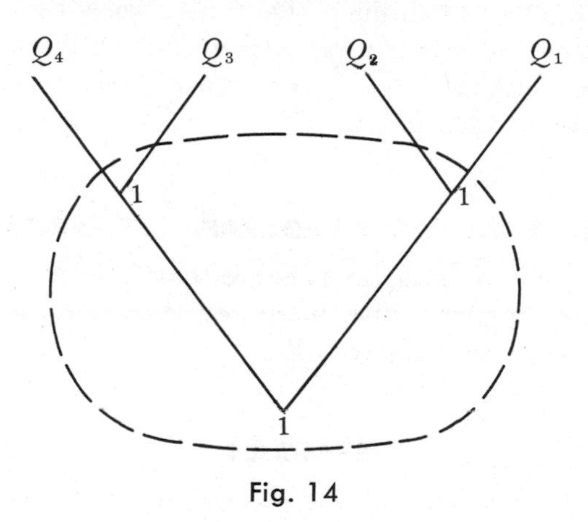

Fig. 14

Although A and B are not allowed to communicate with each other during the play, we nevertheless suppose that they can get together before the game starts to talk about what they should do. They must, then, decide whether both will pick 1, or both will pick 2; or whether A will pick 1 and B will pick 2; or whether A will pick 2 and B will pick 1. Unfortunately for them, however, they do not know whether the umpire will choose A first or B first, and hence they do not know whether the number chosen by A will be called "x" or "y"; and, of course, it is not necessarily true that $M(x, y) = M(y, x)$. As a matter of fact, they do not even know the probability that the umpire will choose A first; hence it is impossible for them to calculate the expected payoff to them in case A chooses x and B chooses y. If α were the probability that the umpire would choose A first, then these choices by A and B would give them an expected payoff of

$$\alpha M(x, y) + (1 - \alpha)M(y, x).$$

It might be thought that this kind of difficulty could be avoided by considering games quite abstractly—i.e., by simply abandoning the assumption, let us say, that a game is necessarily played by people and by supposing that the "players" are some kind of machines without any memories. But even so, if the machine were set to choose the right-hand alternative in the game graphed above, it would choose this alternative both times, and hence end up at Q_1; if it were set to choose the left-hand alternative, it would necessarily end up at Q_4. Thus—and this is a crucial point—the vertices Q_3 and Q_2 can never be reached, and hence they might as well be omitted from the graph.

It therefore appears impossible to find a realization of a game corresponding to Fig. 14 in which the people playing know the consequences to be expected if they behave in one way or another. For this reason, we exclude such graphs from consideration; i.e., we define "game" in such a way that a game cannot have such a graph.

HISTORICAL AND BIBLIOGRAPHICAL REMARKS

A detailed account of games in extensive form is given in von Neumann and Morgenstern [1]. We have based our discussion, however, on the formulation to be found in Kuhn [2].

EXERCISES

1. Find the value and optimal mixed strategies for the game in Example 5.7.

2. Determine the strategy matrices and solutions for the games in Examples 5.4 and 5.8.

3. Determine the strategy matrix for the game in Example 5.9. Find optimal pure strategies for the two players.

4. Draw the graph of a game in which the following moves are made:

Move I. P_1 chooses a number x from $\{1, 2, 3, 4\}$;

Move II. P_2, having been informed whether x is even or odd, chooses a number y from $\{1, 2\}$;

Move III. If $y = 1$, then a chance device chooses a number z from $\{1, 2\}$ in such a manner that the probability of drawing 1 is $1/10$. If $y = 2$, then P_1, knowing x and y, chooses a number z from $\{1, 2\}$;

Move IV. P_1, knowing y, but not z, and having forgotten x, chooses a number w from $\{1, 2\}$.

5. How many strategies are there for the game of ticktacktoe?

6. Find a payoff function M which will make the value of the game in Example 5.5 different from the value of the game in Example 5.2.

7. Show that it is sometimes useful to remember what one has done.

8. We are given the following game: In move I, player P_1 chooses a number x from $\{1, 2\}$. In move II, player P_2, not knowing x, chooses a number y from $\{1, 2\}$. In move III, a chance device picks a number z from $\{1, 2\}$ in such a manner as to assign probability α to 1 and $(1 - \alpha)$ to 2. After x, y, and z are chosen, P_1 is paid an amount $M_1(x, y, z)$ and P_2 is paid an amount $M_2(x, y, z)$. (We are not assuming that we necessarily have, for all x, y, and z, $M_2(x, y, z) = -M_1(x, y, z)$.)

The graph of this game is given in Fig. 15.

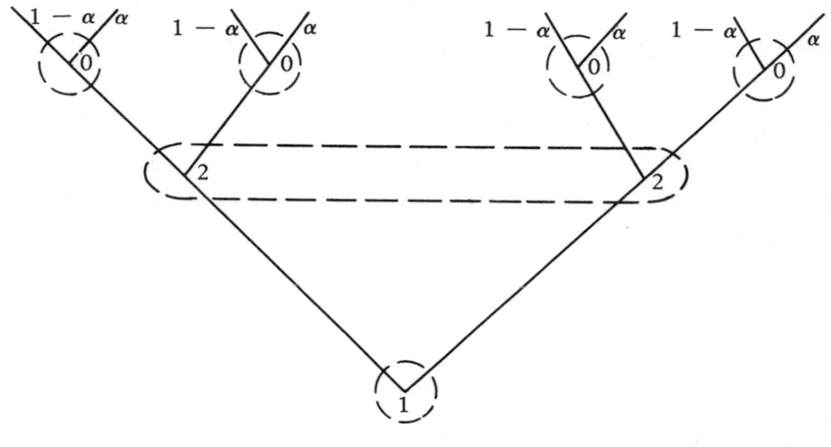

Fig. 15

Enumerate the strategies for this game, find the expected payoff to each of the players for each possible pair of strategies, and find the conditions on α, M_1, and M_2 in order that the game be zero-sum (i.e., in order that the expected payoff to P_2 always be the negative of the expected payoff to P_1). What are the conditions which must be imposed on M_1 and M_2 in order that the game be zero-sum independently of the value of α?

9. What are the conditions on the functions M_1, M_2, M_3, and M_4 in order that the game in Example 5.10 be zero-sum (i.e., in order that, for all strategies, the sum of the expected payoffs to the four players be zero)?

CHAPTER 6

GAMES IN EXTENSIVE FORM—GENERAL THEORY

1. General Definition of Finite Games. In our examples in Chap. 5 we found it very helpful to represent games in extensive form by means of graphs; indeed, it seems that the graphs are often more useful and perspicuous than verbal descriptions. For this reason it is formally convenient simply to identify games with the graphs used to represent them. We shall make such an identification in this chapter, where we are interested in presenting a precise definition of games in extensive form and in proving an important theorem about them.

Thus we shall regard a finite n-person game as a system consisting of:

1. A tree T (in the sense explained in Chap. 5).
2. n real-valued functions $F_1, \cdots, F_n$, which are defined at each of the top points of T; i.e., if t is such a top point, then $F_i(t)$ is the amount to be paid player P_i when the play terminates at t.
3. An assignment, to each branch point of T, of one of the numbers $0, 1, \cdots, n$ to indicate which player moves at the point in question (the number 0 means that a chance device is used at the point).
4. For each branch point q of T to which 0 is assigned by item 3, above, a member $\| x_1 \quad \cdots \quad x_k \|$ of $\mathbf{S}_k$, where k is the number of alternatives of q, i.e., the number of lines rising from q.
5. A partition of the branch points into mutually exclusive and exhaustive sets (the information sets) which satisfy the following assumptions:

   (*a*) All the branch points belonging to a given information set are associated, by item 3, with the same player.

   (*b*) All the branch points belonging to a given information set have the same number of alternatives, which we shall always number from right to left.

   (*c*) If 0 is assigned to a branch point q by item 3, then the information set in which it lies consists of just the one point q.

   (*d*) If S is any play of the game, i.e., if S is a broken line going from the bottom of the tree to one of its top points, and if A is any information set, then there exists, at most, one branch point which belongs both to S and to A.

From the examples in Chap. 5 it should be obvious what functions are served by the various parts of this formal apparatus and what reasons we have for making the four assumptions about information sets. It will be clear, by the way, that assumption (c), above, is quite arbitrary and is made merely to make things definite.

In terms of this general definition of games, we can now define strategies as follows: By a *strategy* for player P_i (for $i = 1, \cdots, n$), we mean a function which is defined for each information set corresponding to P_i and whose value for each such information set is one of the alternatives there available to P_i; thus a strategy tells the player what to do for every possible state of his knowledge. Thus consider a game whose graph is given in Fig. 1 (in this figure we have, for reasons which will appear presently, given names to the various branch points and end points).

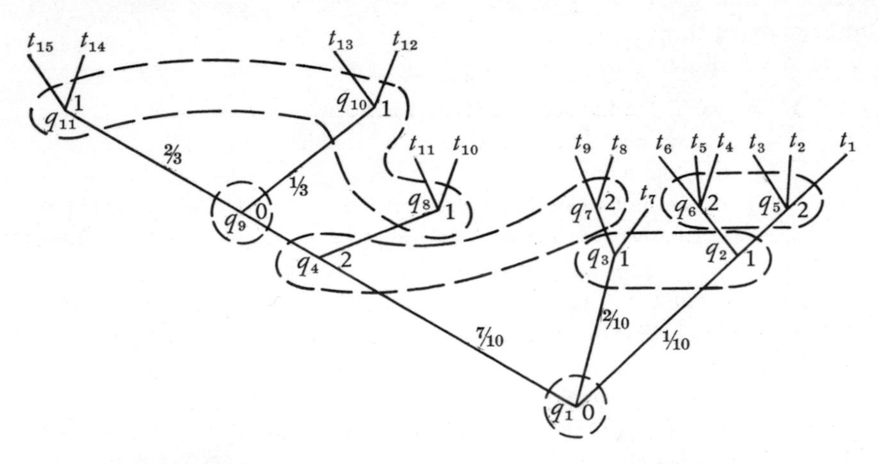

Fig. 1

Here there are just two information sets for P_1, namely, $\{q_2, q_3\}$ and $\{q_8, q_{10}, q_{11}\}$, and there are just two information sets for P_2, namely, $\{q_4, q_7\}$ and $\{q_5, q_6\}$. Hence a strategy for P_1 is any one of the four functions which are defined for the arguments $\{q_2, q_3\}$ and $\{q_8, q_{10}, q_{11}\}$ and assume as values the numbers 1 and 2. A strategy for P_2 is a function F, defined for $\{q_4, q_7\}$ and $\{q_5, q_6\}$, such that

$$F(\{q_4, q_7\}) \in \{1, 2\},$$

$$F(\{q_5, q_6\}) \in \{1, 2, 3\}.$$

To say that P_2 plays the strategy F such that

$$F(\{q_4, q_7\}) = 1,$$
$$F(\{q_5, q_6\}) = 2,$$

for instance, means that whenever, in the course of the play, P_2 finds himself at one of the points q_4 or q_7, he will choose the right-hand alternative, and that whenever he finds himself at one of the points q_5 or q_6, he will choose the middle alternative.

If each player P_i chooses a strategy x_i, then the vector $x = \| x_1 \quad \cdots \quad x_n \|$ determines, for each alternative at each branch point, a probability: the probability that a play which has reached the given branch point will continue along the alternative in question. If the choice at the point is to be made by one of the players,. then this probability is 1 or 0, according as the player's strategy tells him to choose, or not to choose, that alternative. If the choice is to be made by chance, then the probability is determined by item 4 of the definition of the game. We indicate these probabilities by $p(x, q, i)$, where x is the (ordered) set of strategies being used by the players, q is the branch point, and i is the alternative.

Thus, for the game graphed in Fig. 1, suppose that P_1 uses a strategy F such that

$$F(\{q_2, q_3\}) = 1,$$
$$F(\{q_8, q_{10}, q_{11}\}) = 2,$$

and that P_2 uses a strategy G such that

$$G(\{q_4, q_7\}) = 2,$$
$$G(\{q_5, q_6\}) = 3;$$

and let us denote by a the ordered couple $\| F \quad G \|$. Then it is readily verified that we have

$$p(a, q_1, 1) = \frac{1}{10}, \qquad p(a, q_1, 2) = \frac{2}{10}, \qquad p(a, q_1, 3) = \frac{7}{10},$$
$$p(a, q_2, 1) = p(a, q_3, 1) = 1,$$
$$p(a, q_2, 2) = p(a, q_3, 2) = 0,$$
$$p(a, q_5, 1) = p(a, q_6, 1) = 0,$$
$$p(a, q_5, 2) = p(a, q_6, 2) = 0,$$
$$p(a, q_5, 3) = p(a, q_6, 3) = 1,$$
$$p(a, q_4, 1) = p(a, q_7, 1) = 0,$$
$$p(a, q_4, 2) = p(a, q_7, 2) = 1,$$

$$p(a, q_9, 1) = \frac{1}{3}, \qquad p(a, q_9, 2) = \frac{2}{3},$$

$$p(a, q_8, 1) = p(a, q_9, 1) = p(a, q_{11}, 1) = 0,$$

$$p(a, q_8, 2) = p(a, q_9, 2) = p(a, q_{11}, 2) = 1.$$

For some purposes, however, it is preferable to indicate the direction in which a choice leads us, by specifying a top point to which it would lead rather than by specifying the number of the alternative. Thus, in Fig. 1 the first alternative at q_1 could lead to one of the top points t_1, t_2, t_3, t_4, t_5, and t_6; the second alternative, to one of the top points t_7, t_8, and t_9; and the third alternative, to one of the alternatives t_{10}, t_{11}, t_{12}, t_{13}, t_{14}, and t_{15}. The first alternative at q_2 could lead to t_1, t_2, or t_3, and the second alternative, to t_4, t_5, or t_6.

We adopt the notation $p(x, q, t)$, where x is an ordered set of strategies, q is a branch point, and t is a top point, to mean the probability that if the branch point q is reached in a play, then the play will continue along a path that might end at t. Thus, for the game graphed in Fig. 1, instead of writing

$$p(a, q_1, 1) = \frac{1}{10},$$

we might just as well write

$$p(a, q_1, t_1) = \frac{1}{10}$$

or

$$p(a, q_1, t_2) = \frac{1}{10},$$

etc. And instead of writing

$$p(a, q_4, 1) = 0,$$

we could just as well write

$$p(a, q_4, t_{10}) = 0$$

or

$$p(a, q_4, t_{11}) = 0.$$

Using this notation, we shall, of course, also have $p(x, q, t) = 0$ whenever t is not the end point of any play which passes through q. Thus, for instance, for the game graphed in Fig. 1, we have

$$p(a, q_2, t_7) = 0,$$

since a play which goes through q_2 cannot end at t_7.

In terms of this function we can also define the probability, for any fixed strategies for the various players, that the play will terminate at any given top point. Let x be an ordered set of pure strategies for the n players, let t be the top point corresponding to some play, and let $q_1, q_2, \cdots, q_r$ be the branch points of the play which terminates at t. Then, if we set

$$p(x, t) = \prod_{i=1}^{r} p(x, q_i, t),$$

it is clear that $p(x, t)$ is the probability that the play will terminate with t if the players use the ordered set x of strategies. (No confusion can arise from using the same letter "p" for both functions, since one is a function of two variables, and the other, of three.)

Thus, for the game graphed in Fig. 1, and using the same set a of strategies as before, we have

$$p(a, t_1) = p(a, q_1, t_1) \cdot p(a, q_2, t_1) \cdot p(a, q_5, t_1) = \frac{1}{10} \cdot 1 \cdot 0 = 0,$$

$$p(a, t_2) = p(a, q_1, t_2) \cdot p(a, q_2, t_2) \cdot p(a, q_5, t_2) = \frac{1}{10} \cdot 1 \cdot 0 = 0,$$

$$p(a, t_3) = p(a, q_1, t_3) \cdot p(a, q_2, t_3) \cdot p(a, q_5, t_3) = \frac{1}{10} \cdot 1 \cdot 1 = \frac{1}{10}.$$

Similarly, we find

$$\begin{aligned} p(a, t_4) &= 0, & p(a, t_{10}) &= 0, \\ p(a, t_5) &= 0, & p(a, t_{11}) &= 0, \\ p(a, t_6) &= 0, & p(a, t_{12}) &= 0, \\ p(a, t_7) &= \frac{2}{10}, & p(a, t_{13}) &= \frac{7}{30}, \\ p(a, t_8) &= 0, & p(a, t_{14}) &= 0, \\ p(a, t_9) &= 0, & p(a, t_{15}) &= \frac{7}{15}. \end{aligned}$$

As a check, we observe that

$$\sum_{i=1}^{15} p(a, t_i) = 1.$$

In terms of the probabilities just introduced, we can now, in turn, find the expectations of the players for the various choices of pure strategies. Let x be an ordered set of strategies, let $t_1, \cdots, t_s$ be the top points of the graph of the game, and let H_i be the payoff function for player P_i. Then the expectation of P_i, which we denote by $M_i(x)$, is given by the formula

$$M_i(x) = \sum_{j=1}^{s} H_i(t_j) \cdot p(x, t_j).$$

Thus, for instance, suppose that the payoff function, H_1, for the first player of the game of Fig. 1 is as follows:

$$\begin{aligned} H_1(t_1) &= 10, & H_1(t_8) &= 30, \\ H_1(t_2) &= -10, & H_1(t_9) &= 20, \\ H_1(t_3) &= 10, & H_1(t_{10}) &= -30, \\ H_1(t_4) &= 20, & H_1(t_{11}) &= 0, \\ H_1(t_5) &= 30, & H_1(t_{12}) &= 30, \\ H_1(t_6) &= 0, & H_1(t_{13}) &= -30, \\ H_1(t_7) &= -10, & H_1(t_{14}) &= 40, \end{aligned}$$
$$H_1(t_{15}) = 15.$$

Then, for the ordered set a of strategies considered earlier, we have

$$\begin{aligned} M_1(a) &= \sum_{j=1}^{15} H_1(t_j) \cdot p(a, t_j) \\ &= 10 \cdot \frac{1}{10} + (-10) \cdot \frac{2}{10} + (-30) \cdot \frac{7}{30} + 15 \cdot \frac{7}{15} = -1. \end{aligned}$$

We have now shown quite generally that the problem of solving an arbitrary finite game reduces to the problem of solving a game in rectangular form. We find this normalized form of the original game by enumerating all possible strategies for the various players and then calculating the values of the functions $M_1, \cdots, M_n$.

We now have two kinds of payoff functions: (1) the functions $H_1, \cdots, H_n$, which are defined over the set of top points of the graph of the game and which tell what each player will be paid in case the play terminates in a given way; and (2) the functions $M_1, \cdots, M_n$, which are defined over the set of ordered n-tuples of pure strategies and which tell what each player will be paid (on the average) in case the various players use given pure strategies. When it is desirable to distinguish between the two kinds of payoff functions, we shall call the first kind *play payoff functions* and the second kind *strategy payoff functions*. Ordinarily, however, it will be clear from the context which kind of payoff functions we have in mind, so we shall omit the qualifications "play" and "strategy."

It should be noticed that the correct definition of a zero-sum game is in terms of strategy payoff functions rather than play payoff functions. If $M_1, \cdots, M_n$ are the strategy payoff functions, then the game is called *zero-sum* if, for every n-tuple x of strategies for the players $P_1, \cdots, P_n$, we have

$$\sum_{i=1}^{n} M_i(x) = 0.$$

From Exercises 8 and 9 of Chap. 5, we see that this condition can be satisfied in some cases, even when the corresponding equation for play payoff functions does not hold.

2. Games with Perfect Information—Equilibrium Points. We turn now to a more special kind of games—the so-called games with "perfect information." These games are characterized by the fact that at every point in every play the player whose turn it is to move knows exactly what choices have been made previously. So far as the graph of the game is concerned, this means that every information set is a one-element set. Thus, for example, a game whose graph is given in Fig. 2 is a game with perfect information.

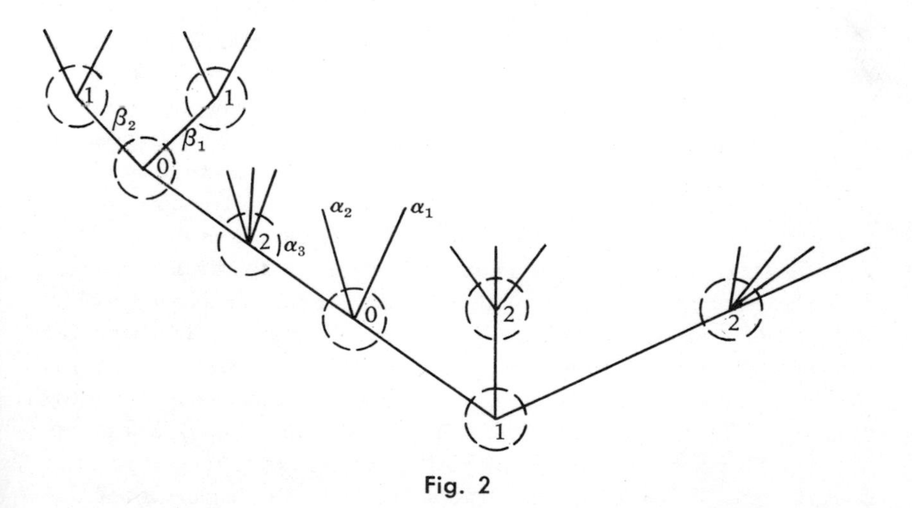

Fig. 2

Since, in the graph of a game with perfect information, each dotted line encloses just one branch point, we can, when it is understood that such a game is under consideration, simply omit the dotted lines altogether. Thus Fig. 2 can be replaced by Fig. 3.

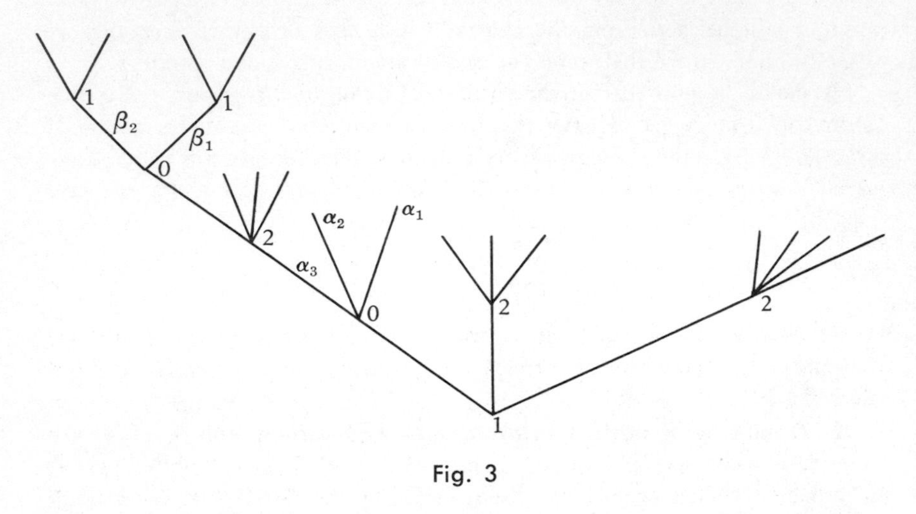

Fig. 3

The student will recognize that some common parlor games are games with perfect information. Ticktacktoe, checkers, chess, and backgammon, for instance, are of this sort. On the other hand, most card games (e.g., bridge, poker, and canasta) are not games with perfect information, since the players do not know what cards have been dealt to the other players.

We are going to show now that the matrix of the normal form of any zero-sum two-person game with perfect information has a saddle-point, i.e., that there are optimal pure strategies for such a game. For the case of so simple a game as ticktacktoe, this result is, of course, known to everyone who has ever played the game a few times: each player of ticktacktoe can play in such a way as to ensure that he will win, if the other player does not play correctly, or that the game will be a draw. This is the reason that adults do not often play ticktacktoe: after optimal strategies for a game are known, it ceases to offer any intellectual challenge, and people stop playing it. It is therefore of some interest to know that there are also optimal pure strategies for chess. In order to find such optimal strategies, it would be necessary merely to enumerate all strategies for the game, to write out the matrix (putting "1," "0," and "-1" for "win," "draw," and "lose"), and to pick out a saddle-point. The number of possible strategies for chess is so great, however, that it hardly appears feasible to make this enumeration; thus people will probably continue to play chess for some time.

Actually, because of the fact that our proof will be by mathematical induction, it turns out to be more convenient to prove a somewhat stronger theorem than the one mentioned above—a theorem which applies to all two-person games with perfect information, rather than merely to zero-sum two-person games with perfect information. In order to formulate this stronger theorem, it is necessary to introduce the notion of an equilibrium point, which is a generalization of the notion of a saddle-point.

Let the strategy payoff functions of an n-person game be $M_1, \cdots, M_n$, let the pure strategies available to player P_i (for $i = 1, \cdots, n$) be A_i, and let A be the Cartesian product of $\mathsf{A}_1, \cdots, \mathsf{A}_n$. Then we say that an element $\| x_1 \quad \cdots \quad x_n \|$ of A is an *equilibrium point*, if, for each i, and for y any element of A_i, we have

$$M_i(x_1, \cdots, x_n) \geq M_i(x_1, \cdots, x_{i-1}, y, x_{i+1}, \cdots, x_n).$$

The intuitive meaning of an equilibrium point is this: It is a way of playing such that, if all the players but one adhere to it, the remaining player cannot do better than adhere to it also.

In the case of a zero-sum two-person game, an equilibrium point is the same as a saddle-point of the matrix of the normal form of the game. For suppose that $\| x_1 \quad x_2 \|$ is an equilibrium point of a zero-sum two-person game with payoff functions M_1 and M_2, and let A_1 and A_2 be the strategies available to players P_1 and P_2. Since $\| x_1 \quad x_2 \|$ is an equilibrium point, we have, for y_1 any member of A_1 and y_2 any member of A_2,

$$M_1(x_1, x_2) \geq M_1(y_1, x_2),$$
$$M_2(x_1, x_2) \geq M_2(x_1, y_2).$$

Since the game is zero-sum, however, we have

$$M_2(x_1, x_2) = -M_1(x_1, x_2),$$
$$M_2(x_1, y_2) = -M_1(x_1, y_2);$$

thus from the second inequality above we conclude that

$$-M_1(x_1, x_2) \geq -M_1(x_1, y_2),$$

or

$$M_1(x_1, x_2) \leq M_1(x_1, y_2).$$

Hence we have

$$M_1(y_1, x_2) \leq M_1(x_1, x_2) \leq M_1(x_1, y_2),$$

which means that $\| x_1 \quad x_2 \|$ is a saddle-point of the matrix of the normal form of the game, as was to be shown. Conversely, it is easily seen that a saddle-point is also an equilibrium point.

Thus, in order to show that every zero-sum two-person game with perfect information has a saddle-point, it will certainly suffice to show that every two-person game with perfect information has an equilibrium point; and this is what we shall do.

In order to carry out our proof, it is convenient to introduce the notion of the *truncations* of a game with perfect information. By this we mean the games which arise from the given game by deleting the first move. Thus suppose that a game has the graph indicated in Fig. 4, and that the payoff

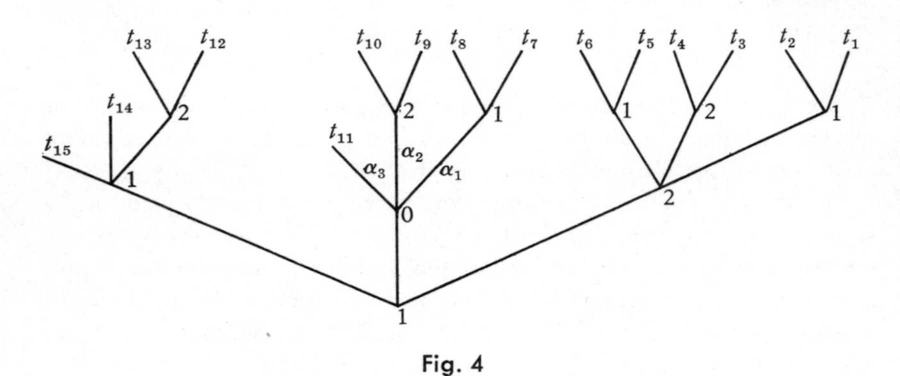

Fig. 4

function for P_i (for $i = 1, 2$) is given by a real-valued function H_i, defined over the points $t_1, \cdots, t_{15}$. Then there are three truncations of this game, corresponding to the three alternatives at the first move. Their graphs are given in Figs. 5, 6, and 7. The payoff functions for these truncated games

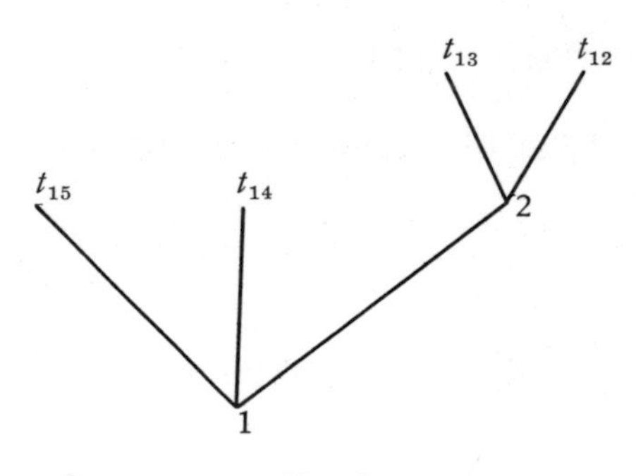

Fig. 5

are the original payoff functions, but with their domains of definition suitably restricted. Thus the payoff function (for P_i) for the game graphed in Fig. 5

is the function K_i, which is defined over the points t_{12}, t_{13}, t_{14}, t_{15}, and is such that

$$K_i(t_j) = H_i(t_j) \qquad \text{for } j = 12, 13, 14, 15.$$

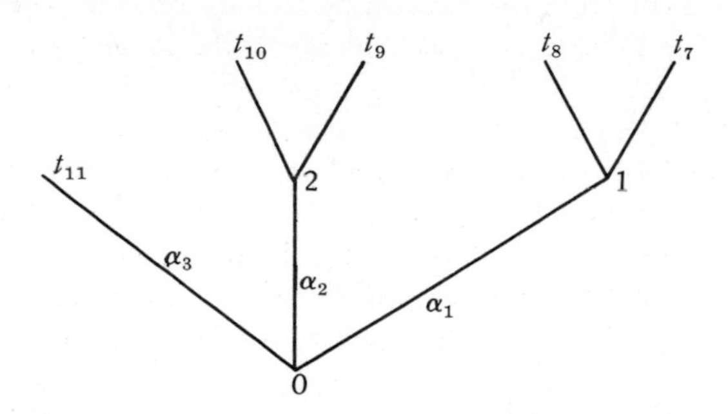

Fig. 6

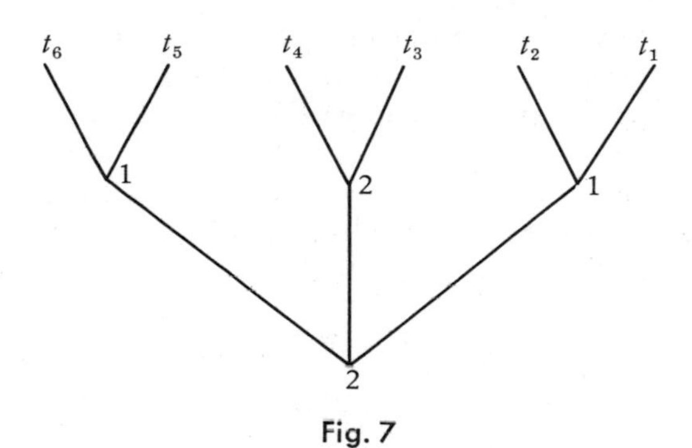

Fig. 7

(It should be noticed that the possibility of forming truncated games in this manner depends on the fact that the game is a game with perfect information. If we were to try to define truncations for such a game as that given in Fig. 1, for example, we should be presented with the problem of splitting in some way the information sets $\{q_2, q_3\}$ and $\{q_4, q_7\}$.)

Since a strategy for a player of a game with perfect information is a function which picks out an alternative at each of the player's moves, we can also consider the truncations of a given strategy corresponding to the various truncations of the game; a *truncation of a strategy* is defined only over the branch points of the corresponding truncation of the game, and it picks out the same alternatives at those branch points as does the original strategy.

THEOREM 6.1. Let $\mathsf{A}^{(1)}$ and $\mathsf{A}^{(2)}$ be the sets of strategies available to P_1 and P_2 in a two-person game with perfect information, and let A be their Cartesian product. Then A has an equilibrium point.

PROOF. We shall prove the theorem by an induction on the *length* of the game, i.e., on the number of branch points in the longest possible play of the game.

For games of length 0 (i.e., games with no moves), the theorem is obvious. For in this case each of the players has just one strategy, which consists in doing nothing, and hence the Cartesian product A also has only one member, which is an equilibrium point by definition.

Hence we suppose that the theorem is true for all games of length less than k, and we let Γ be a game of length k. Suppose that there are r alternatives on the first move, and let $\Gamma_1, \Gamma_2, \cdots, \Gamma_r$ be the r truncations of Γ (they exist because of the fact that Γ is a game with perfect information).

For each of the games Γ_i (for $i = 1, \cdots, r$), let $\mathsf{A}_i^{(1)}$ be the set of pure strategies available to P_1, let $\mathsf{A}_i^{(2)}$ be the set of pure strategies available to P_2, and let A_i be the Cartesian product of $\mathsf{A}_i^{(1)}$ and $\mathsf{A}_i^{(2)}$. As in the statement of the theorem, we let $\mathsf{A}^{(1)}$ and $\mathsf{A}^{(2)}$ be the strategies available to P_1 and P_2 in the original game Γ, and we let A be the Cartesian product of $\mathsf{A}^{(1)}$ and $\mathsf{A}^{(2)}$.

By the induction hypothesis, there is an equilibrium point in each of the sets A_i. For each i, let $\| f_i^* \quad g_i^* \|$ be an equilibrium point of A_i. Then, letting $M_i^{(1)}$ and $M_i^{(2)}$ be the strategy payoff functions in the game Γ_i, we have

$$\left.\begin{aligned} M_i^{(1)}(f_i^*, g_i^*) &\geq M_i^{(1)}(f_i, g_i^*), \\ M_i^{(2)}(f_i^*, g_i^*) &\geq M_i^{(2)}(f_i^*, g_i), \end{aligned}\right\} \tag{1}$$

for $i = 1, \cdots, r$, and for f_i any member of $\mathsf{A}_i^{(1)}$ and for g_i any member of $\mathsf{A}_i^{(2)}$.

We now distinguish three cases: (1) the first move of Γ is a chance move, (2) the first move of Γ is made by P_1, and (3) the first move of Γ is made by P_2.

CASE 1. If q is a branch point of one of the truncated games Γ_i, and corresponds to a move made by P_1, we set

$$f^*(q) = f_i^*(q).$$

Similarly, if q is a branch point of one of the truncated games Γ_i, and corresponds to a move made by P_2, we set

$$g^*(q) = g_i^*(q).$$

Since, by hypothesis, the first move is made by chance (i.e., by neither P_1 nor P_2), it is clear that f^* is defined over every branch point of Γ which corresponds to a move made by P_1, and hence is a member of $\mathsf{A}^{(1)}$. Similarly, g^* is a member of $\mathsf{A}^{(2)}$. We wish to show that $\| f^* \quad g^* \|$, which is thus in A, is an equilibrium point of A.

Let the probabilities assigned to the r alternatives at the first move be $\alpha_1, \alpha_2, \cdots, \alpha_r$, so that $\alpha_i \geq 0$ (for $i = 1, \cdots, r$) and $\sum_{i=1}^{r} \alpha_i = 1$; let $M^{(1)}$ and $M^{(2)}$ be the strategy payoff functions for P_1 and P_2, respectively, in the game Γ. Then it is clear that if f and g are any strategies for P_1 and P_2 in Γ, and if f_i and g_i (for $i = 1, \cdots, n$) are the truncations of these strategies for the truncated game Γ_i, then

$$M^{(1)}(f, g) = \sum_{i=1}^{r} \alpha_i M_i^{(1)}(f_i, g_i),$$

$$M^{(2)}(f, g) = \sum_{i=1}^{r} \alpha_i M_i^{(2)}(f_i, g_i).$$

In particular, since the strategies $f_1^*, \cdots, f_r^*$ are truncations of f^* and $g_1^*, \cdots \; g_r^*$ are truncations of g^*, we have

$$\left. \begin{aligned} M^{(1)}(f, g^*) &= \sum_{i=1}^{r} \alpha_i M_i^{(1)}(f_i, g_i^*), \\ M^{(2)}(f^*, g) &= \sum_{i=1}^{r} \alpha_i M_i^{(2)}(f_i^*, g_i) \end{aligned} \right\} \tag{2}$$

and

$$\left. \begin{aligned} M^{(1)}(f^*, g^*) &= \sum_{i=1}^{r} \alpha_i M_i^{(1)}(f_i^*, g_i^*), \\ M^{(2)}(f^*, g^*) &= \sum_{i=1}^{r} \alpha_i M_i^{(2)}(f_i^*, g_i^*). \end{aligned} \right\} \tag{3}$$

From (1), remembering that $\alpha_1, \cdots, \alpha_r$ are all non-negative, we derive

$$\left. \begin{aligned} \sum_{i=1}^{r} \alpha_i M_i^{(1)}(f_i^*, g_i^*) &\geq \sum_{i=1}^{r} \alpha_i M_i^{(1)}(f_i, g_i^*), \\ \sum_{i=1}^{r} \alpha_i M_i^{(2)}(f_i^*, g_i^*) &\geq \sum_{i=1}^{r} \alpha_i M_i^{(2)}(f_i^*, g_i). \end{aligned} \right\} \tag{4}$$

Substituting (2) and (3) in (4), we obtain

$$M^{(1)}(f^*, g^*) \geq M^{(1)}(f, g^*),$$
$$M^{(2)}(f^*, g^*) \geq M^{(2)}(f^*, g),$$

so that $\| f^* \quad g^* \|$ is indeed an equilibrium point of Γ, as was to be shown.

CASE 2. In this case the first move, q_0, of Γ is made by P_1. The set of numbers

$$\{M_1^{(1)}(f_1^*, g_1^*), M_2^{(1)}(f_2^*, g_2^*), \cdots, M_r^{(1)}(f_r^*, g_r^*)\}$$

is finite, and hence it has a maximum; let μ be an integer such that

$$M_\mu^{(1)}(f_\mu^*, g_\mu^*) = \max_i M_i^{(1)}(f_i^*, g_i^*). \tag{5}$$

We now define a function f^* by setting

$$f^*(q_0) = \mu, \tag{6}$$

and if q is a point of one of the truncated games Γ_i which corresponds to a move made by P_1, then

$$f^*(q) = f_i^*(q).$$

We define a function g^* exactly as in Case 1. That is, if q is a point of one of the truncated games Γ_i which corresponds to a move made by P_2, we set

$$g^*(q) = g_i^*(q).$$

It is clear that f^* and g^* are strategies for P_1 and P_2, respectively, in the game Γ. We want to show that $\| f^* \quad g^* \|$ is an equilibrium point of A.

From (6) we see that if g is any strategy for P_2 in Γ, and if g_μ is its truncation to Γ_μ, then

$$M^{(1)}(f^*, g) = M_\mu^{(1)}(f_\mu^*, g_\mu) \tag{7}$$

and

$$M^{(2)}(f^*, g) = M_\mu^{(2)}(f_\mu^*, g_\mu). \tag{8}$$

From (7) and (8), since g_μ^* is the truncation of g^* to Γ_μ, we have, in particular,

$$M^{(1)}(f^*, g^*) = M_\mu^{(1)}(f_\mu^*, g_\mu^*) \tag{9}$$

and

$$M^{(2)}(f^*, g^*) = M_\mu^{(2)}(f_\mu^*, g_\mu^*). \tag{10}$$

Thus, if g is any strategy for P_2 in Γ, and if g_μ is its truncation to Γ_μ, we see by (10), the second part of (1), and (8), that

$$M^{(2)}(f^*, g^*) = M_\mu^{(2)}(f_\mu^*, g_\mu^*) \geq M_\mu^{(2)}(f_\mu^*, g_\mu) = M^{(2)}(f^*, g). \quad (11)$$

Now let f be any strategy for P_1 in Γ, and suppose that f picks out the ith alternative on the first move, i.e., suppose that

$$f(q_0) = i.$$

Let f_i be the truncation of f to Γ_i. Then we see that, if g is any strategy for P_2 in Γ, and if g_i is its truncation to Γ_i,

$$M^{(1)}(f, g) = M_i^{(1)}(f_i, g_i).$$

In particular,

$$M^{(1)}(f, g^*) = M_i^{(1)}(f_i, g_i^*). \quad (12)$$

From (5) we have

$$M_\mu^{(1)}(f_\mu^*, g_\mu^*) \geq M_i^{(1)}(f_i^*, g_i^*). \quad (13)$$

From (9), (13), the first part of (1), and (12) we conclude that

$$\begin{aligned} M^{(1)}(f^*, g^*) &= M_\mu^{(1)}(f_\mu^*, g_\mu^*) \geq M_i^{(1)}(f_i^*, g_i^*) \\ &\geq M_i^{(1)}(f_i, g_i^*) = M^{(1)}(f, g^*). \end{aligned} \quad (14)$$

From (11) and (14) we see that $\| f^* \quad g^* \|$ is an equilibrium point of Γ, as was to be shown.

CASE 3. This is analogous to Case 2.

This completes the proof of our theorem.

COROLLARY 6.2. The matrix of the normal form of any zero-sum two-person game with perfect information has a saddle-point.

PROOF. From Theorem 6.1, since, as was pointed out earlier in this chapter, an equilibrium point of a zero-sum two-person game is a saddle-point.

REMARK 6.3. Theorem 6.1 could be proved just as easily for n-person games as for two-person games, except that the notation would become slightly more complicated. This extension will be left as an exercise.

We have defined equilibrium points only among the sets of pure strategies. It is clear, however, that one could also define equilibrium points among mixed strategies. It is possible, indeed, to prove that every n-person game has an equilibrium point among the sets of mixed strategies. This fact is considered important by certain people, who cherish the hope of basing the entire theory of the rational manner of playing n-person games on the notion of an equilibrium point. Without necessarily completely accepting this view, we can at least notice the following: When a game is played frequently, and for a fairly long period, by a group of people, it sometimes happens that they fall into the habit of playing an equilibrium point. In case a man is playing the game in question with some members of this group, he cannot do better than to play the equilibrium point himself (and often, of course, he will do worse if he does not play the equilibrium point). Thus an equilibrium point which has been accepted in such a way represents, so to speak, a conventional standard of behavior for the group: one violates the convention at one's peril, unless one succeeds in persuading others to violate it also.

3. Games with Perfect Recall, and Behavior Strategies. An interesting and useful generalization of the notion of games with perfect information is the notion of games with *perfect recall*. Speaking intuitively, by a game with perfect recall we mean one where each of the players always remembers everything he did, or knew, at each of his previous moves. Thus every two-person game which can be played by just two people (rather than by teams) is a game with perfect recall; rummy, for instance, is a game with perfect recall, but bridge is not, since in bridge each player is a pair of people, neither of whom is informed what cards the other has been dealt.

This notion of a game with perfect recall can be made precise in terms of the information sets of the game. A game with perfect recall is one which satisfies the following condition: Let P and Q be any two moves, both of which are made by the same player and such that P precedes Q in some play of the game; let U and V be the information sets containing P and Q, respectively, and suppose that each point of U presents k alternatives; let U_i (for $i = 1, \cdots, k$) be the set of all vertices of the tree (i.e., moves) which can be reached by taking the ith alternatives at some point of U; then, for some i, we have $\mathsf{V} \subseteq \mathsf{U}_i$.

We turn now to the notion of a behavior strategy. It is clear that to play a mixed strategy amounts to using a chance device in order to pick out a pure strategy, i.e., to deciding what will be done at every move of the game. A somewhat analogous procedure would be to use a chance device at each move in order to decide which alternative to choose at that move. Such a system of play is called a *behavior strategy*. Strictly speaking, a behavior strategy for a given player is a function which is defined over the class of his information sets, and which assigns to each information set U a member of S_r, where r is the number of alternatives presented by U. It is clear that, for a given behavior strategy for one player of a game, and for a given pure strategy or mixed strategy or behavior strategy for the other player, one can calculate the expectation of each player.

Thus consider, for example, the game whose graph is given in Fig. 8; here we have indicated the payoff to the first player, for the various possible

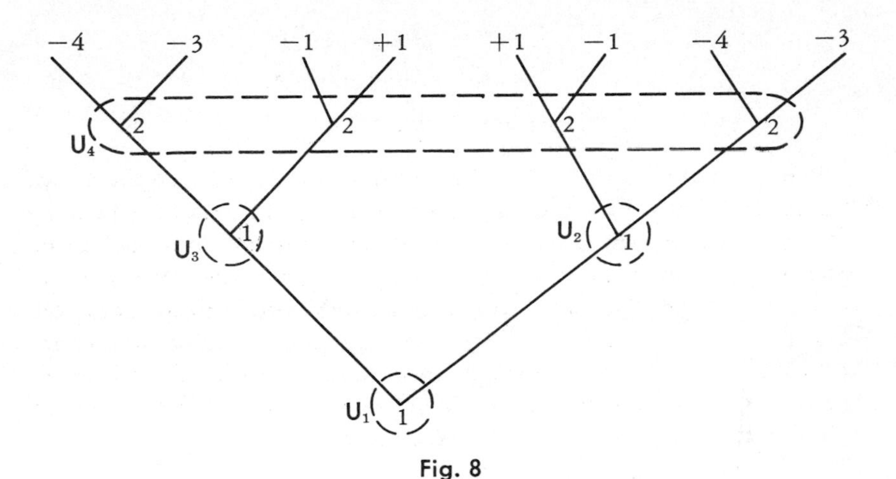

Fig. 8

outcomes, by numerical symbols attached to the top points of the tree. A behavior strategy for the first player is a function which is defined over $\{ \mathsf{U}_1, \mathsf{U}_2, \mathsf{U}_3 \}$ and whose members belong to $\mathbf{S}_2$ (since, for each of these information sets, there are precisely two alternatives from which to choose). A behavior strategy for P_2, on the other hand, is a function which is defined only for U_4, and whose value is in $\mathbf{S}_2$. Thus in this game a behavior strategy for P_2 is essentially the same as a mixed strategy; this always happens, of course, in the case where a player has but one information set.

Now suppose that P_1 uses a behavior strategy f such that

$$f(\mathsf{U}_1) = \| \alpha_1 \quad 1 - \alpha_1 \|,$$
$$f(\mathsf{U}_2) = \| \alpha_2 \quad 1 - \alpha_2 \|,$$
$$f(\mathsf{U}_3) = \| \alpha_3 \quad 1 - \alpha_3 \|,$$

and suppose that P_2 uses a behavior strategy g such that

$$g(\mathsf{U}_4) = \| \beta \quad 1 - \beta \|;$$

then the expectation $E(f, g)$ of P_1, or (as we shall sometimes write it) $E(\alpha_1, \alpha_2, \alpha_3; \beta)$, is given by the formula:

$$
\begin{aligned}
E(\alpha_1, \alpha_2, \alpha_3; \beta) = & - 4(\alpha_1)(\alpha_3)(\beta) - 3(\alpha_1)(\alpha_3)(1 - \beta) \\
& - (\alpha_1)(1 - \alpha_3)(\beta) + (\alpha_1)(1 - \alpha_3)(1 - \beta) \\
& + (1 - \alpha_1)(\alpha_2)(\beta) \\
& - (1 - \alpha_1)(\alpha_2)(1 - \beta) - 4(1 - \alpha_1)(1 - \alpha_2)(\beta) \\
& - 3(1 - \alpha_1)(1 - \alpha_2)(1 - \beta) \\
= & \ \alpha_1\alpha_3\beta - 3\alpha_1\alpha_2\beta - 4\alpha_1\alpha_3 - 2\alpha_1\alpha_2 \\
& - \alpha_1\beta + 3\alpha_2\beta + 4\alpha_1 + 2\alpha_2 - \beta - 3.
\end{aligned}
$$

It is intuitively clear, and it is not difficult to prove in a formal way, that one can always do at least as well by using mixed strategies as by using behavior strategies; but it is easy to construct games in which one can do better with mixed strategies than with any possible behavior strategy. It can be shown, on the other hand, that a game with perfect recall always has optimal behavior strategies: if E is the expectation function of a game with perfect recall, then there is a behavior strategy f for P_1 and a behavior strategy g for P_2 such that, if α and β are any mixed strategies (or any behavior strategies) for P_1 and P_2, respectively,

$$E(\alpha, g) \leq E(f, g) \leq E(f, \beta).$$

(A reference to a proof of this theorem is given in the Historical and Bibliographical Remarks at the end of the chapter.)

EXAMPLE 6.4. Consider the game whose graph is given in Fig. 8. Since this is a game with perfect recall, we see that there are optimal behavior strategies for the two players. As before, we have

$$
\begin{aligned}
E(\alpha_1, \alpha_2, \alpha_3; \beta) = & \ \alpha_1\alpha_3\beta - 3\alpha_1\alpha_2\beta - 4\alpha_1\alpha_3 - 2\alpha_1\alpha_2 \\
& - \alpha_1\beta + 3\alpha_2\beta + 4\alpha_1 + 2\alpha_2 - \beta - 3.
\end{aligned}
$$

The coefficient of α_3 in this equation is $(\alpha_1\beta - 4\alpha_1) = \alpha_1(\beta - 4)$; since this is never positive, it is clear that P_1, who wants to make $E(\alpha_1, \alpha_2, \alpha_3; \beta)$ large, cannot do better than to take

$$\alpha_3 = 0.$$

Similarly, the coefficient of α_2 is

$$-3\alpha_1\beta - 2\alpha_1 + 3\beta + 2 = (3\beta + 2)(-\alpha_1 + 1),$$

which is never negative; and hence P_1 cannot do better than to take

$$\alpha_2 = 1.$$

Since, finally,

$$E(\alpha_1, 1, 0; \beta) = -4\alpha_1\beta + 2\alpha_1 + 2\beta - 1 = -(2\alpha_1 - 1)(2\beta - 1),$$

we conclude that P_1 cannot do better than to take

$$\alpha_1 = \frac{1}{2},$$

and, similarly, that P_2 cannot do better than to take

$$\beta = \frac{1}{2}.$$

Thus for P_1 an optimal system of playing is to toss a coin to decide which alternative to take at U_1, to choose always the left-hand alternative at U_2, and to choose always the right-hand alternative at U_3.

REMARK 6.5. It should be observed that, in general, the number of parameters to be determined in calculating optimal behavior strategies is much smaller than the number of parameters involved in calculating optimal mixed strategies. Thus in Example 6.4 we had to deal with only three parameters in calculating an optimal behavior strategy for P_1; since P_1 has eight pure strategies in this game, on the other hand, a mixed strategy is an element of S_8, and hence the calculation of optimal mixed strategies involves seven independent parameters.

This advantage of behavior strategies, however, is at least partially balanced by the fact that the expectation function is usually of a more manageable type when expressed in terms of mixed strategies.

HISTORICAL AND BIBLIOGRAPHICAL REMARKS

The formal definition, given above, of games in extensive form is very similar to that to be found in Kuhn [2].

Our Theorem 6.1 is also proved in Kuhn [2].

The notion of an equilibrium point was introduced in Nash [2]. The proof that every n-person game has an equilibrium point among its mixed strategies is also due to Nash.

A proof of the theorem that a game with perfect recall has optimal behavior strategies will be found in Kuhn [2].

EXERCISES

1. Formulate and prove a generalization of Theorem 6.1 for the case of n-person games.

2. Find the equilibrium points (there are eight of them) for the game with perfect information whose graph is shown in Fig. 9 and whose play payoff functions are given by Table. 1.

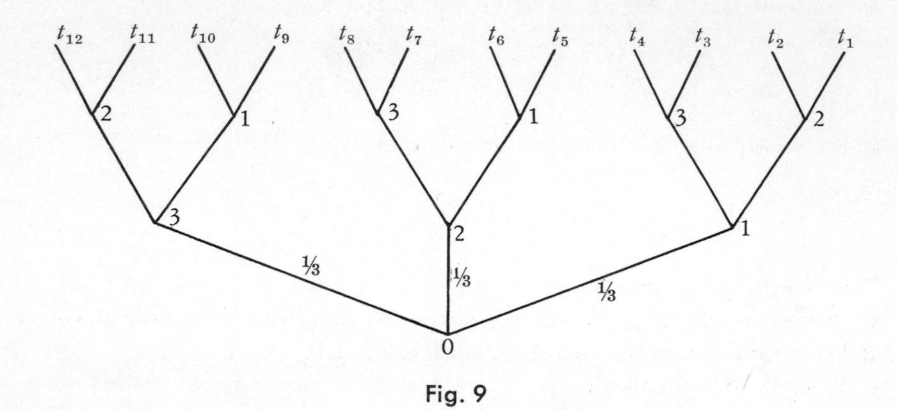

Fig. 9

Table 1

t	$H_1(t)$	$H_2(t)$	$H_3(t)$
t_1	1	5	3
t_2	2	6	4
t_3	3	7	1
t_4	4	8	2
t_5	5	1	9
t_6	6	2	10
t_7	7	3	11
t_8	8	4	12
t_9	9	11	5
t_{10}	10	12	6
t_{11}	11	9	7
t_{12}	12	10	8

3. In the case of a zero-sum two-person game, show that a saddle-point is also an equilibrium point.

4. Define an equilibrium point for the set of all n-tuples of mixed strategies for a normalized n-person game.

5. Making use of the notion of an equilibrium point for mixed strategies introduced in Exercise 4, find an equilibrium point for the two-person rectangular game where the payoff matrices for the first and second player are given by the matrices:

$$\begin{Vmatrix} 3 & 5 \\ 4 & 2 \end{Vmatrix}, \quad \begin{Vmatrix} 6 & -2 \\ -3 & 8 \end{Vmatrix}.$$

6. Show that every pair of mixed strategies is an equilibrium point for the game whose payoff matrices are

$$\begin{Vmatrix} 2 & 3 \\ 2 & 3 \end{Vmatrix}, \quad \begin{Vmatrix} -1 & -1 \\ 0 & 0 \end{Vmatrix}.$$

7. Find a payoff function for a zero-sum game having the graph shown in Fig. 10, and not having a saddle-point (in pure strategies).

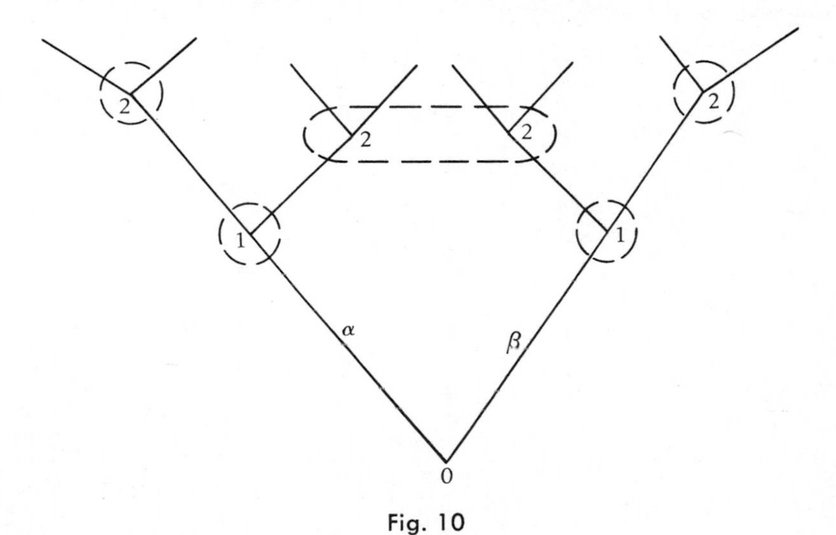

Fig. 10

8. Find optimal behavior strategies for the game whose graph is the same as Fig. 8, except that the payoffs, reading from left to right, are

$$-10, -9, -1, +1, +3, -2, -10, -9,$$

instead of

$$-4, -3, -1, +1, +1, -1, -4, -3.$$

9. Construct an example of a game for which you can show that there is no optimal behavior strategy for P_1.

10. There are three similar decks of n cards, each set labeled with symbols for positive numbers $x_1, \cdots, x_n$. Each player holds one set, and the third is randomly arranged in a stack placed face down. There are n stages of this game, each stage consisting of the following:

1. The top card of the stack is turned up.
2. The players simultaneously bid by each playing one card.
3. The player showing the higher card wins from his opponent the face value of the card turned up from the stack (there is no payoff if they play cards numbered alike).

The game clearly has value zero, since it is symmetrical.

Construct the graph of this game for the case where $n = 2$ and show that there are optimal pure strategies for this case. Show, moreover, that there are optimal pure strategies for the case where $n = 3$, if we have $x_1 = 1$, $x_2 = 2$, and $x_3 = 3$. Find optimal strategies for the case where $n = 3$, if we have $x_1 = 2$, $x_2 = 3$, and $x_3 = 4$.

CHAPTER 7

GAMES WITH INFINITELY MANY STRATEGIES

Up to this point we have confined our attention to finite games. However, in some situations of practical importance—and situations which it is convenient to study from the point of view of game theory—the choices of the people involved are made from infinite sets. Thus suppose, for instance, that a manufacturer is faced with the problem of how much soap to put into a package which he proposes to sell for ten cents. He would like to put in enough soap to be able to compete favorably with other manufacturers and thus to sell many packages, but, of course, he does not want to put in so much that he will lose money on each package sold. Since he must try somehow to take into account the actions of the other manufacturers, whose interests oppose his own, the situation is similar to a game. And, since there are infinitely many possible weights for a cake of soap (or, at least, so large a finite number that it is convenient to regard it as infinite), we are confronted with a gamelike situation in which the players make their choices from infinite sets.

Thus it appears profitable to extend our notion of a game so as to include also *infinite* games, i.e., games in which at least some of the choices are made from infinite sets. We shall not treat this subject, however, with anything like the generality with which we discussed finite games.

In the first place, we shall consider only games with one move for each player, and with no information given to either player about the choice of the other player. Thus we shall deal only with games in which, as in the rectangular games defined in Chap. 1, player P_1 chooses an element x from a set A and player P_2 chooses an element y from a set B. As with rectangular games, we shall suppose, moreover, that there is a real-valued function M of two variables, such that if P_1 chooses x and P_2 chooses y, then P_2 pays P_1 the amount $M(x, y)$. Thus we generalize the notion of a rectangular game only to the extent of allowing the sets A and B to be infinite.

Our attention will be almost entirely confined, moreover, to the case where both A and B are the closed interval $[0, 1]$; we shall call such games *continuous* games. The name derives from the fact that a closed interval (say of points or real numbers) is sometimes called a continuum; it should not be understood as imposing any kind of continuity properties on the function M. It should be noticed that this restriction of **A** and **B** is not quite so severe as might appear at first glance; if **A** and **B** are any finite closed intervals, then a mere relabeling of the elements of **A** and **B** will give us a game where the choices are both made from the closed interval $[0, 1]$. Thus a game is a continuous game, or can be reduced to a continuous game, if it is like a rectangular game except that the players make their choices from finite closed intervals.

EXAMPLE 7.1. Consider, for instance, the game where P_1 chooses a number x from the interval $[0, 1]$ and P_2 independently chooses a number y from the same interval, and where the payoff to P_1 is

$$M(x, y) = 2x^2 - y^2.$$

This game is, in fact, so simple that we can immediately recognize the players' optimal strategies: P_1, wishing to maximize $M(x, y)$, will choose $x = 1$, and P_2, in order to minimize $M(x, y)$, will also choose $y = 1$; the resulting payoff will then be 1.

EXAMPLE 7.2. As an example of a slightly more sophisticated continuous game, consider the following situation: A colonel, let us name him Blotto, wishes to attack two equally valuable positions A and B by sending a fraction x of his regiment to A and the remainder, $1 - x$, to B. His opponent, the defender of these positions, has only a fraction α of a regiment at his disposal, where $0 < \alpha < 1$, to be distributed between A and B. Suppose that the defender assigns the fraction y of his force to A and $1 - y$ to B, so that, in terms of regiments, the defending forces are αy and $\alpha(1 - y)$, respectively. Assume that each of the ensuing battles is fought to the death and that equal numbers of men are lost on both sides—namely, a number equal to the smaller of the opposing forces—resulting in the position's falling to the larger force. For each position, the payoff to its winner is taken to be the position itself (valued at one regiment) plus the number of survivors; if there is no winner at a position, the payoff there is 0. Therefore we may write the payoff to Blotto as follows:

$$\begin{aligned} M(x, y) = \operatorname{sgn} [x - \alpha y] &+ x - \alpha y \\ &+ \operatorname{sgn} [1 - x - \alpha(1 - y)] + 1 - x - \alpha(1 - y), \end{aligned}$$

where $\operatorname{sgn} z$ is defined as follows:

$$\operatorname{sgn} z = \begin{cases} -1 & \text{if } z < 0, \\ 0 & \text{if } z = 0, \\ 1 & \text{if } z > 0. \end{cases}$$

To resume our general discussion, let us now consider a game whose payoff function is M, and suppose that

$$\max_{0\le x\le 1} \min_{0\le y\le 1} M(x, y) = v_1$$

and

$$\min_{0\le y\le 1} \max_{0\le x\le 1} M(x, y) = v_2$$

both exist. Then, by reasoning as we did in Chap. 1, we see that P_1 can make a choice which will ensure that he gets at least v_1; and P_2 can see to it that P_1 gets at most v_2. If $v_1 = v_2$, we see by Theorem 1.5 that M possesses a saddle-point $\| x_0 \quad y_0 \|$, and that

$$M(x_0, y_0) = v_1 = v_2 .$$

In this case it is natural to call x_0 and y_0 optimal ways for P_1 and P_2 to play and to call v_1 (and hence also v_2) the value of the game.

For instance, in Example 7.1 we had

$$M(x, y) = 2x^2 - y^2,$$

which has a saddle-point at $\| 1 \quad 1 \|$; hence

$$M(1, 1) = v_1 = v_2 = 1 .$$

In Example 7.2, on the other hand, it can be shown that no saddle-point exists.

We are left with two cases which are more difficult: (1) the case in which v_1 and v_2 do not both exist; and (2) the case in which they both exist, but are unequal. The second case will be discussed in Chap. 10, after we have introduced (in Chaps. 8 and 9) some necessary concepts of a purely mathematical nature. So far as regards the first case, we shall confine ourselves to showing that this situation can actually arise.

Often, when v_1 and v_2 do not exist, we are at a loss as to how to define the value of the game or the optimal strategies for the two players.

EXAMPLE 7.3. The payoff function of a continuous game is defined as follows:

$$\begin{aligned} M(x, y) &= \frac{1}{x} - \frac{1}{y} \qquad && \text{if } x \neq 0 \text{ and } y \neq 0, \\ M(0, y) &= -\frac{1}{y} && \text{if } y \neq 0, \\ M(x, 0) &= \frac{1}{x} && \text{if } x \neq 0, \\ M(0, 0) &= 0. \end{aligned}$$

It is clear that P_1 should not always pick 0, for then P_2 could win large amounts by taking y positive, but small. Moreover, P_1 should not always pick the same positive number, for then P_2 could win by choosing y positive, but smaller than x. Hence it would seem that P_1 should choose x by a chance device, but in such a way that small positive values are chosen more frequently than large. Not even such a procedure as this can be optimal, however; for P_1 can always increase his expectation by using a new chance device which makes him choose still smaller numbers with the given frequencies. Thus it does not appear that we can assign a value to this game or show that there exist optimal ways of playing it.

It should not be thought, however, that the same situation arises in every case in which v_1 and v_2 do not exist. Indeed, it can even happen that a game has a saddle-point although v_1 and v_2 do not exist. Consider, for example, the continuous game whose payoff function M is defined as follows:

$$M(x, y) = \frac{1}{x} - \frac{1}{y} \qquad \text{if } x \neq 0 \text{ and } y \neq 0,$$
$$M(x, 0) = M(0, y) = 0.$$

It is easily verified that, for $x \neq 0$, $\min_y M(x, y)$ does not exist. Thus $\max_x \min_y M(x, y)$ does not exist; and, similarly, $\min_y \max_x M(x, y)$ does not exist. On the other hand, the function has a saddle-point at $\| 0 \quad 0 \|$. Thus there is a way for P_1 to play (namely, to take $x = 0$) which will ensure that he will get at least $M(0, y) = 0$; and there is a way for P_2 to play (namely, to take $y = 0$) which will ensure that P_1 will get at most $M(x, 0) = 0$. Hence it is reasonable to call 0 the value of this game and to say that an optimal way for P_1 to play is to take $x = 0$, and, similarly, for P_2 to take $y = 0$.

REMARK 7.4. It is easy to find a function M of two variables such that

$$\max_{0 \leq x \leq 1} \min_{0 \leq y \leq 1} M(x, y) \tag{1}$$

and

$$\min_{0 \leq y \leq 1} \max_{0 \leq x \leq 1} M(x, y) \tag{2}$$

do not exist, while the quantities

$$\sup_{0 \leq x \leq 1} \inf_{0 \leq y \leq 1} M(x, y) \tag{3}$$

and

$$\inf_{0 \leq y \leq 1} \sup_{0 \leq x \leq 1} M(x, y) \tag{4}$$

both exist and are equal. If a continuous game has such a function as a payoff matrix, then, although the game has no saddle-point, we nevertheless sometimes speak of the common value of (3) and (4) as the value, v, of the game. In such a case, although P_1 cannot in general pick a strategy x_0 which will ensure that, for every choice of a strategy y by P_2,

$$M(x_0, y) \geq v,$$

nevertheless he can, for every positive ε, choose a strategy x_ε such that, for every choice of a y by P_2,

$$M(x_\varepsilon, y) \geq v - \varepsilon.$$

Thus P_1 can come as close as he pleases to an optimal method of play; and, similarly, P_2 can come arbitrarily close to an optimal method of play.

EXAMPLE 7.5. The payoff function, M, of a continuous game is defined by the conditions:

$$M(x, y) = x + y \qquad \text{if } x \neq 1 \text{ and } y \neq 0,$$

$$M(1, y) = \frac{1}{2} + y \qquad \text{if } y \neq 0,$$

$$M(x, 0) = \frac{1}{2} + x \qquad \text{if } x \neq 1,$$

$$M(1, 0) = 2.$$

In this case we readily verify that the terms (1) and (2) do not exist. On the other hand, we have

$$\inf_y \sup_x M(x, y) = 1 = \sup_x \inf_y M(x, y).$$

Thus we call 1 the value of this game. By choosing $x_\varepsilon = 1 - \varepsilon$, player P_1 can be sure that, for every y chosen by P_2,

$$M(x_\varepsilon, y) \geq 1 - \varepsilon.$$

Similarly, by choosing $y_\varepsilon = \varepsilon$, player P_2 can be sure that, for every x chosen by P_1,

$$M(x, y_\varepsilon) \leq 1 + \varepsilon.$$

HISTORICAL AND BIBLIOGRAPHICAL REMARK

The earliest general treatment of infinite games is to be found in Ville [1].

EXERCISES

1. A continuous game has a payoff function M defined as follows:

$$M(x, y) = xy - \frac{1}{3}x - \frac{1}{2}y.$$

Show that this function has a saddle-point at $\| \frac{1}{2} \quad \frac{1}{3} \|$. That is to say, referring to the definition in Chap. 1 of a saddle-point, show that

$$M\left(x, \frac{1}{3}\right) \leq M\left(\frac{1}{2}, \frac{1}{3}\right) \leq M\left(\frac{1}{2}, y\right)$$

holds for all x and y in the closed interval $[0, 1]$. Hence conclude that the value of the game (to P_1) is $-\frac{1}{6}$ and that an optimal way for P_1 to play is to take $x = \frac{1}{2}$ and an optimal way for P_2 to play is to take $y = \frac{1}{3}$.

2. A continuous game has a payoff function M defined as follows:

$$M(x, y) = \left[\left(x - \frac{1}{2}\right)^4 - \left(x - \frac{1}{2}\right)^2\right] \cdot \left[1 + \left(y - \frac{1}{3}\right)G(x, y)\right] + \left[\left(y - \frac{1}{3}\right)^2 - \left(y - \frac{1}{3}\right)^3\right]\left[1 + \left(x - \frac{1}{2}\right)H(x, y)\right],$$

where G and H are functions defined over the closed unit square. Show that the value of the game is 0, that an optimal strategy for P_1 is $x = \frac{1}{2}$, and that an optimal strategy for P_2 is $y = \frac{1}{3}$. That is to say, show that

$$M\left(x, \frac{1}{3}\right) \leq M\left(\frac{1}{2}, \frac{1}{3}\right) = 0 \leq M\left(\frac{1}{2}, y\right)$$

holds for all x and y in the closed interval $[0, 1]$.

3. Let M be the payoff function of a continuous game, and suppose that $\partial M/\partial x$ and $\partial M/\partial y$ exist at all points in the open unit square; i.e., for all x and y such that

$$0 < x < 1,$$
$$0 < y < 1.$$

Show that if $\| x_0 \quad y_0 \|$ is a saddle-point of the game and lies in the open unit square, then

$$\left.\frac{\partial M}{\partial x}\right|_{\substack{x=x_0 \\ y=y_0}} = 0$$

and

$$\left.\frac{\partial M}{\partial y}\right|_{\substack{x=x_0\\y=y_0}} = 0.$$

4. Use the principle established in Exercise 3 to find a saddle-point, and hence the value and optimal strategies, of the continuous game whose payoff function is

$$M(x, y) = 15xy - 3x - 5y + 2.$$

5. Find a value of k such that the game whose payoff function is

$$M(x, y) = 10xy - x - y + k$$

will have 0 for its value.

6. Show that the game whose payoff function is

$$M(x, y) = xy + x + y$$

has no saddle-point within the open unit square. Show that the game has no saddle-point of the form $\| 0 \quad y_0 \|$. Find a number x_0 such that $\| x_0 \quad 0 \|$ is a saddle-point.

7. Find the inequalities which must be satisfied by the constants α and β in order that the game whose payoff function is

$$M(x, y) = xy - \alpha x - \beta y + \gamma$$

shall have a saddle-point in the open unit square.

8. Show that every continuous game whose payoff function is of the form

$$M(x, y) = xy - \alpha x - \beta y + \gamma$$

has a saddle-point (which may, however, lie upon the boundary of the unit square). Hint: Make use of the fact that every polynomial of the given type can be represented in the form

$$(x + a) \cdot (y + b) + c,$$

where a, b, and c are real numbers.

9. Show that the continuous game whose payoff function is

$$M(x, y) = (x - y)^2$$

has no saddle-point at all, i.e., neither within the unit square nor upon its boundary.

10. Show that the continuous game whose payoff function is

$$M(x, y) = \frac{1}{1 + 2(x - y)^2}$$

has no saddle-point at all.

11. Let F_1 and F_2 be continuous increasing functions (i.e., such that $u > v$ implies $F_1(u) > F_1(v)$ and $F_2(u) > F_2(v)$) satisfying

$$F_1(0) = F_2(0) = 0,$$
$$F_1(1) = F_2(1) = 1,$$

and let the function M be defined as follows:

$$\begin{aligned} M(x, y) &= 2F_1(x) - 1 && \text{if } x < y, \\ M(x, y) &= F_1(x) - F_2(y) && \text{if } x = y, \\ M(x, y) &= 1 - 2F_2(y) && \text{if } x > y. \end{aligned}$$

Show that the game whose payoff function is M has a saddle-point. Hint: From the hypothesis about F_1 and F_2, it follows that there exists a unique point u in $[0, 1]$ such that

$$F_1(u) + F_2(u) = 1.$$

12. Show that the game with payoff sgn $(x - y)$ has a saddle-point. (See Example 7.2 for the definition of "sgn.")

13. Compute v_1 and v_2 for the game described in Example 7.2.

14. Show that the game described in Example 7.2 has a value if we assume that $0 < \alpha < ½$. Find this value and a pair of optimal strategies.

15. Reformulate Example 7.2 for the case where Position A is worth 1 regiment and Position B is worth c regiments ($c > 1$). Show that, for sufficiently large c, this game has a saddle-point.

16. Show that if M is a bounded function which is defined over the unit square and has a saddle-point, then

$$\sup_{0 \leq x \leq 1} \inf_{0 \leq y \leq 1} M(x, y)$$

and

$$\inf_{0 \leq y \leq 1} \sup_{0 \leq x \leq 1} M(x, y)$$

exist and are equal.

17. The payoff function of a game is given by

$$M(x, y) = 10xy - y - 5x \qquad \text{for } x \neq \frac{1}{10},$$

$$M\left(\frac{1}{10}, y\right) = -y.$$

Find the value, v, of the game. What strategy for P_1 will ensure that he gets at least $v - \varepsilon$?

CHAPTER 8

DISTRIBUTION FUNCTIONS

1. Intuitive Considerations. It will be recalled that in order to find optimal ways of playing rectangular games without saddle-points it became necessary to consider mixed strategies—i.e., ways of making choices at random, and only with certain probabilities. It is clear that something similar must be done in the case of continuous games without saddle-points (examples of such games were given in Exercises 9 and 10 of Chap. 7).

It is desirable first to clear our minds of the misconception that a mixed strategy for playing a continuous game is simply a rule which ascribes a probability to each number in the closed interval $[0, 1]$. For it can happen that a random way of choosing a number assigns the probability 0 to each number in the interval and, indeed, that two such random ways can be different, even though both of them assign probability 0 to each particular number.

Thus suppose that we mark the numbers x, such that $0 \leq x \leq 1$, along a circle (see Fig. 1) whose circumference is of unit length, spin a pointer, and choose the number corresponding to the place where the pointer stops.

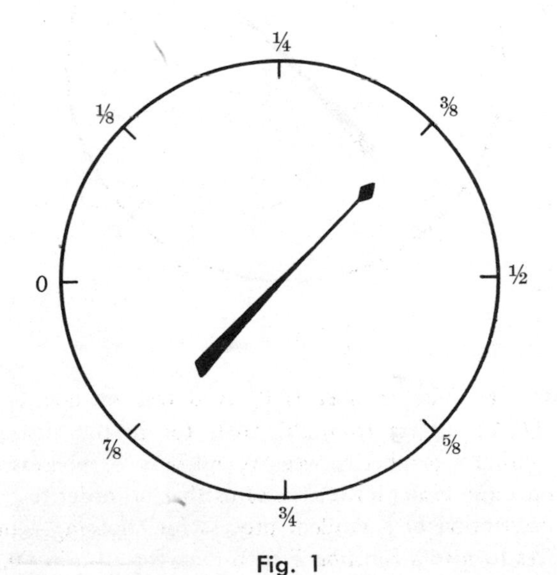

Fig. 1

If we suppose that the pivot of the pointer is well oiled and that the whole apparatus is properly constructed, then the probability (in the sense of relative frequency) that the pointer will come to rest within any interval is simply equal to the length of that interval; the probability that it will come to rest, for instance, within the interval $[\frac{3}{8}, \frac{5}{8}]$ is $\frac{1}{4}$. In general, the probability that the number chosen will fall in the interval $[\frac{1}{2} - (\varepsilon/2), \frac{1}{2} + (\varepsilon/2)]$ is ε; since the probability, $p(\frac{1}{2})$, that the pointer will come to rest at $\frac{1}{2}$ is certainly not greater than the probability that it will come to rest in the interval $[\frac{1}{2} - (\varepsilon/2), \frac{1}{2} + (\varepsilon/2)]$, we see that $p(\frac{1}{2}) \leq \varepsilon$ for all positive ε, and hence $p(\frac{1}{2}) = 0$. Since this argument depends on no special property of the point $\frac{1}{2}$, we have here a random way of choosing a number from $[0, 1]$ which assigns probability 0 to each particular number in the interval.

Now suppose that the machine is changed slightly by marking the numbers from 0 to $\frac{1}{2}$ along the first fourth of the circumference of the circle and by marking the numbers from $\frac{1}{2}$ to 1 along the other three-fourths of the circumference, as shown in Fig. 2. It is clear that again the probability

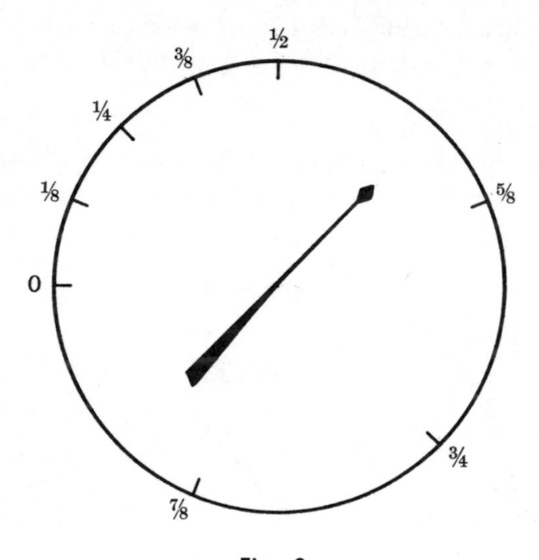

Fig. 2

of choosing any particular number is 0. And this method of choosing a number from $[0, 1]$ differs from the first; for in the first method, the probability of getting a number between 0 and $\frac{1}{2}$ is $\frac{1}{2}$, whereas now it is $\frac{1}{4}$.

Reflection on these examples convinces us that, in order to give a formal mathematical description of a random process for choosing a number from $[0, 1]$, it suffices to give a function F such that, for all a in $[0, 1]$, $F(a)$ is the probability that the number chosen will be at most equal to a. For mathematical convenience in the theory we shall develop later, however, it turns out to be better to modify this definition slightly for the case where $a = 0$. Thus we consider functions F such that $F(a)$, for $a \neq 0$, is the probability that the number chosen will be at most equal to a and such that $F(0) = 0$ (thus $F(0)$ is the probability that the number chosen will be actually less than 0, not that it will be at most 0). If a and b are two numbers from $[0, 1]$ such that $0 < a < b$, then we see that $F(b) - F(a)$ will be the probability that the number x will be chosen in the interval $a < x \leq b$ and that $F(b) - F(0)$ will be the probability that the number x will be chosen in the interval $0 \leq x \leq b$. Such a function F is called a *cumulative distribution function*, or, for brevity, simply a *distribution function.*

The distribution function for the first example given above is the function F such that, for all x in $[0, 1]$,

$$F(x) = x.$$

The graph of this function is shown in Fig. 3. (The function is, of course, not defined except for $0 \leq x \leq 1$.)

The distribution function for the second example is the function F such that

$$F(x) = \frac{1}{2}x \qquad \text{for } 0 \leq x \leq \frac{1}{2},$$

$$F(x) = \frac{3}{2}x - \frac{1}{2} \qquad \text{for } \frac{1}{2} < x \leq 1.$$

The graph of this function is shown in Fig. 4.

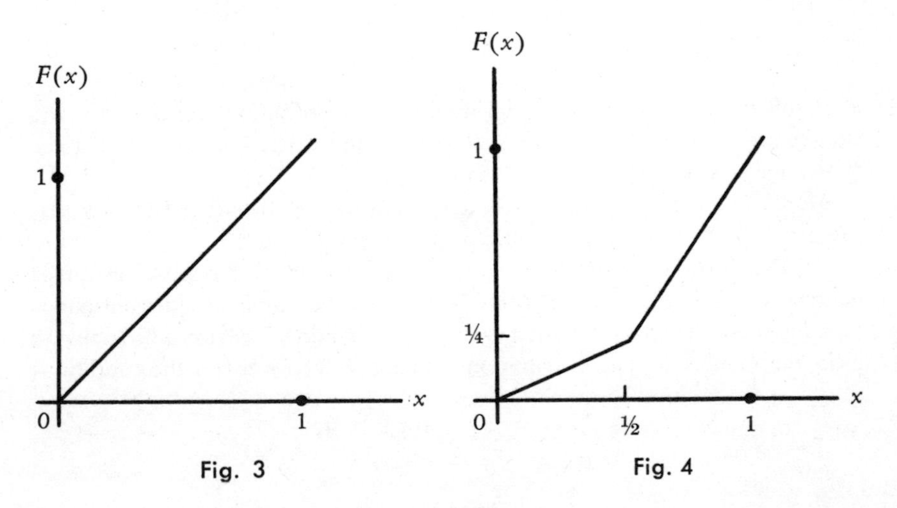

Fig. 3 Fig. 4

We notice that both the above functions are *nondecreasing*; i.e., that $F(x_1) \leq F(x_2)$ whenever $x_1 \leq x_2$. (Graphically this means, of course, that the curves slope upward to the right.) This is immediately seen to be true of every distribution function, for if $x_1 \leq x_2$, then the probability that the number chosen lies in $[0, x_2]$ must be at least as great as the probability that it lies in $[0, x_1]$.

Our two special functions, as a matter of fact, are what is called *increasing*; i.e., $F(x_1) < F(x_2)$ holds whenever $x_1 < x_2$. This is not always true of distribution functions, however. Thus, for instance, a distribution function can have a graph such as that shown in Fig. 5. The flat interval on this graph means that

$$F\left(\frac{3}{4}\right) - F\left(\frac{1}{4}\right) = 0;$$

hence the probability that the number chosen will lie in the interval $\frac{1}{4} < x \leq \frac{3}{4}$ is 0. It is readily seen how a random device could be constructed that would generate this distribution function: it would only be necessary to make a machine like those described above, but leaving out the interval $[\frac{1}{4}, \frac{3}{4}]$.

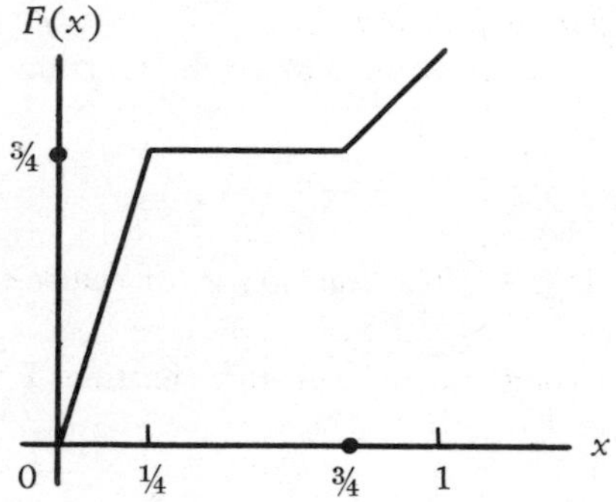

Fig. 5

We notice that all the graphs pass through the point $\|1 \quad 1\|$. This must be so for all distribution functions; i.e., for every distribution function F, we have

$$F(1) = 1.$$

This follows directly from the definition of a distribution function; for the choices are made within the interval $[0, 1]$, and hence it is certain that the choice made will not be greater than 1.

Moreover, it is specified in the definition of a distribution function that the graph of such a function must pass through $\|0 \quad 0\|$.

All the distribution functions so far considered are continuous, but this is not true of all distribution functions. The simplest example of a discontinuous distribution function is obtained by using a "random" device which always picks the number 0. The distribution function F then satisfies the conditions

$$F(x) = 1 \qquad \text{for } x > 0,$$
$$F(0) = 0,$$

and the graph for it is shown in Fig. 6. We can obtain a less trivially discontinuous distribution function by means of the following random process: we toss an unbiased coin. If it shows heads, we pick ¼; if it shows tails, we use the machine shown in Fig. 1 to pick a number from the interval $[0, 1]$. The distribution function F is then seen to satisfy the conditions

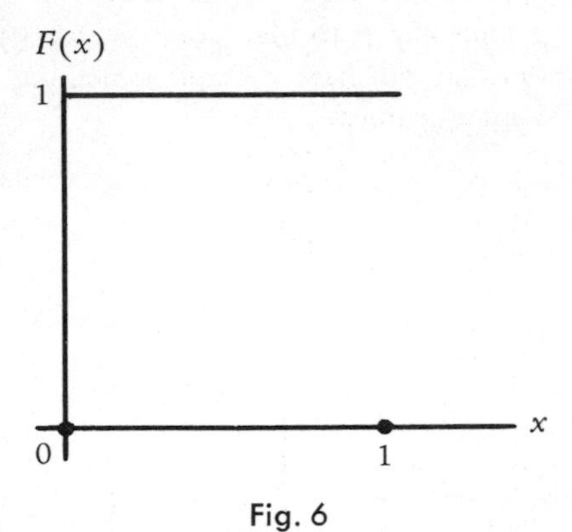

Fig. 6

$$F(x) = \frac{1}{2}x \qquad \text{for } x < \frac{1}{4},$$

$$F(x) = \frac{1}{2}(1 + x) \qquad \text{for } x \geq \frac{1}{4}.$$

The graph of this function is shown in Fig. 7. (The heavy dot at the left end of the upper segment indicates that $F(¼)$ is ⅝, not ⅛.)

The graph of Fig. 7 has a jump at $x = ¼$, which is a point to which finite (non-zero) probability is assigned by the distribution function. It is easily seen that the points to which finite probability is assigned will always correspond to points of discontinuity of the graph of the distribution function.

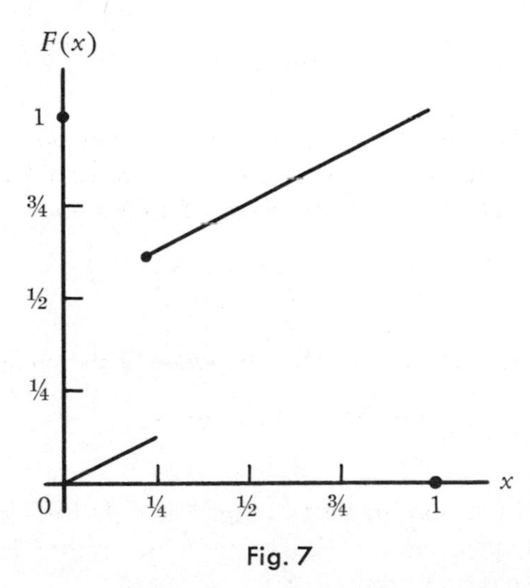

Fig. 7

All the distribution functions we have considered so far have graphs that consist of either a straight line-segment, or of sets of such segments. That this is not always the case, however, can be seen from the following example. Let us suppose that, in the machine of Fig. 1, the tip of the pointer is made of steel and that a small magnet is placed in the plane of the circle, somewhat to the right of the point marked "½." Then the pointer will more frequently come to rest in the neighborhood of ½ than elsewhere (but the probability of getting any particular point is still 0). It is then seen that the distribution function will have a graph something like that indicated in Fig. 8 (thus, not a straight line).

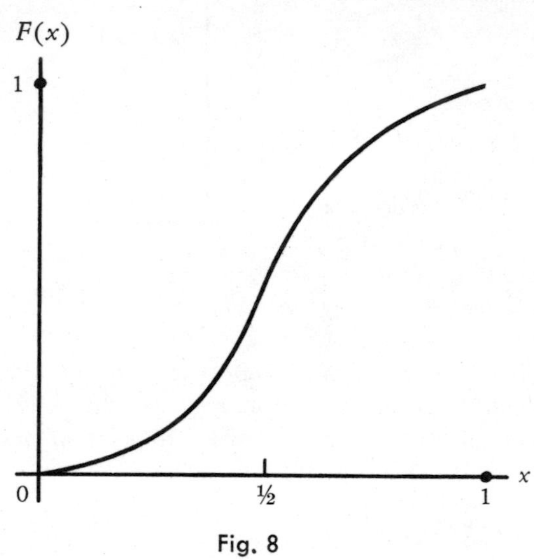

Fig. 8

We wish finally to consider a slightly less elementary property of distribution functions. In order to do this, it is necessary to introduce a definition. A function F is called *right-hand continuous* at a point x if

$$\lim_{\substack{\varepsilon \to 0 \\ \varepsilon > 0}} F(x + \varepsilon) = F(x).$$

Similarly, we call a function *left-hand continuous* at the point x if

$$\lim_{\substack{\varepsilon \to 0 \\ \varepsilon > 0}} F(x - \varepsilon) = F(x).$$

We notice that the function graphed in Fig. 7 is right-hand continuous, but not left-hand continuous, at $x = \frac{1}{4}$. The function graphed in Fig. 6 is not right-hand continuous at $x = 0$.

We now have the following theorem:

THEOREM 8.1. Every distribution function is right-hand continuous at all points in the open interval $(0, 1)$.

PROOF. Let F be a distribution function, and let $0 < x < 1$. We want to show that

$$\lim_{\substack{\varepsilon \to 0 \\ \varepsilon > 0}} [F(x + \varepsilon) - F(x)] = 0 .$$

Now the quantity

$$F(x + \varepsilon) - F(x),$$

by the definition of a distribution function, equals the probability that a number z, chosen according to this distribution function, satisfies the inequality

$$x < z \leq x + \varepsilon .$$

However, for any given z it is possible to find a positive ε which is sufficiently small so that the above inequality cannot hold. Hence the probability that z will satisfy this inequality approaches 0 as ε approaches 0, which completes the proof.

2. Formal Development. Thus we see that a distribution function F always satisfies the following conditions: (i) for any x in $[0, 1]$, $F(x)$ is a non-negative real number; (ii) $F(0) = 0$ and $F(1) = 1$; (iii) F is a nondecreasing function within the interval $[0, 1]$; and (iv) F is right-hand continuous in the open interval $(0, 1)$.

From the standpoint of our intuitions about probability, moreover, it appears fairly plausible that any function which satisfies conditions (i), (ii), (iii), and (iv), above, is a distribution function; i.e., if F is any function which satisfies these conditions, then there exists some way of choosing numbers from the interval $[0, 1]$ which has F for its distribution function. In any case, we shall now make the notion of a distribution function mathematically precise by using the term hereafter merely to mean a function which satisfies the above four conditions. Thus, to verify that a function is a distribution function, we need not describe the actual random process which generates it; we need only verify that it is defined over $[0, 1]$, that it never assumes negative values, that it is nondecreasing, and so on.

We shall use the letter **D** to stand for the set of all distribution functions, i.e., for the set of all functions which satisfy the above four conditions.

We now have a theorem which will be useful later.

THEOREM 8.2. Every sequence of distribution functions contains a subsequence which converges to a distribution function F at every point of continuity of F.

PROOF. It is well known that the proper fractions can be ordered into an

At disc. of F, the seq. may for instance oscillate.

infinite sequence. We can do this, for example, by first writing down the only proper fraction (namely, ½) whose denominator is 2; then, in order of magnitude, the proper fractions whose denominators are 3; then those whose denominators are 4 (but omitting those already listed); and so on. Thus we obtain

$$\frac{1}{2}, \frac{1}{3}, \frac{2}{3}, \frac{1}{4}, \frac{3}{4}, \frac{1}{5}, \frac{2}{5}, \frac{3}{5}, \frac{4}{5}, \frac{1}{6}, \frac{5}{6}, \frac{1}{7}, \cdots;$$

it is clear that every proper fraction will eventually be reached in this way.

Now let

$$r_1, r_2, r_3, \cdots, r_n, \cdots$$

be any such enumeration of the proper fractions, and let

$$G_1, G_2, \cdots, G_n, \cdots \tag{1}$$

be an arbitrary sequence of distribution functions. We consider first the sequence of numbers

$$G_1(r_1), G_2(r_1), \cdots, G_n(r_1), \cdots;$$

since this sequence is infinite and bounded, it contains a convergent subsequence

$$G_{i_1}(r_1), G_{i_2}(r_1), \cdots, G_{i_n}(r_1), \cdots.$$

Setting

$$F_n^{(1)}(x) = G_{i_n}(x) \qquad \text{for } n = 1, 2, \cdots,$$

we see that the sequence

$$F_1^{(1)}, F_2^{(1)}, \cdots, F_n^{(1)}, \cdots$$

is a subsequence of (1) which converges to a limit at the point r_1.

Next we consider the sequence of numbers

$$F_1^{(1)}(r_2), F_2^{(1)}(r_2), \cdots, F_n^{(1)}(r_2), \cdots;$$

as before, we see that this sequence contains a subsequence

$$F_{i_1}^{(1)}(r_2), F_{i_2}^{(1)}(r_2), \cdots, F_{i_n}^{(1)}(r_2), \cdots$$

which is convergent. Now setting

$$F_n^{(2)}(x) = F_{i_n}^{(1)}(x) \qquad \text{for } n = 1, 2, \cdots,$$

we conclude that the sequence

$$F_1^{(2)}, F_2^{(2)}, \cdots, F_n^{(2)}, \cdots$$

is a subsequence of (1), which converges to a limit at r_2 as well as at r_1.

Continuing in this way we obtain a sequence of sequences

$$\begin{array}{ccccc} F_1^{(1)}, & F_2^{(1)}, & F_3^{(1)}, & \cdots, & F_n^{(1)}, \cdots \\ F_1^{(2)}, & F_2^{(2)}, & F_3^{(2)}, & \cdots, & F_n^{(2)}, \cdots \\ F_1^{(3)}, & F_2^{(3)}, & F_3^{(3)}, & \cdots, & F_n^{(3)}, \cdots \\ \vdots & \vdots & \vdots & & \vdots \\ F_1^{(m)}, & F_2^{(m)}, & F_3^{(m)}, & \cdots, & F_n^{(m)}, \cdots \\ \vdots & \vdots & \vdots & & \vdots \end{array}$$

each of which is a subsequence of all the preceding ones, and hence also of the initial sequence (1), and is such that the mth sequence converges to a limit for $r_1, \cdots, r_m$.

From this sequence of sequences we now form the "diagonal sequence"

$$F_1, F_2, \cdots, F_n, \cdots \tag{2}$$

by setting

$$F_n = F_n^{(n)} \qquad \text{for } n = 1, 2, \cdots.$$

It is clear that (2) is a subsequence of (1). Moreover, for each m, all but a finite number of the elements of (2) belong to the sequence

$$F_1^{(m)}, F_2^{(m)}, \cdots, F_n^{(m)}, \cdots,$$

and hence the sequence (2) converges to a limit for all proper fractions.

We next define a function ϕ, over the proper fractions, by setting, for r any proper fraction,

$$\phi(r) = \lim_{n\to\infty} F_n(r); \tag{3}$$

and then define a function F, over all numbers in the open unit interval, by setting for x any such number

$$F(x) = \inf_{r>x} \phi(r); \tag{4}$$

we also set

$$\left.\begin{aligned} F(0) &= 0, \\ F(1) &= 1, \end{aligned}\right\} \tag{5}$$

so that F is defined over the whole unit interval (including the end points). In order to prove our theorem it will suffice to show that F is a distribution function and that (2) converges to F at every point of continuity of F.

To show that F is a distribution function it suffices, by (5), to show that it is nondecreasing and right-hand continuous. If x and y are any numbers such that $x < y$, however, then the set A_y of all rationals greater than y is a subset of the set A_x of all rationals greater than x; hence

$$\inf_{r \in \mathsf{A}_x} \phi(r) \leq \inf_{r \in \mathsf{A}_y} \phi(r),$$

or, by (4),

$$F(x) \leq F(y).$$

Thus F is nondecreasing.

In order to show that F is right-hand continuous, we note first that the function ϕ, by its definition (3), is nondecreasing. Now suppose that

$$x_1, x_2, \cdots, x_n, \cdots$$

is a decreasing sequence of numbers whose limit is a point x of $(0, 1)$; we are to show that

$$F(x) = \lim_{n \to \infty} F(x_n).$$

By (4) we see that, for every $\varepsilon > 0$, there exists a rational number r such that

$$r > x, \tag{6}$$

$$\phi(r) - F(x) < \varepsilon. \tag{7}$$

Now if we let N be an integer so large that $x < x_N < r$, we see that, for all $n > N$,

$$x < x_n < r,$$

and hence, by the fact that F is nondecreasing, and from (4), that

$$F(x) \leq F(x_n) \leq \phi(r). \tag{8}$$

From (7) and (8) we conclude that, for $n > N$,

$$0 \leq F(x_n) - F(x) \leq \varepsilon,$$

so that

$$\lim_{n \to \infty} [F(x_n) - F(x)] = 0,$$

as was to be shown.

Before going to the last part of the proof, it is convenient to note the following consequence of (4): for r rational, x arbitrary, and $r < x$, we have

$$\phi(r) \leq F(x). \tag{9}$$

We shall now show, finally, that the diagonal sequence (2) converges to F at every point of continuity of F. Let x be a point of continuity of F, and let ε be an arbitrary positive number. From (4) we see immediately that there exists a rational r such that $r > x$ and

$$0 \leq \phi(r) - F(x) \leq \frac{\varepsilon}{6}. \tag{10}$$

From the continuity of F at x, we see that there exists some y such that $y < x$ and

$$0 \leq F(x) - F(y) \leq \frac{\varepsilon}{6}. \tag{11}$$

Now let s be any rational such that

$$y < s < x. \tag{12}$$

From (12) we conclude, by means of (4) and (9), that

$$F(y) \leq \phi(s) \leq F(x); \tag{13}$$

and then, by (11) and (13), we obtain

$$0 \leq F(x) - \phi(s) \leq \frac{\varepsilon}{6}. \tag{14}$$

From (3) we see that there is an integer N such that, for all $n > N$,

$$|\phi(r) - F_n(r)| \leq \frac{\varepsilon}{6} \tag{15}$$

and

$$|\phi(s) - F_n(s)| \leq \frac{\varepsilon}{6}. \tag{16}$$

Moreover, since F_n is a distribution function, and hence nondecreasing,

$$F_n(s) \leq F_n(x) \leq F_n(r). \tag{17}$$

From (16), (14), (10), and (15) we now obtain

$$\begin{aligned} |F_n(s) - F_n(r)| &\leq |F_n(s) - \phi(s)| \\ &\quad + |\phi(s) - F(x)| + |F(x) - \phi(r)| \\ &\quad + |\phi(r) - F_n(r)| \\ &\leq 4\left(\frac{\varepsilon}{6}\right) = \frac{2}{3}\varepsilon. \end{aligned} \tag{18}$$

From (17) and (18), it follows that

$$|F_n(x) - F_n(s)| \leq \frac{2}{3}\varepsilon. \tag{19}$$

From (14), (16), and (19) we conclude finally that, for all $n > N$,

$$\begin{aligned} |F(x) - F_n(x)| &\leq |F(x) - \phi(s)| + |\phi(s) - F_n(s)| + |F_n(s) - F_n(x)| \\ &\leq \frac{\varepsilon}{6} + \frac{\varepsilon}{6} + \frac{2}{3}\varepsilon = \varepsilon, \end{aligned} \tag{20}$$

and hence that

$$F(x) = \lim_{n \to \infty} F_n(x),$$

which completes the proof of our theorem.

The following theorem gives us a useful method for constructing new distribution functions from given ones.

THEOREM 8.3. Let $F_1, \cdots, F_n$ be a set of distribution functions, let $\| a_1 \quad \cdots \quad a_n \|$ be any member of $\mathbf{S}_n$, and let the function F be defined, for all x in $[0, 1]$, by the equation

$$F(x) = a_1F_1(x) + \cdots + a_nF_n(x).$$

Then F is also a distribution function.

PROOF. Since $F_i \in \mathbf{D}$ (for $i = 1, \cdots, n$), F_i satisfies the four conditions stated above. In order to prove that $F \in \mathbf{D}$, it is only necessary to prove that F also satisfies the four given conditions.

First, we have

$$F(x) = a_1F_1(x) + \cdots + a_nF_n(x) \geq 0,$$

since $a_i \geqq 0$ and $F_i(x) \geqq 0$ by assumption.

Secondly, since $F_i(1) = 1$, we have

$$F(1) = a_1F_1(1) + \cdots + a_nF_n(1) = a_1 + \cdots + a_n = 1.$$

The proof that $F(0) = 0$ is similar.

Thirdly, to show that F is nondecreasing, let u and v be any numbers in $[0, 1]$ such that $u \geq v$; we are to show that $F(u) \geq F(v)$. Since each F_i is nondecreasing,

$$F_i(u) \geq F_i(v),$$

and hence, since each a_i is non-negative,

$$a_i F_i(u) \geq a_i F_i(v).$$

Adding these inequalities, we obtain

$$a_1 F_1(u) + \cdots + a_n F_n(u) \geq a_1 F_1(v) + \cdots + a_n F_n(v),$$

or

$$F(u) \geq F(v),$$

as was to be shown.

Fourthly, to complete our proof, it remains only to show that F is right-hand continuous in $(0, 1)$. Suppose that x is an arbitrary positive number in $(0,1)$. Then, making use of the fact that each of the functions $F_1, \cdots, F_n$ is right-hand continuous at x, we have

$$\begin{aligned} \lim_{\substack{\varepsilon \to 0 \\ \varepsilon > 0}} F(x + \varepsilon) &= \lim_{\substack{\varepsilon \to 0 \\ \varepsilon > 0}} [a_1 F_1(x + \varepsilon) + \cdots + a_n F_n(x + \varepsilon)] \\ &= a_i \lim_{\substack{\varepsilon \to 0 \\ \varepsilon > 0}} F_1(x + \varepsilon) + \cdots + a_n \lim_{\substack{\varepsilon \to 0 \\ \varepsilon > 0}} F_n(x + \varepsilon) \\ &= a_1 F_1(x) + \cdots + a_n F_n(x) = F(x), \end{aligned}$$

as was to be shown.

REMARK 8.4. The above theorem can be generalized, in that we can construct a distribution function F from an infinite sequence $F_1, F_2, \cdots, F_n, \cdots$ of distribution functions, together with an infinite sequence $a_1, a_2, \cdots, a_n, \cdots$ of non-negative real numbers whose sum is 1, by setting

$$F(x) = \sum_{i=1} a_i F_i(x).$$

Since we shall not have to make use of the theorem in this form, however, we have confined ourselves to proving it for the case of a finite number of distribution functions. The proof for infinite sequences of distribution functions will be left as an exercise.

It is useful, finally, to single out a special class of distribution functions to which we shall often refer hereafter. Suppose there is a finite increasing sequence of points of $[0, 1]$,

$$x_1 < x_2 < x_3 < \cdots < x_n,$$

such that a distribution function F has discontinuities at each of these n points but is constant elsewhere; i.e., $F(u) = F(v)$ if, for some i, $x_i < u < x_{i+1}$ and $x_i < v < x_{i+1}$. Then F is called a *step-function with n steps.* We denote the class of all step-functions with n steps by D_n.

The graph of a step-function always consists of a finite number of horizontal line-segments, and the breaks occur at the points of discontinuity; thus, for example, the distribution function in Fig. 9 is a member of D_4, which has discontinuities at 0, ¼, ¾, and 1.

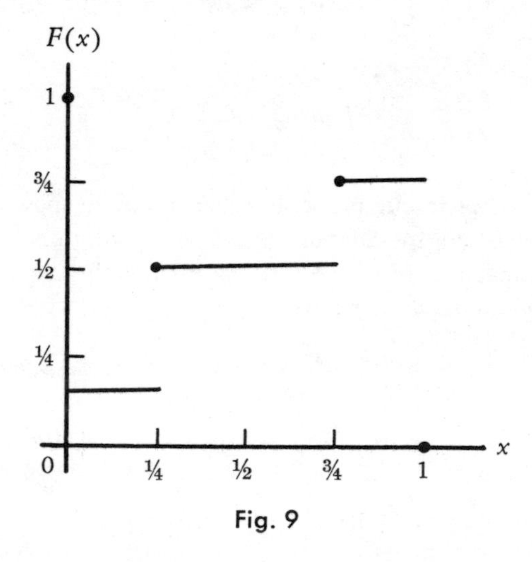

Fig. 9

It is easily seen that if $F_1, \cdots, F_n$ are step-functions, and if $a_1, \cdots, a_n$ are non-negative real numbers whose sum is 1, then the function F, defined by the equation

$$F(x) = a_1F_1(x) + \cdots + a_nF_n(x),$$

is also a step-function. The number of discontinuities of F is not greater than the sum of the numbers of discontinuities of $F_1, \cdots, F_n$; if none of the functions $F_1, \cdots, F_n$ have a common point of discontinuity, and $a_i \neq 0$ (for $i = 1, \cdots, n$), then the number of discontinuities of F is equal to this sum.

The members of D_1 have only one step, which must therefore be one unit. Thus to specify a member of D_1 it suffices to tell where this one step occurs; we write I_a to mean the member of D_1 which has its step at $x = a$. Thus

I_a, for $a \neq 0$, satisfies the equations

$$I_a(x) = 0 \qquad \text{for } x < a,$$
$$I_a(x) = 1 \qquad \text{for } x \geq a.$$

The function I_0 satisfies the equations

$$I_0(0) = 0,$$
$$I_0(x) = 1 \qquad \text{for } x > 0.$$

HISTORICAL AND BIBLIOGRAPHICAL REMARK

A more detailed discussion of distribution functions can be found in Cramér [1].

EXERCISES

1. Find the distribution function corresponding to the following random process for choosing a number from $[0, 1]$: (1) two coins are tossed; (2) if both coins show heads, then the number ¼ is chosen; (3) if both coins show tails, then a number is chosen from $[0, 1]$ by means of the machine pictured in Fig. 1; and (4) if one of the coins shows heads, while the other shows tails, then a number is chosen from $[0, 1]$ by means of the machine pictured in Fig. 2.

2. Find the distribution function for the following random process: roll a die, and take the reciprocal of the number which appears.

3. Show that if a function is both right-hand continuous and left-hand continuous at a given point, then it is continuous at the point.

4. Show that, for an arbitrary n, every member of D_n can be constructed, by the method established in Theorem 8.3, from members of D_1.

5. Prove the following generalization of Theorem 8.3: Let $F_1, \cdots, F_n, \cdots$ be an infinite sequence of distribution functions, let $a_1, \cdots, a_n, \cdots$ be an infinite sequence of non-negative real numbers such that

$$a_1 + \cdots + a_n + \cdots = 1,$$

and let the function F be defined by the equation

$$F(x) = a_1F_1(x) + \cdots + a_nF_n(x) + \cdots;$$

then F is also a distribution function.

6. Show that a distribution function cannot have more than a countable infinity of discontinuities.

7. If $0 \leq a \leq b \leq 1$, show that

$$I_a(x) \cdot I_b(x) = I_b(x).$$

8. Show that the product of two distribution functions is always a distribution function; i.e., if F_1 and F_2 are in D, then the function F defined by the equation

$$F(x) = F_1(x) \cdot F_2(x) \qquad \text{for all } x \text{ in } [0, 1]$$

is also in D.

9. Show that the product of two step-functions is a step-function. Show that, if $F_1 \in \mathsf{D}_r$ and $F_2 \in \mathsf{D}_s$, then there exists a number $t \leq r + s - 1$ such that the product of F_1 and F_2 is in D_t.

CHAPTER 9
STIELTJES INTEGRALS

Suppose that we are given a continuous game with a payoff function M, and suppose that P_2 picks the number y_0. Then, for each x which can be chosen by P_1, the amount $G(x)$, determined by the equation

$$G(x) = M(x, y_0),$$

will be what P_1 receives.

Now suppose that P_1, instead of always picking the same number x, decides to use a random device to which there corresponds a distribution function F. It is clearly important to P_1, if he is to decide among the various possible distribution functions he might use, to be able to find his mathematical expectation if he uses F.

Thus we have the following problem: For $0 \leq x \leq 1$, let $G(x)$ be the amount P_1 will receive if he chooses x; if P_1 makes his choices by means of the distribution function F, then what will be the mathematical expectation of P_1?

It is clear that we cannot say that this expectation is simply the sum of all products $G(x) \cdot P(x)$, where $P(x)$ is the probability that x will be chosen; for, as we saw in the last chapter, $P(x)$ may be 0 for all x.

On the other hand, we can make an approximation to this expectation in the following way. We take three values of x, say $x = 0$, $x = \frac{1}{2}$, and $x = 1$, and form the following expression:

$$G\left(\frac{1}{2}\right)\left[F\left(\frac{1}{2}\right) - F(0)\right] + G(1)\left[F(1) - F\left(\frac{1}{2}\right)\right].$$

The first term of this sum is the amount P_1 will get if he chooses $\frac{1}{2}$, multiplied by the probability that he will choose an x in the interval $0 \leq x \leq \frac{1}{2}$; the second term is the amount he will get if he chooses 1, multiplied by the probability that he will choose an x in the interval $\frac{1}{2} < x \leq 1$. Another approximation can be obtained from the given subdivision of the interval by writing

$$G(x_1)\left[F\left(\frac{1}{2}\right) - F(0)\right] + G(x_2)\left[F(1) - F\left(\frac{1}{2}\right)\right],$$

where x_1 and x_2 are any numbers such that $0 \leq x_1 \leq \frac{1}{2}$ and $\frac{1}{2} < x_2 \leq 1$. If G does not change very much within $[0, \frac{1}{2}]$ or within $[\frac{1}{2}, 1]$, then these two approximations will not differ much from each other. If we suppose that the minimum value of G in the interval $0 \leq x \leq \frac{1}{2}$ is assumed at $\bar{z}_1$ and that the minimum value of G in the interval $\frac{1}{2} < x \leq 1$ is assumed at $\bar{z}_2$, then our intuitive notion of the nature of mathematical expectation is sufficiently clear that we are willing to agree that the expectation of P_1 is at least

$$G(\bar{z}_1)\left[F\left(\frac{1}{2}\right) - F(0)\right] + G(\bar{z}_2)\left[F(1) - F\left(\frac{1}{2}\right)\right].$$

Similarly, if the maximum values of G in these two intervals are assumed at $\bar{\bar{z}}_1$ and $\bar{\bar{z}}_2$, respectively, then the expectation of P_1 is at most

$$G(\bar{\bar{z}}_1)\left[F\left(\frac{1}{2}\right) - F(0)\right] + G(\bar{\bar{z}}_2)\left[F(1) - F\left(\frac{1}{2}\right)\right].$$

It appears, however, that, in general, we could get better upper and lower bounds for the expectation of P_1 by taking four points of division; still better by taking five; and so on. Thus we are led to consider a subdivision of the interval by the points $0 = x_0 < x_1 < \cdots < x_n = 1$, and to form the sum

$$\sum_{i=1}^{n} G(z_i)[F(x_i) - F(x_{i-1})],$$

where $x_{i-1} < z_i \leq x_i$ (for $i = 1, \cdots, n$). Intuitively, it now seems clear that if G is a "reasonable" kind of function, then we ought to be able to get arbitrarily close to the expectation of P_1 by taking n large enough and seeing to it that all the differences $x_i - x_{i-1}$ approach 0 as n becomes large. (If G is not sufficiently "reasonable" that this limit exists, then our intuition fails us, and we are no longer quite sure exactly what we would mean by the expectation of P_1.)

The sum

$$\sum_{i=1}^{n} G(z_i)[F(x_i) - F(x_{i-1})] \tag{1}$$

has a certain resemblance to the sum

$$\sum_{i=1}^{n} G(z_i)[x_i - x_{i-1}] \tag{2}$$

used in defining the ordinary (Riemann) integral; indeed, (2) is just a special case of (1), where we take F to be the function such that, for all x,

$$F(x) = x.$$

For this reason it is natural to regard the limit of (1) also as an integral. We now proceed to give a formal definition of this kind of integral, which is called a *Stieltjes integral.*

DEFINITION 9.1. Let Δ denote a subdivision of the interval $[a, b]$ by the points $x_0, x_1, \cdots, x_n$, where

$$a = x_0 < x_1 < \cdots < x_{n-1} < x_n = b.$$

Let the maximum of the differences

$$x_1 - x_0, x_2 - x_1, \cdots, x_n - x_{n-1}$$

be denoted by $\| \Delta \|$. If

$$\lim_{\substack{n \to \infty \\ \|\Delta\| \to 0}} \sum_{i=1}^{n} G(z_i)[F(x_i) - F(x_{i-1})] \tag{3}$$

(where $x_{i-1} \leq z_i \leq x_i$, for $i = 1, \cdots, n$) exists and is independent of the choice of the z_i's, then this limit is called the *Stieltjes integral of G with respect to F from a to b*, and is denoted by

$$\int_a^b G(x)\, dF(x).$$

Using this definition and notation, we can now say that if $G(x)$ is the payoff to P_1 when he chooses x, and if he chooses x by means of the distribution function F, then his expectation will be

$$\int_0^1 G(x)\, dF(x).$$

Since the Stieltjes integral is defined by means of a complicated limiting process, it should not be an occasion for surprise that it does not always exist. In particular, it is easily shown that it does not exist when G and F have a common point of discontinuity.

For example, let

$$F(x) = G(x) = 0 \qquad \text{for } 0 \leq x \leq 1,$$

and let

$$F(x) = G(x) = 1 \qquad \text{for } 1 < x \leq 2.$$

Here exactly one of the differences $F(x_i) - F(x_{i-1})$ is different from 0, and this one, call it $F(x_k) - F(x_{k-1})$, is equal to 1. Then $G(z_k)$ can be either 0 or 1, depending on the choice of z_k. Hence in this case the sum (1) is either 0 or 1, depending on whether $G(z_k)$ is 0 or 1. Thus the limit (3) is not independent of the choice of the z_i's, and hence

$$\int_0^2 G(x)\, dF(x)$$

does not exist.

We shall now state and prove a theorem which gives sufficient conditions for the existence of the Stieltjes integral.

> THEOREM 9.2. If G is continuous in the interval $[a, b]$, and F is a nondecreasing function, then $\int_a^b G(x)\, dF(x)$ exists.

REMARK 9.3. In the special case where $F(x) \equiv x$, Theorem 9.2 reduces to the familiar theorem of integral calculus which states that if G is continuous over $[a, b]$, then the Riemann integral $\int_a^b G(x)\, dx$ exists. In view of this relationship, it is not surprising that the proof of Theorem 9.2 is very similar to, though slightly more complicated than, the proof of the latter theorem. We shall give the proof of Theorem 9.2, but thereafter we shall omit proofs of theorems which are closely analogous to the proofs of corresponding theorems concerning Riemann integrals.

We shall first introduce some further notations which will be used in the proof of Theorem 9.2 and of some lemmas which lead up to this proof. We assume throughout this discussion that Δ, Δ_1, Δ_2, etc., are subdivisions of $[a, b]$, that G is continuous over $[a, b]$, and that F is nondecreasing over $[a, b]$. We define

$$M_k = \max_{x_{k-1} \leq x \leq x_k} G(x),$$

$$m_k = \min_{x_{k-1} \leq x \leq x_k} G(x),$$

$$S_\Delta = \sum_{i=1}^{n} M_i [F(x_i) - F(x_{i-1})],$$

$$s_\Delta = \sum_{i=1}^{n} m_i [F(x_i) - F(x_{i-1})].$$

Clearly, $m_k \leq M_k$; and, since F is nondecreasing, $s_\Delta \leq S_\Delta$.

The points $x_0 = a, x_1, \cdots, x_{n-1}, x_n = b$ which determine the subdivision Δ are called *points of division* of Δ. If every point of division of Δ_1 is also a point of division of Δ_2, then the subdivision Δ_2 is called a *refinement* of Δ_1. Clearly, if Δ_2 is a refinement of Δ_1, and Δ_1 is a refinement of Δ_2, then Δ_1 and Δ_2 are identical.

LEMMA 9.4. If Δ_0 is a refinement of Δ, then $s_{\Delta_0} \geq s_\Delta$.

PROOF. Let the points of division of Δ be $x_0, x_1, \cdots, x_n$, and suppose, to begin with, that Δ_0 has exactly one point of division, $\bar{x}$, which is not a point of division of Δ. Let $x_{j-1} < \bar{x} < x_j$. Let

$$m'_j = \min_{x_{j-1} \leq x \leq \bar{x}} G(x)$$

and

$$m''_j = \min_{\bar{x} \leq x \leq x_j} G(x).$$

Clearly, $m'_j \geq m_j$ and $m''_j \geq m_j$. Hence

$$m'_j[F(\bar{x}) - F(x_{j-1})] + m''_j[F(x_j) - F(\bar{x})] \geq m_j[F(x_j) - F(x_{j-1})],$$

while, for $i = 1, 2, \cdots, j-1, j+1, j+2, \cdots, n$, the term $m_i[F(x_i) - F(x_{i-1})]$ occurs in both the expression for s_Δ and in the expression for s_{Δ_0}. Consequently, in this special case, $s_\Delta \leq s_{\Delta_0}$.

Now if Δ_0 is any refinement of Δ other than Δ itself, then there is a finite chain of subdivisions beginning with Δ and ending with Δ_0, and having the property that each subdivision of the chain (except Δ) is a refinement of its predecessor which contains exactly one point of division which is not also a point of division of its predecessor. Hence the lemma follows in this case by repeated applications of the argument above.

LEMMA 9.5. If Δ_0 is a refinement of Δ, then $S_{\Delta_0} \leq S_\Delta$.

PROOF. Similar to the proof of Lemma 9.4.

LEMMA 9.6. If Δ_1 and Δ_2 are subdivisions of $[a, b]$, then $s_{\Delta_1} \leq S_{\Delta_2}$.

PROOF. Let Δ_3 be the subdivision whose points of division are exactly those points each of which is a point of division of either Δ_1 or Δ_2. Then Δ_3 is a refinement of Δ_1 and also of Δ_2. Hence, by Lemmas 9.4 and 9.5, respectively, we have $s_{\Delta_1} \leq s_{\Delta_3}$ and $S_{\Delta_3} \leq S_{\Delta_2}$. Since $s_{\Delta_3} \leq S_{\Delta_3}$, therefore, we obtain the desired conclusion.

It follows from this result that if Δ_1 is any subdivision, then S_{Δ_1} is an upper bound of all the numbers s_Δ. Hence the numbers s_Δ have a least upper bound, which we denote by s. Similarly the numbers S_Δ have a greatest lower bound, which we denote by S.

LEMMA 9.7. $\lim_{||\Delta|| \to 0} S_\Delta = \lim_{||\Delta|| \to 0} s_\Delta = S = s.$

PROOF. Since G is continuous over the closed interval $[a, b]$, it is uniformly continuous over $[a, b]$; i.e., given any $\varepsilon > 0$, there exists a $\delta > 0$ such that if $|z_1 - z_2| < \delta$, then $|G(z_1) - G(z_2)| < \varepsilon$. This means that if $\|\Delta\| < \delta$, then $|M_k - m_k| < \varepsilon$, for $k = 1, 2, \cdots, n$. Therefore,

$$\begin{aligned} 0 \leq S_\Delta - s_\Delta &= \sum_{i=1}^{n} M_i[F(x_i) - F(x_{i-1})] \\ &\quad - \sum_{i=1}^{n} m_i[F(x_i) - F(x_{i-1})] \\ &= \sum_{i=1}^{n} (M_i - m_i)[F(x_i) - F(x_{i-1})] \\ &< \varepsilon \sum_{i=1}^{n} [F(x_i) - F(x_{i-1})] \\ &= \varepsilon[F(b) - F(a)]. \end{aligned}$$

Hence we can write

$$0 \leq S_\Delta - s_\Delta = (S_\Delta - S) + (S - s) + (s - s_\Delta) \leq \varepsilon[F(b) - F(a)].$$

Each of the three terms in parentheses is non-negative. Therefore, since $F(b) - F(a)$ is fixed, and ε can be made arbitrarily small, it follows that each of these three terms must approach 0 as $\|\Delta\|$ approaches 0; and this completes the proof.

PROOF OF THEOREM 9.2. It is clear from the definitions of m_k, M_k, s_Δ, and S_Δ that for any subdivision Δ and any choice of the z_i's in (3), we have

$$s_\Delta \leq \sum_{i=1}^{n} G(z_i)[F(x_i) - F(x_{i-1})] \leq S_\Delta.$$

From this, together with Lemma 9.7, it follows that the limit (3) exists and is equal to $s = S$.

REMARK 9.8. It is easy to show that the conditions which have been imposed on G and F are not necessary for the existence of $\int_a^b G(x)\, dF(x)$. In particular, it is obvious that the condition that F be nondecreasing could be replaced by the condition that F be nonincreasing. Other generalizations are possible, but are not needed for our purposes.

The following nine theorems can be established by rather simple proofs, several of which are similar to those usually given for analogous theorems about Riemann integrals. We shall omit these proofs.

Theorem 9.9. If the integrals involved exist, and if $a < b < c$, then

$$\int_a^c G(x)\,dF(x) = \int_a^b G(x)\,dF(x) + \int_b^c G(x)\,dF(x).$$

Theorem 9.10. If the integrals involved exist, then

$$\int_a^b [G(x) + H(x)]\,dF(x) = \int_a^b G(x)\,dF(x) + \int_a^b H(x)\,dF(x).$$

Theorem 9.11. If the integrals involved exist, then

$$\int_a^b G(x)\,d[F(x) + H(x)] = \int_a^b G(x)\,dF(x) + \int_a^b G(x)\,dH(x).$$

Theorem 9.12. If the integrals involved exist, and k is any real number, then

$$\int_a^b kG(x)\,dF(x) = k\int_a^b G(x)\,dF(x).$$

Theorem 9.13. If the integrals involved exist, and k is any real number, then

$$\int_a^b G(x)\,d[kF(x)] = k\int_a^b G(x)\,dF(x).$$

Theorem 9.14. If the integrals involved exist, and if F is nondecreasing, then

$$\left|\int_a^b G(x)\,dF(x)\right| \le \int_a^b |G(x)|\,dF(x).$$

Theorem 9.15. If the integrals involved exist, if F is nondecreasing, and if, for all x in $[a, b]$, $G(x) \le H(x)$, then

$$\int_a^b G(x)\,dF(x) \le \int_a^b H(x)\,dF(x).$$

Moreover, if F is not constant over $[a, b]$, if G and H are continuous over $[a, b]$, and if, for all x in $[a, b]$, $G(x) < H(x)$,

then

$$\int_a^b G(x)\,dF(x) < \int_a^b H(x)\,dF(x).$$

THEOREM 9.16. If F is any distribution function, then

$$\int_0^1 1 \cdot dF(x) = F(1) - F(0) = 1.$$

THEOREM 9.17. If the integrals involved exist, then

$$\int_a^b G(x)\,dF(x) = G(b)F(b) - G(a)F(a) - \int_a^b F(x)\,dG(x).$$

REMARK 9.18. Theorem 9.17 corresponds to the familiar process of "integration by parts" which is used in finding ordinary indefinite integrals.

We now prove a theorem which enables us in certain cases to reduce the problem of evaluating a Stieltjes integral to the problem of evaluating an ordinary (Riemann) integral.

THEOREM 9.19. If the integrals involved exist, and if the function F possesses a derivative F' at every point in $[a, b]$, then

$$\int_a^b G(x)\,dF(x) = \int_a^b G(x)F'(x)\,dx.$$

PROOF. Let $x_0, x_1, \cdots, x_n$ be the points of division of any subdivision Δ of $[a, b]$. Since F possesses a derivative in $[a, b]$, it follows by the theorem of the mean that there are numbers $y_1, \cdots, y_n$ such that $x_{i-1} \le y_i \le x_i$, for $i = 1, \cdots, n$, and

$$F(x_i) - F(x_{i-1}) = F'(y_i)[x_i - x_{i-1}]. \tag{4}$$

Since the Riemann integral $\int_a^b G(x)F'(x)\,dx$ exists, the limit

$$\lim_{\substack{n\to\infty \\ \|\Delta\|\to 0}} \sum_{i=1}^{n} G(y_i)F'(y_i)[x_i - x_{i-1}]$$

exists and is independent of the choice of the y_i's. But, by (4),

$$\lim_{\substack{n\to\infty \\ \|\Delta\|\to 0}} \sum_{i=1}^{n} G(y_i)F'(y_i)[x_i - x_{i-1}] = \lim_{\substack{n\to\infty \\ \|\Delta\|\to 0}} \sum_{i=1}^{n} G(y_i)[F(x_i) - F(x_{i-1})].$$

Our result now follows from this equation, together with the hypothesis that the two integrals mentioned in the theorem exist.

We next prove a theorem which enables us to evaluate the Stieltjes integral of a function with respect to a step-function.

THEOREM 9.20. Let a, b, and $a_1, a_2, \cdots, a_n$ be real numbers satisfying

$$a \leq a_1 \leq \cdots \leq a_n \leq b,$$

let $b_1, \cdots, b_n$ be real numbers, and let the step-function F be defined as follows:

$$F(x) = b_1 I_{a_1}(x) + b_2 I_{a_2}(x) + \cdots + b_n I_{a_n}(x).$$

Let G be any function which is defined over the interval $[a, b]$ and is continuous at the points $a_1, a_2, \cdots, a_n$. Then

$$\int_a^b G(x)\, dF(x) = b_1 G(a_1) + \cdots + b_n G(a_n).$$

REMARK 9.21. Note that if the condition that G be continuous at the points $a_1, a_2, \cdots, a_n$ is not satisfied, then the integral does not exist. Consequently, the theorem enables us to evaluate the integral $\int_a^b G(x)\, dF(x)$ whenever F is a step-function and the integral exists.

We recall that $I_a(x)$ was defined in the last paragraph of Chap. 8.

PROOF OF THEOREM 9.20. In view of Theorems 9.11 and 9.13, it is clearly sufficient to prove our present theorem for the case in which

$$F(x) = I_{a_1}(x).$$

The function G is continuous at $x = a_1$; therefore, given any positive number ε, there exists a positive number δ such that if $|x - a_1| < \delta$, then $|G(x) - G(a_1)| < \varepsilon$. Let Δ be any subdivision of $[a, b]$, for which $\|\Delta\| < \delta$, and let $x_0, x_1, \cdots, x_n$ be the points of division of Δ. Let $x_{j-1} < a_1 \leq x_j$. Then $F(x_j) - F(x_{j-1}) = 1$, while $F(x_i) - F(x_{i-1}) = 0$, for $i \neq j$. Hence

$$\begin{aligned} \sum_{i=1}^{n} G(z_i)[F(x_i) - F(x_{i-1})] &= G(z_j)[F(x_j) - F(x_{j-1})] \\ &= G(z_j). \end{aligned}$$

Since $x_{j-1} < a_1 \leq x_j$, $x_{j-1} \leq z_j \leq x_j$, and $|x_{j-1} - x_j| < \delta$, it follows that $|z_j - a_1| < \delta$. Hence

$$\left| \sum_{i=1}^{n} G(z_i)[F(x_i) - F(x_{i-1})] - G(a_1) \right| = |G(z_j) - G(a_1)| < \varepsilon.$$

It now follows at once that

$$\lim_{\substack{n\to\infty \\ ||\Delta||\to 0}} \sum_{i=1}^{n} G(z_i)[F(x_i) - F(x_{i-1})] = G(a_1),$$

and the theorem is proved.

We shall conclude this chapter by proving three special theorems which will be useful in later chapters.

THEOREM 9.22. If G is a continuous function in the closed interval $[0, 1]$, then

$$\min_{F \in \mathbf{D}} \int_0^1 G(x)\, dF(x)$$

exists, and

$$\min_{F \in \mathbf{D}} \int_0^1 G(x)\, dF(x) = \min_{0\le x\le 1} G(x);$$

moreover,

$$\max_{F \in \mathbf{D}} \int_0^1 G(x)\, dF(x)$$

exists, and

$$\max_{F \in \mathbf{D}} \int_0^1 G(x)\, dF(x) = \max_{0\le x\le 1} G(x).$$

PROOF. We recall that $\mathbf{D}$ is the set of all distribution functions.

Since G is continuous in the closed interval, $\min_{0\le x\le 1} G(x)$ exists. Let a be a number in $[0, 1]$ such that

$$G(a) = \min_{0\le x\le 1} G(x).$$

Now since, for all x in $[0, 1]$,

$$G(a) \le G(x),$$

we see, by Theorems 9.16, 9.12, and 9.15, that for any distribution function F,

$$G(a) = G(a) \cdot \int_0^1 1 \cdot dF(x) = \int_0^1 G(a)\, dF(x) \leq \int_0^1 G(x)\, dF(x),$$

and hence that

$$G(a) \leq \inf_{F \in \mathrm{D}} \int_0^1 G(x)\, dF(x). \tag{5}$$

Moreover, we have, using Theorem 9.20,

$$\inf_{F \in \mathrm{D}} \int_0^1 G(x)\, dF(x) \leq \int_0^1 G(x)\, dI_a(x) = G(a). \tag{6}$$

From (5) and (6) we conclude that

$$\inf_{F \in \mathrm{D}} \int_0^1 G(x)\, dF(x) = G(a) = \int_0^1 G(x)\, dI_a(x).$$

Thus the greatest lower bound, with respect to F, of the numbers

$$\int_0^1 G(x)\, dF(x)$$

is actually assumed (namely, for $F = I_a$), which means that

$$\min_{F \in \mathrm{D}} \int_0^1 G(x)\, dF(x)$$

exists, and that

$$\min_{F \in \mathrm{D}} \int_0^1 G(x)\, dF(x) = G(a). \tag{7}$$

This completes the proof of the first part of our theorem. The second part is established in a similar way.

THEOREM 9.23. If G is a continuous function over the closed interval $[0, 1]$, and if $F_1, F_2, \cdots$ is a sequence of distribution functions converging to the distribution function F (at every point of continuity of F), then

$$\lim_{t \to \infty} \int_0^1 G(x)\, dF_t(x) = \int_0^1 G(x)\, dF(x).$$

REMARK 9.24. In the proof of this theorem we make use of the notation introduced just before Lemma 9.4, specializing it to the case in which $[a, b]$ is $[0, 1]$. We also use the following lemma, the proof of which is easy and is omitted (see Exercise 6 of Chap. 8).

LEMMA 9.25. If F is a distribution function and δ is a positive number, then there exist points

$$0 = x_0 < x_1 < \cdots < x_{n-1} < x_n = 1$$

such that F is continuous at x_i (for $i = 1, \cdots, n - 1$), and $|x_i - x_{i-1}| < \delta$ (for $i = 1, \cdots, n$).

PROOF OF THEOREM 9.23. Let the positive number ε be given, and let δ be a positive number such that if Δ_1 is any subdivision of $[0, 1]$, for which $\| \Delta_1 \| < \delta$, then $S_{\Delta_1} - s_{\Delta_1} < \varepsilon$. Choose a subdivision Δ of $[0, 1]$, for which $\| \Delta \| < \delta$, such that the points of division $x_0, x_1, \cdots, x_n$ of Δ, except possibly x_0 and x_n, are points of continuity of F. (The lemma assures us that such a subdivision exists.)

Since $F_t(x_i)$ converges to $F(x_i)$ (for $i = 0, 1, \cdots, n$), it follows that there exists an integer T such that if $t \geq T$, then

$$| F_t(x_i) - F(x_i) | < \frac{\varepsilon}{n} \qquad \text{for } i = 0, 1, \cdots, n. \tag{8}$$

Since $S_\Delta - s_\Delta < \varepsilon$, it follows that

$$\left| \sum_{i=1}^{n} M_i[F(x_i) - F(x_{i-1})] - \int_0^1 G(x)\, dF(x) \right| < \varepsilon \tag{9}$$

and

$$\left| \sum_{i=1}^{n} m_i[F(x_i) - F(x_{i-1})] - \int_0^1 G(x)\, dF(x) \right| < \varepsilon. \tag{10}$$

Moreover,

$$\sum_{i=1}^{n} M_i[F_t(x_i) - F_t(x_{i-1})] \geq \int_0^1 G(x)\, dF_t(x) \tag{11}$$

and

$$\int_0^1 G(x)\, dF_t(x) \geq \sum_{i=1}^{n} m_i[F_t(x_i) - F_t(x_{i-1})]. \tag{12}$$

Let

$$A = \max_{0 \leq x \leq 1} | G(x) |.$$

Then $M_i \leq A$ (for $i = 1, \cdots, n$). Now if $t \geq T$, we have

$$\left| \sum_{i=1}^{n} M_i[F(x_i) - F(x_{i-1})] - \sum_{i=1}^{n} M_i[F_t(x_i) - F_t(x_{i-1})] \right|$$

$$= \left| \sum_{i=1}^{n} M_i[F(x_i) - F_t(x_i) + F_t(x_{i-1}) - F(x_{i-1})] \right|$$

$$\leq A\left[\sum_{i=1}^{n} | F(x_i) - F_t(x_i) | + \sum_{i=1}^{n} | F_t(x_{i-1}) - F(x_{i-1}) | \right]$$

$$\leq 2A\varepsilon, \tag{13}$$

where the last step follows from (8).

From (9) and (13) we have

$$\left| \int_0^1 G(x)\, dF(x) - \sum_{i=1}^{n} M_i[F_t(x_i) - F_t(x_{i-1})] \right| \leq (1 + 2A)\varepsilon. \tag{14}$$

Similarly we obtain, making use of (10) instead of (9),

$$\left| \int_0^1 G(x)\, dF(x) - \sum_{i=1}^{n} m_i[F_t(x_i) - F_t(x_{i-1})] \right| \leq (1 + 2A)\varepsilon. \tag{15}$$

From (14) and (15) we see, by means of (11) and (12), that

$$\left| \int_0^1 G(x)\, dF(x) - \int_0^1 G(x)\, dF_t(x) \right| \leq (1 + 2A)\varepsilon. \tag{16}$$

Since $2A + 1$ is fixed and ε is arbitrary, it follows from (16) that

$$\lim_{t \to \infty} \int_0^1 G(x)\, dF_t(x) = \int_0^1 G(x)\, dF(x),$$

as was to be shown.

Before proving the final theorem of this chapter we introduce a definition.

DEFINITION 9.26. Let F be a function defined over the interval $[a - h, a]$, where h is a positive number. Suppose that for every decreasing sequence $\varepsilon_1, \varepsilon_2, \cdots$ of non-negative numbers which converges to 0 and for which $\varepsilon_1 \leq h$ we have

$$\lim_{n \to \infty} \left[\frac{F(a) - F(a - \varepsilon_n)}{\varepsilon_n} \right] = c.$$

Then we say that c is the *left-hand derivative of* F *at* a.

Similarly, let F be defined over the interval $[a, a + h]$, where h is any positive number, and suppose that for every decreasing sequence $\varepsilon_1, \varepsilon_2, \cdots$ of non-negative numbers which converges to 0 and for which $\varepsilon_1 \leq h$ we have

$$\lim_{n\to\infty}\left[\frac{F(a+\varepsilon_n) - F(a)}{\varepsilon_n}\right] = d.$$

Then we say that d is the *right-hand derivative of* F *at* a.

Clearly, if the derivative of F exists at a point, then the left-hand derivative and the right-hand derivative both exist and are equal to the derivative. Conversely, if the left-hand derivative and the right-hand derivative both exist at a point and are equal, then the derivative also exists. Since a distribution function is nondecreasing, it is clear that it cannot have a negative left-hand derivative nor a negative right-hand derivative.

THEOREM 9.27. Let H be a continuous function over $[0, 1]$; let v be a real number such that, for all x in $[0, 1]$, $H(x) \leq v$; and let F be a distribution function such that

$$\int_0^1 H(x)\, dF(x) = v.$$

Then if $\bar{x}$ is any point such that $0 < \bar{x} < 1$, and the left-hand derivative of F is not 0 at $\bar{x}$ (i.e., it is positive or does not exist at $\bar{x}$), we have

$$H(\bar{x}) = v.$$

PROOF. Suppose, if possible, that the theorem is false. Then $H(\bar{x}) = v - 2\varepsilon$, where ε is positive. Since H is continuous, there exists a positive number δ such that $H(x) \leq v - \varepsilon$ whenever $|x - \bar{x}| \leq \delta$. We now have

$$\begin{aligned}
v &= \int_0^1 H(x)\, dF(x) \\
&= \int_0^{\bar{x}-\delta} H(x)\, dF(x) + \int_{\bar{x}-\delta}^{\bar{x}} H(x)\, dF(x) + \int_{\bar{x}}^1 H(x)\, dF(x) \\
&\leq v \int_0^{\bar{x}-\delta} 1 \cdot dF(x) + (v - \varepsilon) \int_{\bar{x}-\delta}^{\bar{x}} 1 \cdot dF(x) + v \int_{\bar{x}}^1 1 \cdot dF(x) \\
&= v \int_0^1 1 \cdot dF(x) - \varepsilon \int_{\bar{x}-\delta}^{\bar{x}} 1 \cdot dF(x) \\
&= v - \varepsilon \int_{\bar{x}-\delta}^{\bar{x}} 1 \cdot dF(x). \qquad (17)
\end{aligned}$$

Now it is easy to prove that

$$\int_{\bar{x}-\delta}^{\bar{x}} 1 \cdot dF(x) = F(\bar{x}) - F(\bar{x} - \delta). \tag{18}$$

Since F is nondecreasing, and the left-hand derivative of F is different from 0 at $\bar{x}$, it follows that

$$F(\bar{x}) - F(\bar{x} - \delta) > 0,$$

so that

$$\varepsilon \int_{\bar{x}-\delta}^{\bar{x}} 1 \cdot dF(x) > 0.$$

Since this contradicts (17), we conclude that our theorem is true after all.

BIBLIOGRAPHICAL REMARK

A more detailed treatment of Stieltjes integrals will be found in Widder [1] and [2], on which the discussion in this chapter is largely based.

EXERCISES

1. Let the function F be defined as follows:

$$F(x) = 0 \qquad \text{for } x < \frac{1}{2},$$

$$F(x) = x^2 \qquad \text{for } x \geq \frac{1}{2}.$$

Evaluate the integrals

$$\int_0^1 x^3 \, dF(x),$$

$$\int_0^1 \sin x \, dF(x).$$

2. Let the function G be defined as follows:

$$G(x) = 0 \qquad \text{for } x < \frac{1}{2},$$

$$G(x) = \frac{1}{10} \qquad \text{for } \frac{1}{2} \leq x < \frac{3}{4},$$

$$G(x) = x^3 \qquad \text{for } \frac{3}{4} \leq x \leq 1.$$

Evaluate the integral

$$\int_0^1 e^x \, dG(x).$$

3. Evaluate the integral

$$\int_0^1 x^2 \, dh(x),$$

where

$$h(x) = \frac{1}{n+1} \sum_{k=0}^{n} I_{k/n}(x).$$

4. If F is the function defined in Exercise 1 and G is the function defined in Exercise 2, show that the integral

$$\int_0^1 G(x) \, dF(x)$$

does not exist.

5. Prove Theorem 9.9.

6. Prove Theorem 9.10.

7. Prove Theorem 9.14.

8. Prove Lemma 9.25.

9. Prove Eq. (18).

10. Show that the conclusion of Theorem 9.14 is no longer true if we drop the hypothesis that F is nondecreasing.

11. Show that the function F defined by the equation

$$F(x) = |x|$$

has a left-hand derivative and a right-hand derivative at $x = 0$.

12. Let the function F be defined as follows:

$$F(x) = x \sin \frac{1}{x} \qquad \text{if } x \neq 0,$$

$$F(0) = 0.$$

Show that F has neither a left-hand derivative nor a right-hand derivative at $x = 0$.

13. Construct a function which will have a right-hand derivative, but no left-hand derivative, at $x = 0$.

CHAPTER 10

THE FUNDAMENTAL THEOREM FOR CONTINUOUS GAMES

1. The Value of a Continuous Game. We turn now to the problem of defining the value of a continuous game and optimal mixed strategies for the two players. We shall prove that these entities always exist in case the payoff function is continuous.

Suppose that the payoff function for a continuous game is M, and suppose that P_1 chooses x from $[0, 1]$ by means of the distribution function F and that P_2 chooses y from $[0, 1]$ by means of the distribution function G. For any given y, the expectation of P_1 will then be

$$\int_0^1 M(x, y)\, dF(x),$$

and hence, making use of the fact that y is chosen by the distribution function G, the total expectation of P_1 will be

$$\int_0^1 \left[\int_0^1 M(x, y)\, dF(x) \right] dG(y),$$

which is usually written simply

$$\int_0^1 \int_0^1 M(x, y)\, dF(x)\, dG(y).$$

Setting

$$E(F, G) = \int_0^1 \int_0^1 M(x, y)\, dF(x)\, dG(y),$$

we therefore say that the expectation of P_1, if P_1 uses the distribution function F and if P_2 uses the distribution function G, is $E(F, G)$. Since the game is zero-sum, the expectation of P_2 is $-E(F, G)$.

It can be shown (see Bray [1]) that if M is continuous, then

$$\int_0^1 \int_0^1 M(x, y)\, dG(y)\, dF(x) = \int_0^1 \int_0^1 M(x, y)\, dF(x)\, dG(y).$$

Hence, when M is continuous, we could just as well define $E(F, G)$ by the equation

$$E(F, G) = \int_0^1 \int_0^1 M(x, y) \, dG(y) \, dF(x).$$

If it happens that

$$v_1 = \max_{F \in D} \min_{G \in D} E(F, G)$$

and

$$v_2 = \min_{G \in D} \max_{F \in D} E(F, G)$$

both exist, then we see (by an argument similar to the one used in Chap. 1 in connection with rectangular games) that P_1 can choose a distribution function so as to be sure of getting at least v_1 and that P_2 can choose a distribution function which will keep P_1 from getting more than v_2. If v_1 and v_2 are equal, then P_1 can get exactly v_1 ($= v_2$) and cannot hope to get more, unless P_2 behaves stupidly. Hence the question of when v_1 and v_2 exist and are equal is very important from the point of view of the theory of games.

When the two quantities v_1 and v_2 exist and are equal, we call their common value the *value* of the game (to P_1). In this case, as was shown in Theorem 1.5, there exists a saddle-point $\| F_0 \quad G_0 \|$ of the function $E(F, G)$; i.e., there is a pair of distribution functions F_0 and G_0 such that, for all distribution functions F and G,

$$E(F, G_0) \leq E(F_0, G_0) \leq E(F_0, G).$$

Such an F_0, or such a G_0, is called an *optimal mixed strategy* for P_1, or for P_2, respectively. We sometimes call an ordered pair $\| F_0 \quad G_0 \|$ of optimal strategies for the two players a *solution* of the game.

2. Two Algebraic Lemmas. In order to prove the fundamental theorem about continuous games, it is convenient first to establish two algebraic lemmas.

LEMMA 10.1. Let

$$\sum_{j=1}^{n} a_{1j}x_j = a_{11}x_1 + a_{12}x_2 + \cdots + a_{1n}x_n,$$

$$\sum_{j=1}^{n} a_{2j}x_j = a_{21}x_1 + a_{22}x_2 + \cdots + a_{2n}x_n,$$

$$\vdots$$

$$\sum_{j=1}^{n} a_{mj}x_j = a_{m1}x_1 + a_{m2}x_2 + \cdots + a_{mn}x_n$$

be m homogeneous linear forms in n unknowns (with real coefficients) and let v be a real number such that, for every $\| x_1 \quad \cdots \quad x_n \|$ in $\mathbf{S}_n$, there exists an integer $k \leq m$ for which

$$\sum_{j=1}^{n} a_{kj}x_j \leq v.$$

Then there exists an element $\| \bar{y}_1 \quad \cdots \quad \bar{y}_m \|$ of $\mathbf{S}_m$ such that, for every element $\| x_1 \quad \cdots \quad x_n \|$ of $\mathbf{S}_n$,

$$\sum_{i=1}^{m} \sum_{j=1}^{n} a_{ij}\bar{y}_i x_j \leq v.$$

PROOF. Let $\| \bar{x}_1 \quad \cdots \quad \bar{x}_n \|$ and $\| \bar{y}_1 \quad \cdots \quad \bar{y}_m \|$ be optimal strategies for P_1 and P_2, respectively, in the rectangular game whose matrix is

$$\begin{Vmatrix} a_{11} & a_{21} & \cdots & a_{m1} \\ a_{12} & a_{22} & \cdots & a_{m2} \\ \vdots & \vdots & & \vdots \\ a_{1n} & a_{2n} & \cdots & a_{mn} \end{Vmatrix}.$$

(These optimal strategies exist by Theorem 2.6.) By the hypothesis of our lemma, there is an integer $k_0 \leq m$ such that

$$\sum_{j=1}^{n} a_{k_0 j}\bar{x}_j \leq v.$$

Let $\| y_1^* \quad y_2^* \quad \cdots \quad y_m^* \|$ be the element of $\mathbf{S}_m$ such that $y_i = 1$ if $i = k_0$ and $y_i = 0$ if $i \neq k_0$.

Now if $\| x_1 \quad \cdots \quad x_n \|$ is any member of $\mathbf{S}_n$, we have, making use of Theorem 2.6,

$$\begin{aligned} \sum_{i=1}^{m} \sum_{j=1}^{n} a_{ij}\bar{y}_i x_j &\leq \sum_{i=1}^{m} \sum_{j=1}^{n} a_{ij}\bar{y}_i \bar{x}_j \\ &\leq \sum_{i=1}^{m} \sum_{j=1}^{n} a_{ij} y_i^* \bar{x}_j \\ &= \sum_{j=1}^{n} a_{k_0 j}\bar{x}_j \leq v, \end{aligned}$$

as was to be shown.

The following lemma can be proved in a similar fashion.

LEMMA 10.2. Let

$$\sum_{j=1}^{n} a_{1j}x_j = a_{11}x_1 + a_{12}x_2 + \cdots + a_{1n}x_n,$$

$$\sum_{j=1}^{n} a_{2j}x_j = a_{21}x_1 + a_{22}x_2 + \cdots + a_{2n}x_n,$$

$$\vdots$$

$$\sum_{j=1}^{n} a_{mj}x_j = a_{m1}x_1 + a_{m2}x_2 + \cdots + a_{mn}x_n$$

be m homogeneous linear forms in n unknowns (with real coefficients) and let v be a real number such that, for every $\| x_1 \quad \cdots \quad x_n \|$ in S_n, there exists an integer $k \leq m$ for which

$$\sum_{j=1}^{n} a_{kj}x_j \geq v.$$

Then there exists an element $\| \bar{y}_1 \quad \cdots \quad \bar{y}_m \|$ of S_m such that, for every element $\| x_1 \quad \cdots \quad x_n \|$ of S_n, we have

$$\sum_{i=1}^{m} \sum_{j=1}^{n} a_{ij}\bar{y}_i x_j \geq v.$$

REMARK 10.3. It is interesting to note that although Lemmas 10.1 and 10.2 are of a purely algebraic nature, we have proved them by means of a theorem concerning games (Theorem 2.6). It is also easy, on the other hand, to prove Theorem 2.6 if we assume these algebraic lemmas.

3. The Fundamental Theorem.

THEOREM 10.4. If M is a continuous function of two variables in the closed unit square, then the quantities

$$\max_{F \in \mathsf{D}} \min_{G \in \mathsf{D}} \int_0^1 \int_0^1 M(x, y)\, dF(x)\, dG(y)$$

and

$$\min_{G \in \mathsf{D}} \max_{F \in \mathsf{D}} \int_0^1 \int_0^1 M(x, y)\, dF(x)\, dG(y)$$

exist and are equal.

PROOF. Since $M(x, y)$ is continuous in x and y, we conclude that, for any distribution function G,

$$\int_0^1 M(x, y)\, dG(y)$$

is a continuous function of x in the closed interval $[0, 1]$. Hence by Theorem 9.22, we see that

$$\max_{F \in \mathrm{D}} \int_0^1 \int_0^1 M(x, y)\, dG(y)\, dF(x)$$

exists, and that

$$\max_{F \in \mathrm{D}} \int_0^1 \int_0^1 M(x, y)\, dG(y)\, dF(x) = \max_x \int_0^1 M(x, y)\, dG(y). \tag{1}$$

Letting x_G be a value of x which maximizes the integral on the right side of (1), we have

$$\max_x \int_0^1 M(x, y)\, dG(y) = \int_0^1 M(x_G, y)\, dG(y). \tag{2}$$

Since $M(x_G, y) \geq \min_x \min_y M(x, y)$, we see from (1) and (2), by means of Theorem 9.15, that

$$\begin{aligned}
\max_{F \in \mathrm{D}} \int_0^1 \int_0^1 M(x, y)\, dG(y)\, dF(x) &= \int_0^1 M(x_G, y)\, dG(y) \\
&\geq \int_0^1 [\min_x \min_y M(x, y)]\, dG(y) \\
&= [\min_x \min_y M(x, y)] \int_0^1 dG(y) \\
&= \min_x \min_y M(x, y).
\end{aligned}$$

Since this inequality holds for every G, and since the right member does not involve G, we conclude that

$$\max_{F \in \mathrm{D}} \int_0^1 \int_0^1 M(x, y)\, dG(y)\, dF(x)$$

has a lower bound, and hence a greatest lower bound. We set

$$\mu = \inf_{G \in \mathrm{D}} \max_{F \in \mathrm{D}} \int_0^1 \int_0^1 M(x, y)\, dG(y)\, dF(x). \tag{3}$$

By the definition of a greatest lower bound, we see that there exists a sequence $G_1, G_2, G_3, \cdots$ of distribution functions such that

$$\mu = \lim_{n\to\infty} \max_{F\in D} \int_0^1 \int_0^1 M(x, y)\, dG_n(y)\, dF(x). \tag{4}$$

By Theorem 8.2 we can suppose that the sequence $G_1, G_2, G_3, \cdots$ is chosen so as to converge to a distribution function G_0 at all points of continuity of G_0.

Let $\bar{x}$ be a value of x such that

$$\max_x \int_0^1 M(x, y)\, dG_0(y) = \int_0^1 M(\bar{x}, y)\, dG_0(y). \tag{5}$$

By Theorem 9.23 we have,

$$\lim_{n\to\infty} \int_0^1 M(\bar{x}, y)\, dG_n(y) = \int_0^1 M(\bar{x}, y)\, dG_0(y). \tag{6}$$

Since, for each n,

$$\int_0^1 M(\bar{x}, y)\, dG_n(y) \le \max_x \int_0^1 M(x, y)\, dG_n(y),$$

we see that

$$\lim_{n\to\infty} \int_0^1 M(\bar{x}, y)\, dG_n(y) \le \lim_{n\to\infty} \max_x \int_0^1 M(x, y)\, dG_n(y). \tag{7}$$

From (5), (6), and (7) we obtain

$$\max_x \int_0^1 M(x, y)\, dG_0(y) \le \lim_{n\to\infty} \max_x \int_0^1 M(x, y)\, dG_n(y). \tag{8}$$

Making use of Theorem 9.22, we now conclude from (8) that

$$\max_{F\in D} \int_0^1 \int_0^1 M(x, y)\, dG_0(y)\, dF(x)$$

$$\le \lim_{n\to\infty} \max_{F\in D} \int_0^1 \int_0^1 M(x, y)\, dG_n(y)\, dF(x), \tag{9}$$

and hence by (3) and (4) that

$$\max_{F\in D} \int_0^1 \int_0^1 M(x, y)\, dG_0(y)\, dF(x)$$

$$\le \inf_{G\in D} \max_{F\in D} \int_0^1 \int_0^1 M(x, y)\, dG(y)\, dF(x). \tag{10}$$

On the other hand, it is obvious from the definition of a greatest lower bound that

$$\inf_{G \in D} \max_{F \in D} \int_0^1 \int_0^1 M(x, y)\, dG(y)\, dF(x)$$
$$\leq \max_{F \in D} \int_0^1 \int_0^1 M(x, y)\, dG_0(y)\, dF(x); \tag{11}$$

hence we have

$$\inf_{G \in D} \max_{F \in D} \int_0^1 \int_0^1 M(x, y)\, dG(y)\, dF(x)$$
$$= \max_{F \in D} \int_0^1 \int_0^1 M(x, y)\, dG_0(y)\, dF(x). \tag{12}$$

Equation (12) means that the greatest lower bound of

$$\max_{F \in D} \int_0^1 \int_0^1 M(x, y)\, dG(y)\, dF(x)$$

is assumed at $G = G_0$ and hence that the above expression has a minimum. Thus we can write

$$\mu = \min_{G \in D} \max_{F \in D} \int_0^1 \int_0^1 M(x, y)\, dG(y)\, dF(x). \tag{13}$$

The proof of the existence of

$$\nu = \max_{F \in D} \min_{G \in D} \int_0^1 \int_0^1 M(x, y)\, dG(y)\, dF(x) \tag{14}$$

is similar.

It remains to be shown that $\mu = \nu$. If G is any member of D, then we see by (1) and (2) that the number x_G satisfies the following condition:

$$\int_0^1 M(x_G, y)\, dG(y) = \max_x \int_0^1 M(x, y)\, dG(y)$$
$$= \max_{F \in D} \int_0^1 \int_0^1 M(x, y)\, dG(y)\, dF(x)$$
$$\geq \min_{G \in D} \max_{F \in D} \int_0^1 \int_0^1 M(x, y)\, dG(y)\, dF(x) = \mu.$$

Hence we conclude: For every G in D, there exists an $\bar{x}$ in $[0, 1]$ such that

$$\int_0^1 M(\bar{x}, y)\, dG(y) \geq \mu. \tag{15}$$

Now let ε be any positive number. By the continuity of M we see that there exists an n such that, whenever $|x' - x''| < 1/n$ and $|y' - y''| < 1/n$, we have $|M(x', y') - M(x'', y'')| < \varepsilon$. If $\pi = \| p_1 \quad \cdots \quad p_n \|$ is any member of $\mathbf{S}_n$, let G_π be the step-function defined as follows:

$$\begin{aligned}
G_\pi(y) &= 0 && \text{for } y < \frac{1}{n}, \\
G_\pi(y) &= p_1 && \text{for } \frac{1}{n} \le y < \frac{2}{n}, \\
G_\pi(y) &= p_1 + p_2 && \text{for } \frac{2}{n} \le y < \frac{3}{n}, \\
&\vdots \\
G_\pi(y) &= p_1 + \cdots + p_{n-1} && \text{for } \frac{n-1}{n} \le y < 1, \\
G_\pi(y) &= p_1 + \cdots + p_n = 1 && \text{for } y = 1.
\end{aligned}$$

It is clear that, for any x in $[0, 1]$, there exists an $i \le n$ such that $|x - i/n| < 1/n$, and hence, by the continuity of M, such that

$$\begin{aligned}
&\left| \int_0^1 M(x, y)\, dG_\pi(y) - \int_0^1 M\left(\frac{i}{n}, y\right) dG_\pi(y) \right| \\
&= \left| \int_0^1 \left[M(x, y) - M\left(\frac{i}{n}, y\right) \right] dG_\pi(y) \right| \\
&\le \int_0^1 \varepsilon\, dG_\pi(y) = \varepsilon .
\end{aligned}$$

Thus for every x in $[0, 1]$ there exists an $i \le n$ such that

$$\int_0^1 M\left(\frac{i}{n}, y\right) dG_\pi(y) \ge \int_0^1 M(x, y)\, dG_\pi(y) - \varepsilon . \tag{16}$$

From (16) and (15) we conclude that there exists an $i \le n$ such that

$$\int_0^1 M\left(\frac{i}{n}, y\right) dG_\pi(y) \ge \mu - \varepsilon . \tag{17}$$

Evaluating the left member of (17) by means of Theorem 9.20, we conclude that there exists an i such that

$$\sum_{j=1}^{n} M\left(\frac{i}{n}, \frac{j}{n}\right) p_j \ge \mu - \varepsilon .$$

Thus we have shown the following: For every positive ε, there exists an

n such that, for every element $\| p_1 \quad \cdots \quad p_n \|$ of $\mathbf{S}_n$, there exists an i such that

$$\sum_{j=1}^{n} M\left(\frac{i}{n}, \frac{j}{n}\right) p_j \geq \mu - \varepsilon. \tag{18}$$

For a fixed n, the quantities $M(1/n, 1/n)$, $M(1/n, 2/n)$, $M(2/n, 1/n)$, etc., are constants. Thus we have n linear forms

$$\sum_{j=1}^{n} M\left(\frac{i}{n}, \frac{j}{n}\right) p_j \qquad \text{for } i = 1, \cdots, n$$

in the n unknowns $p_1, \cdots, p_n$; and (18) means that these forms satisfy the hypothesis of Lemma 10.2. Hence by means of Lemma 10.2 we conclude: For every positive ε, there exists an n and an element $\| q_1 \quad \cdots \quad q_n \|$ of $\mathbf{S}_n$ such that, for every element $\| p_1 \quad \cdots \quad p_n \|$ of $\mathbf{S}_n$,

$$\sum_{i=1}^{n} \sum_{j=1}^{n} M\left(\frac{i}{n}, \frac{j}{n}\right) p_j q_i \geq \mu - \varepsilon. \tag{19}$$

We now define a step-function F_0 by setting

$$\begin{aligned}
F_0(x) &= 0 && \text{for } 0 \leq x < \frac{1}{n}, \\
F_0(x) &= q_1 && \text{for } \frac{1}{n} \leq x < \frac{2}{n}, \\
F_0(x) &= q_1 + q_2 && \text{for } \frac{2}{n} \leq x < \frac{3}{n}, \\
&\vdots \\
F_0(x) &= q_1 + \cdots + q_{n-1} && \text{for } \frac{n-1}{n} \leq x < 1, \\
F_0(x) &= q_1 + \cdots + q_n = 1 && \text{for } x = 1.
\end{aligned}$$

Then, from Theorem 9.20, we see that, for all y,

$$\int_0^1 M(x, y)\, dF_0(x) = \sum_{i=1}^{n} M\left(\frac{i}{n}, y\right) q_i.$$

In particular we have

$$\int_0^1 M\left(x, \frac{j}{n}\right) dF_0(x) = \sum_{i=1}^{n} M\left(\frac{i}{n}, \frac{j}{n}\right) q_i \qquad \text{for } j = 1, \cdots, n. \tag{20}$$

Multiplying the jth equation of (20) by p_j and summing we obtain by means of (19)

$$\begin{aligned}\sum_{j=1}^{n}\left[p_j \int_0^1 M\left(x, \frac{j}{n}\right) dF_0(x)\right] &= \sum_{j=1}^{n}\left[p_j \sum_{i=1}^{n} M\left(\frac{i}{n}, \frac{j}{n}\right)q_i\right] \\ &= \sum_{i=1}^{n}\sum_{j=1}^{n} M\left(\frac{i}{n}, \frac{j}{n}\right)p_j q_i \\ &\geq \mu - \varepsilon. \qquad (21)\end{aligned}$$

Since (21) holds for all $\| p_1 \quad \cdots \quad p_n \|$ in $\mathbf{S}_n$, in particular we can take $p_j = 1$ and $p_k = 0$, for $k \neq j$; hence we conclude that

$$\int_0^1 M\left(x, \frac{j}{n}\right) dF_0(x) \geq \mu - \varepsilon \qquad \text{for } j = 1, \cdots, n. \qquad (22)$$

From the continuity of M, it follows that for every y there exists some j such that

$$\int_0^1 M(x, y)\, dF_0(x) \geq \int_0^1 M\left(x, \frac{j}{n}\right) dF_0(x) - \varepsilon. \qquad (23)$$

From (22) and (23) we conclude that, for every y,

$$\int_0^1 M(x, y)\, dF_0(x) \geq \mu - 2\varepsilon. \qquad (24)$$

Applying Theorem 9.15 to (24), we see that for every distribution function G,

$$\begin{aligned}\int_0^1 \int_0^1 M(x, y)\, dF_0(x)\, dG(y) &\geq \int_0^1 (\mu - 2\varepsilon)\, dG(y) \\ &= \mu - 2\varepsilon,\end{aligned}$$

and hence that

$$\min_{G \in \mathbf{D}} \int_0^1 \int_0^1 M(x, y)\, dF_0(x)\, dG(y) \geq \mu - 2\varepsilon. \qquad (25)$$

From (14), (25), and the definition of a maximum, we now obtain

$$\begin{aligned}v &= \max_{F \in \mathbf{D}} \min_{G \in \mathbf{D}} \int_0^1 \int_0^1 M(x, y)\, dF(x)\, dG(y) \\ &\geq \min_{G \in \mathbf{D}} \int_0^1 \int_0^1 M(x, y)\, dF_0(x)\, dG(y) \\ &\geq \mu - 2\varepsilon.\end{aligned}$$

Since we thus have

$$\nu \geq \mu - 2\varepsilon$$

for all positive ε, we conclude that

$$\nu \geq \mu .$$

On the other hand, by (13), (14), and Theorem 1.1, we have

$$\nu \leq \mu ,$$

so that

$$\mu = \nu ,$$

as was to be shown.

4. Devices for Computing and Verifying Solutions. Though we are thus assured that continuous games with continuous payoff functions have solutions (i.e., have values and optimal strategies), we are not yet provided with any method of computing the solutions for given payoff functions. This problem is, in general, a difficult one, since in order to solve it a Stieltjes integral must be maximized with respect to the set of all distribution functions. No completely general methods are known for dealing with the problem, but in the next two chapters we shall consider two special cases in which solutions can be obtained. We are now going to prove some theorems which are useful in deciding whether given strategies are actually optimal.

THEOREM 10.5. *Let M be the continuous payoff function of a continuous game, and let F_0 and G_0 be distribution functions. Then the following conditions are all equivalent:*

(i) *F_0 is an optimal strategy for P_1 and G_0 is an optimal strategy for P_2.*

(ii) *If F and G are any distribution functions, then*

$$\int_0^1 \int_0^1 M(x, y)\, dF(x)\, dG_0(y)$$
$$\leq \int_0^1 \int_0^1 M(x, y)\, dF_0(x)\, dG_0(y)$$
$$\leq \int_0^1 \int_0^1 M(x, y)\, dF_0(x)\, dG(y) .$$

(iii) *If z and w are any points of $[0, 1]$, then*

$$\int_0^1 M(z, y)\, dG_0(y)$$
$$\leq \int_0^1 \int_0^1 M(x, y)\, dF_0(x)\, dG_0(y)$$
$$\leq \int_0^1 M(x, w)\, dF_0(x).$$

PROOF. The equivalence of (i) and (ii) follows directly from Theorem 10.4, together with the definition of a saddle-point. To see that (ii) implies (iii), we take $F(x) = I_z(x)$ and $G(y) = I_w(y)$ and apply Theorem 9.20. To see that (iii) implies (ii), suppose that, for all z and w in $[0, 1]$, we have

$$\int_0^1 M(z, y)\, dG_0(y) \leq \int_0^1 \int_0^1 M(x, y)\, dF_0(x)\, dG_0(y)$$
$$\leq \int_0^1 M(x, w)\, dF_0(x).$$

Replacing z by x in the first part of this inequality, we obtain by Theorem 9.15, for any distribution function F,

$$\int_0^1 \int_0^1 M(x, y)\, dG_0(y)\, dF(x)$$
$$\leq \int_0^1 \left[\int_0^1 \int_0^1 M(x, y)\, dF_0(x)\, dG_0(y) \right] dF(x)$$
$$= \int_0^1 \int_0^1 M(x, y)\, dF_0(x)\, dG_0(y).$$

In a similar way we can obtain the second part of the inequality of (ii) from the second part of the inequality of (iii), which completes the proof.

THEOREM 10.6. Let M be the continuous payoff function of a continuous game whose value is v. Then a distribution function F_0 is an optimal strategy for the first player if, and only if, for every y in $[0, 1]$,

$$v \leq \int_0^1 M(x, y)\, dF_0(x).$$

A distribution function G_0 is an optimal strategy for the second player if, and only if, for every x in $[0, 1]$,

$$\int_0^1 M(x, y)\, dG_0(y) \leq v.$$

PROOF. If F_0 is an optimal strategy for the first player, then we see by Theorem 10.5 that, for all y,

$$v \leq \int_0^1 M(x, y)\, dF_0(x). \tag{26}$$

Now suppose that F_0 is a distribution function which satisfies (26) for all y. Let $\overline{G}$ be an optimal strategy for the second player. By Theorems 9.16 and 9.15 we conclude from (26) that

$$v = \int_0^1 v\, d\overline{G}(y) \leq \int_0^1 \int_0^1 M(x, y)\, dF_0(x)\, d\overline{G}(y). \tag{27}$$

Since $\overline{G}$ is optimal, we see by Theorem 10.5 that, for any distribution function F,

$$\int_0^1 \int_0^1 M(x, y)\, dF(x)\, d\overline{G}(y) \leq v;$$

in particular,

$$\int_0^1 \int_0^1 M(x, y)\, dF_0(x)\, d\overline{G}(y) \leq v. \tag{28}$$

From (27) and (28) we have

$$v = \int_0^1 \int_0^1 M(x, y)\, dF_0(x)\, d\overline{G}(y), \tag{29}$$

and hence from (26), for all y,

$$\int_0^1 \int_0^1 M(x, y)\, dF_0(x)\, d\overline{G}(y) \leq \int_0^1 M(x, y)\, dF_0(x). \tag{30}$$

Since $\overline{G}$ is optimal, we see from Theorem 10.5 that, for all x,

$$\int_0^1 M(x, y)\, d\overline{G}(y) \leq v,$$

and hence from (29) that

$$\int_0^1 M(x, y)\, d\overline{G}(y) \leq \int_0^1 \int_0^1 M(x, y)\, dF_0(x)\, d\overline{G}(y). \tag{31}$$

From (30) and (31) we conclude by Theorem 10.5 that F_0 is an optimal strategy for the first player, as was to be shown.

The proof of the second part of the theorem is similar.

THEOREM 10.7. Let M be the continuous payoff function of a continuous game, and suppose that ν is a real number and that F_0 and G_0 are distribution functions such that, for all x and y in $[0, 1]$,

$$\int_0^1 M(x, y)\, dG_0(y) \leq \nu \leq \int_0^1 M(x, y)\, dF_0(x).$$

Then ν is the value of the game, and F_0 and G_0 are optimal strategies for the first and second players, respectively.

PROOF. From the hypothesis we see, as in the proof of Theorem 10.5, that, for all distribution functions F and G,

$$\int_0^1 \int_0^1 M(x, y)\, dF(x)\, dG_0(y) \leq \nu \leq \int_0^1 \int_0^1 M(x, y)\, dF_0(x)\, dG(y);$$

hence

$$\begin{aligned} v &= \min_G \max_F \int_0^1 \int_0^1 M(x, y)\, dF(x)\, dG(y) \\ &\leq \max_F \int_0^1 \int_0^1 M(x, y)\, dF(x)\, dG_0(y) \\ &\leq \nu \\ &\leq \min_G \int_0^1 \int_0^1 M(x, y)\, dF_0(x)\, dG(y) \\ &\leq \max_F \min_G \int_0^1 \int_0^1 M(x, y)\, dF(x)\, dG(y) = v, \end{aligned}$$

so that $\nu = v$. The fact that F_0 and G_0 are optimal strategies now follows from the hypothesis of our theorem by means of Theorem 10.6.

The theorems just proved provide methods of checking a suggested solution of a game. This will now be illustrated by an example.

EXAMPLE 10.8. The payoff function of a continuous game is

$$M(x, y) = \frac{1}{1 + (x - y)^2};$$

we wish to show that a solution of the game is given by

$$\begin{aligned} v &= \frac{4}{5}, \\ F_0(x) &= I_{1/2}(x), \\ G_0(y) &= \frac{1}{2} I_0(y) + \frac{1}{2} I_1(y). \end{aligned}$$

By Theorem 10.7 it is only necessary to show that

$$\frac{4}{5} \leq \int_0^1 \left[\frac{1}{1 + (x - y)^2}\right] dI_{1/2}(x) \qquad \text{for } 0 \leq y \leq 1,$$

and that

$$\int_0^1 \left[\frac{1}{1 + (x - y)^2}\right] d\left[\frac{1}{2} I_0(y) + \frac{1}{2} I_1(y)\right] \leq \frac{4}{5} \qquad \text{for } 0 \leq x \leq 1.$$

By Theorems 9.11 and 9.20, this is equivalent to showing that

$$\frac{4}{5} \leq \frac{1}{1 + \left(\frac{1}{2} - y\right)^2} \qquad \text{for } 0 \leq y \leq 1, \tag{32}$$

$$\frac{1}{2}\left[\frac{1}{1 + x^2} + \frac{1}{1 + (x - 1)^2}\right] \leq \frac{4}{5} \qquad \text{for } 0 \leq x \leq 1. \tag{33}$$

Multiplying both sides of (32) by the positive quantity

$$5\left[1 + \left(\frac{1}{2} - y\right)^2\right],$$

we obtain the equivalent inequality

$$4 + 4\left(\frac{1}{2} - y\right)^2 \leq 5,$$

which, in turn, is equivalent to

$$\left(\frac{1}{2} - y\right)^2 \leq \frac{1}{4};$$

and the latter is clearly satisfied for all y in $[0, 1]$.

In a similar way, multiplying both sides of (33) by the product of the denominators of the various fractions involved (this product is positive) and making obvious simplifications, we arrive at the equivalent inequality

$$8x^4 - 16x^3 + 14x^2 - 6x + 1 \geq 0,$$

which can also be written

$$(2x - 1)^2[x^2 + (x - 1)^2] \geq 0.$$

Since the last inequality is obviously satisfied for all x, we conclude that (33) is satisfied and hence that the proposed entities indeed constitute a solution to the given game.

THEOREM 10.9. Let M be the continuous payoff function of a continuous game whose value is v, and let F_0 and G_0 be optimal strategies for the first and second player, respectively. If we set

$$H(x) = \int_0^1 M(x, y)\, dG_0(y)$$

and

$$K(y) = \int_0^1 M(x, y)\, dF_0(x),$$

then

$$v = \max_x H(x) = \min_y K(y).$$

PROOF. By Theorem 10.6 we have, for all x,

$$H(x) \leq v$$

and hence

$$\max_x H(x) \leq v.$$

Now suppose, if possible, that

$$\max_x H(x) < v.$$

Then, for all x,

$$H(x) < v,$$

and hence,

$$v = \int_0^1 H(x)\, dF_0(x) < \int_0^1 v\, dF_0(x) = v,$$

which is absurd. Hence

$$\max_x H(x) = v,$$

as was to be shown. The proof that

$$\min_y K(y) = v$$

is similar.

In the following theorem we give a sufficient condition that a pure strategy yield the value of the game when used against an optimal mixed strategy of the opponent.

THEOREM 10.10. Let M be the continuous payoff function of a continuous game whose value is v, let F_0 and G_0 be optimal mixed strategies for the first and second players, respectively, and let us set

$$H(x) = \int_0^1 M(x, y)\, dG_0(y)$$

and

$$K(y) = \int_0^1 M(x, y)\, dF_0(x).$$

Then if $\bar{x}$ is any point of $[0, 1]$ at which the left-hand derivative of F_0 is not 0 (i.e., is positive or does not exist), then

$$H(\bar{x}) = v = \max_x H(x);$$

similarly, if $\bar{y}$ is any point of $[0, 1]$ at which the left-hand derivative of G_0 is not 0, then

$$K(\bar{y}) = v = \min_y K(y).$$

In particular, if F_0 is a step-function, so that

$$F_0(x) = \sum_{i=1}^{m} \alpha_i I_{a_i}(x), \qquad \text{where } \alpha_i \neq 0,$$

then

$$H(a_1) = H(a_2) = \cdots = H(a_m) = v = \max_x H(x).$$

And if G_0 is a step-function so that

$$G_0(y) = \sum_{i=1}^{n} \beta_i I_{b_i}(y), \qquad \text{where } \beta_i \neq 0,$$

then

$$K(b_1) = K(b_2) = \cdots = K(b_n) = v = \min_y K(y).$$

PROOF. By Theorem 10.6 we see that, for all x in $[0, 1]$,

$$H(x) \leq v.$$

Moreover, by the definition of the value of a game, we have

$$\int_0^1 H(x)\, dF_0(x) = v.$$

Hence if the left-hand derivative of F_0 is not 0 at $x = \bar{x}$, we conclude by means of Theorems 9.27 and 10.9 that

$$H(\bar{x}) = v = \max_x H(x).$$

In particular, if a_i is a point at which F_0 has a step, then the left-hand derivative of F_0 is infinite at a_i, and hence

$$H(a_i) = v = \max_x H(x).$$

The proof of the other part of the theorem is similar.

The following example illustrates a fairly general method for actually finding solutions of a given game. The method consists essentially in trying to find a solution of a given form (in this case, step-functions). If the method fails for a given form of solution, we can continue by trying functions of different forms (e.g., we can assume more steps in the step-functions).

EXAMPLE 10.11. The payoff function of a continuous game is

$$M(x, y) = \frac{1}{1 + \frac{5}{4}(x - y)^2}.$$

We wish, if possible, to find for the two players optimal strategies having the form

$$\left.\begin{aligned} F_0(x) &= I_a(x), \\ G_0(y) &= \beta I_b(y) + (1 - \beta) I_c(y). \end{aligned}\right\} \qquad (34)$$

Supposing that there is a solution of the form (34), we have

$$\begin{aligned} v &= \int_0^1 \int_0^1 \frac{1}{1 + \frac{5}{4}(x - y)^2}\, dF_0(x)\, dG_0(y) \\ &= \frac{\beta}{1 + \frac{5}{4}(a - b)^2} + \frac{1 - \beta}{1 + \frac{5}{4}(a - c)^2}; \end{aligned}$$

hence by Theorem 10.6, for all y in $[0, 1]$,

$$\frac{\beta}{1+\frac{5}{4}(a-b)^2}+\frac{1-\beta}{1+\frac{5}{4}(a-c)^2} \leq \int_0^1 \frac{1}{1+\frac{5}{4}(x-y)^2}\,dF_0(x)$$

$$= \frac{1}{1+\frac{5}{4}(a-y)^2}. \tag{35}$$

Since $c \in [0, 1]$, we obtain from (35), in particular,

$$\frac{\beta}{1+\frac{5}{4}(a-b)^2}+\frac{1-\beta}{1+\frac{5}{4}(a-c)^2} \leq \frac{1}{1+\frac{5}{4}(a-c)^2},$$

or

$$\frac{\beta}{1+\frac{5}{4}(a-b)^2} \leq \frac{\beta}{1+\frac{5}{4}(a-c)^2}. \tag{36}$$

Applying the methods of Chap. 7, we can show that the function $M(x, y)$ has no saddle-point in the unit square; hence our game cannot have solutions of the form (34) where $\beta = 0$ or $1 - \beta = 0$. Thus from (36) we conclude that

$$\frac{1}{1+\frac{5}{4}(a-b)^2} \leq \frac{1}{1+\frac{5}{4}(a-c)^2}.$$

In a similar fashion, by substituting b for y in (35), we obtain

$$\frac{1}{1+\frac{5}{4}(a-c)^2} \leq \frac{1}{1+\frac{5}{4}(a-b)^2},$$

and hence

$$\frac{1}{1+\frac{5}{4}(a-c)^2} = \frac{1}{1+\frac{5}{4}(a-b)^2};$$

from this we conclude by means of (35) that, for all y in $[0, 1]$,

$$\frac{\beta}{1+\frac{5}{4}(a-b)^2}+\frac{1-\beta}{1+\frac{5}{4}(a-b)^2} \leq \frac{1}{1+\frac{5}{4}(a-y)^2},$$

or

$$\frac{1}{1+\frac{5}{4}(a-b)^2} \leq \frac{1}{1+\frac{5}{4}(a-y)^2}, \tag{37}$$

and hence

$$(a-b)^2 \geq (a-y)^2. \tag{38}$$

From (38) it follows that we must have either $b = 0$ or $b = 1$. In a similar way we conclude that either $c = 0$ or $c = 1$. Moreover, we cannot have $b = c$, since M has no saddle-point. Without loss of generality we can take

$$\left.\begin{aligned} b &= 0, \\ c &= 1. \end{aligned}\right\} \tag{39}$$

From (38) and (39) we obtain

$$a^2 \geq (a-1)^2,$$

and hence

$$a \geq \frac{1}{2}.$$

Similarly, we obtain

$$a \leq \frac{1}{2},$$

and hence

$$a = \frac{1}{2}. \tag{40}$$

Thus our functions F_0 and G_0 become

$$\begin{aligned} F_0(x) &= I_{1/2}(x), \\ G_0(y) &= \beta I_0(x) + (1-\beta)I_1(x), \end{aligned}$$

and we have

$$\begin{aligned} v &= \frac{\beta}{1+\frac{5}{4}(a-b)^2} + \frac{1-\beta}{1+\frac{5}{4}(a-c)^2} \\ &= \frac{\beta}{1+\frac{5}{4}\cdot\frac{1}{4}} + \frac{1-\beta}{1+\frac{5}{4}\cdot\frac{1}{4}} = \frac{16}{21}. \end{aligned} \tag{41}$$

Setting

$$H(x) = \int_0^1 M(x, y)\, dG_0(y)$$
$$= \frac{\beta}{1 + \frac{5}{4}x^2} + \frac{1 - \beta}{1 + \frac{5}{4}(x - 1)^2},$$

we see by Theorem 10.10 that

$$H\left(\frac{1}{2}\right) = \max_x H(x). \tag{42}$$

Since H is differentiable in the open interval $(0, 1)$, and since ½ is in this interval, we conclude from (42) that

$$\frac{d}{dx} H(x)$$

must be 0 when $x = ½$. Since

$$\frac{d}{dx} H(x) = \frac{-\frac{5}{2}\beta x}{\left[1 + \frac{5}{4}x^2\right]^2} + \frac{-\frac{5}{2}(1 - \beta)(x - 1)}{\left[1 + \frac{5}{4}(x - 1)^2\right]^2},$$

we therefore obtain

$$\frac{-\frac{5}{2}\beta\left(\frac{1}{2}\right)}{\left[1 + \frac{5}{4}\cdot\frac{1}{4}\right]^2} + \frac{-\frac{5}{2}(1 - \beta)\left(\frac{1}{2} - 1\right)}{\left[1 + \frac{5}{4}\cdot\frac{1}{4}\right]^2} = 0,$$

and hence

$$\beta = \frac{1}{2}. \tag{43}$$

Thus, from (39), (40), (41), and (43), we see that if there exist for our game any optimal strategies having the form (34), then we must have

$$\left.\begin{aligned} F_0(x) &= I_{½}(x), \\ G_0(y) &= \frac{1}{2}I_0(y) + \frac{1}{2}I_1(y), \\ v &= \frac{16}{21}. \end{aligned}\right\} \tag{44}$$

Checking as in Example 10.8, we now readily verify that (44) is indeed a solution.

The following example shows how one can verify that a given set of continuous distribution functions constitute optimal strategies for a given game.

EXAMPLE 10.12. The payoff function of a continuous game is

$$M(x, y) = \frac{(1 + x)(1 + y)}{(1 + xy)^2};$$

we wish to show that a solution of the game is given by

$$v = \frac{1}{\log 2},$$

$$F_0(x) = \frac{\log (1 + x)}{\log 2},$$

$$G_0(y) = \frac{\log (1 + y)}{\log 2}.$$

Since F_0 is differentiable in the interval $[0, 1]$, we see by Theorem 9.19 that, for all y in $[0, 1]$,

$$\begin{aligned}\int_0^1 M(x, y)\, dF_0(x) &= \int_0^1 M(x, y) \frac{d}{dx} F_0(x)\, dx \\ &= \int_0^1 \frac{(1 + x)(1 + y)}{(1 + xy)^2} \left[\frac{1}{(x + 1) \log 2}\right] dx \\ &= \frac{1}{\log 2}.\end{aligned} \tag{45}$$

Similarly, for all x in $[0, 1]$,

$$\int_0^1 M(x, y)\, dG_0(y) = \frac{1}{\log 2}. \tag{46}$$

From (45) and (46), we see by Theorem 10.7 that the given strategies are indeed optimal and that

$$v = \frac{1}{\log 2}.$$

The following theorem enables us, in certain cases, to find the value of a continuous game by inspection.

THEOREM 10.13. *Let M be the payoff function of a continuous game which possesses a solution, and suppose that, for all x and*

y in the interval $[0, 1]$,

$$M(x, y) = -M(y, x).$$

Then the value of the game is 0, and every optimal strategy for one player is also an optimal strategy for the other.

PROOF. Let v be the value of the game, and suppose that F_0 is an optimal strategy for the first player and that G_0 is an optimal strategy for the second player. Then by Theorem 10.5 we see that, for all distribution functions F and G,

$$\int_0^1 \int_0^1 M(x, y)\, dF(x)\, dG_0(y) \leq \int_0^1 \int_0^1 M(x, y)\, dF_0(x)\, dG_0(y) \leq \int_0^1 \int_0^1 M(x, y)\, dF_0(x)\, dG(y). \quad (47)$$

Applying to (47) the hypothesis

$$M(x, y) \equiv -M(y, x),$$

we obtain

$$\int_0^1 \int_0^1 -M(y, x)\, dF(x)\, dG_0(y) \leq \int_0^1 \int_0^1 M(x, y)\, dF_0(x)\, dG_0(y) \leq \int_0^1 \int_0^1 -M(y, x)\, dF_0(x)\, dG(y);$$

hence, multiplying through by -1 and reversing the order,

$$\int_0^1 \int_0^1 M(y, x)\, dF_0(x)\, dG(y) \leq -\int_0^1 \int_0^1 M(x, y)\, dF_0(x)\, dG_0(y) \leq \int_0^1 \int_0^1 M(y, x)\, dF(x)\, dG_0(y). \quad (48)$$

Changing the symbols used for the variables of integration in (48), we get

$$\int_0^1 \int_0^1 M(x, y)\, dG(x)\, dF_0(y) \leq -\int_0^1 \int_0^1 M(x, y)\, dF_0(x)\, dG_0(y) \leq \int_0^1 \int_0^1 M(x, y)\, dG_0(x)\, dF(y). \quad (49)$$

From (49) we conclude by means of Theorem 10.5 that

$$v = -\int_0^1 \int_0^1 M(x, y)\, dF_0(x)\, dG_0(y) = -v$$

and hence that $v = 0$. Moreover, it follows from (49) and Theorem 10.7 that F_0 is an optimal strategy for the second player (as well as for the first) and that G_0 is an optimal strategy for the first player (as well as for the second).

In the following example, as in Example 10.11, we solve the problem of finding optimal strategies of given forms—but in this case we confine our search to continuous distribution functions of a given form rather than to step-functions.

EXAMPLE 10.14. The payoff function of a continuous game is

$$M(x, y) = \sin 2\pi(x - y);$$

we wish, if possible, to find for the two players optimal strategies having the forms

$$\left.\begin{aligned} F_0(0) &= 0, \qquad F_0(x) = a_1 + a_2 x \qquad \text{for } x \neq 0, \\ G_0(0) &= 0, \qquad G_0(y) = b_1 + b_2 y \qquad \text{for } y \neq 0. \end{aligned}\right\} \tag{50}$$

Since F_0 is a distribution function, we must have

$$1 = F_0(1) = a_1 + a_2,$$

and hence $a_2 = 1 - a_1$. Thus, dropping the subscript, we can write

$$F_0(x) = a + (1 - a)x \qquad \text{for } x \neq 0$$

and, similarly,

$$G_0(y) = b + (1 - b)y \qquad \text{for } y \neq 0.$$

We notice that

$$M(y, x) = \sin 2\pi(y - x) = -\sin 2\pi(x - y) = -M(x, y),$$

so that, by Theorem 10.13,

$$v = 0,$$

and every optimal strategy for the first player is also an optimal strategy for the second player; i.e., we can set

$$\left.\begin{aligned} F_0(x) &= a + (1 - a)x \qquad \text{for } x \neq 0, \\ G_0(y) &= a + (1 - a)y \qquad \text{for } y \neq 0. \end{aligned}\right\} \tag{51}$$

Since the point $\| 0 \quad 0 \|$ is not a saddle-point of the function M, we cannot have $a = 1$ in (51). Therefore, since the function F_0 has a positive derivative (and a fortiori a positive left-hand derivative) at every point x in the open interval $(0, 1)$, we conclude by Theorem 10.10 that, for every x in $(0, 1)$,

$$\int_0^1 M(x, y)\, dG_0(y) = 0;$$

i.e.,

$$\int_0^1 \sin 2\pi(x - y)\, dG_0(y) = 0.$$

Since G_0 has a step of magnitude a at $y = 0$ and is continuous everywhere else, we easily see that

$$\begin{aligned}\int_0^1 \sin 2\pi(x - y)\, dG_0(y) &= a \sin 2\pi(x - 0) \\ &\quad + (1 - a) \int_0^1 \sin 2\pi(x - y)\, dy \\ &= a \sin 2\pi x - (1 - a) \frac{1}{2\pi} \{\cos(-2\pi x) \\ &\quad - \cos[2\pi(1 - x)]\} \\ &= a \sin 2\pi x + (1 - a)\left(\frac{1}{2\pi}\right)(0) \\ &= a \sin 2\pi x.\end{aligned}$$

Thus we have

$$a \sin 2\pi x = 0$$

for all x in the open interval $(0, 1)$, and hence we conclude that

$$a = 0.$$

Therefore we obtain

$$\left.\begin{aligned} v &= 0, \\ F_0(x) &= x, \\ G_0(y) &= y. \end{aligned}\right\} \tag{52}$$

We now readily verify, as in Example 10.12, that (52) is indeed a solution of the given game.

As a final example we shall solve the problem of finding a solution for a certain game with a discontinuous payoff function. In solving this problem we make use of theorems such as 10.10, even when they are not known to apply to discontinuous payoff functions; no error can arise from such a procedure, of course, since we eventually check our answer in any case. In this example we succeed in reducing our problem to the solution of a certain differential equation; this method, though highly ingenious, unfortunately appears to be applicable only to very special games.

EXAMPLE 10.15. The payoff function of a continuous game is M, where

$$\begin{aligned} M(x, y) &= x - y + xy && \text{for } x < y, \\ M(x, y) &= 0 && \text{for } x = y, \\ M(x, y) &= x - y - xy && \text{for } x > y. \end{aligned}$$

If possible, we wish to find, for the first player, a continuous optimal strategy F_0 which will satisfy the following conditions (where α is a fixed constant):

$$F_0(x) = 0 \qquad \text{for } 0 \leq x \leq \alpha, \tag{53}$$

$$F_0(x) \text{ is twice differentiable} \qquad \text{for } x \neq \alpha, \tag{54}$$

$$\frac{d}{dx} F_0(x) \neq 0 \qquad \text{for } x > \alpha. \tag{55}$$

Since $M(x, y) = -M(y, x)$, we see by Theorem 10.13 that

$$v = 0$$

and that F_0 is also an optimal strategy for the second player.

By Theorem 10.10 we conclude from (55) that, for $x > \alpha$,

$$\int_0^1 M(x, y)\, dF_0(y) = v = 0$$

and thus that

$$\int_0^x [x - y - xy]\, dF_0(y) + \int_x^1 [x - y + xy]\, dF_0(y) = 0. \tag{56}$$

Since, by hypothesis, F_0 is differentiable, we set

$$P_0(y) = \frac{d}{dy} F_0(y);$$

then, using Theorem 9.19, we obtain from (56), for any $x > \alpha$,

$$\int_0^x [x - y - xy] P_0(y)\, dy + \int_x^1 [x - y + xy] P_0(y)\, dy = 0,$$

which easily reduces to

$$x - \int_0^1 y P_0(y)\, dy - x \int_0^x y P_0(y)\, dy + x \int_x^1 y P_0(y)\, dy = 0. \tag{57}$$

Making use of the rule for differentiating under an integral sign, we obtain from (57), by differentiating with respect to x,

$$1 - \int_0^x yP_0(y)\,dy - x \cdot 1 \cdot xP_0(x) + \int_x^1 yP_0(y)\,dy - x \cdot 1 \cdot xP_0(x) = 0,$$

or

$$-1 + 2x^2P_0(x) + \int_0^x yP_0(y)\,dy - \int_x^1 yP_0(y)\,dy = 0. \tag{58}$$

Differentiating (58) again with respect to x gives us

$$4xP_0(x) + 2x^2\frac{d}{dx}P_0(x) + xP_0(x) + xP_0(x) = 0,$$

and hence

$$x^2\frac{d}{dx}P_0(x) + 3xP_0(x) = 0.$$

Therefore, for $x > \alpha$, we must have

$$x\frac{d}{dx}P_0(x) = -3P_0(x).$$

Solving this differential equation, we obtain

$$P_0(x) = \frac{k}{x^3}, \tag{59}$$

where k is some constant, and hence

$$F_0(x) = \frac{-k}{2x^2} + h, \tag{60}$$

where h is also a constant.

Thus we have

$$\left.\begin{aligned} F_0(x) &= 0 && \text{for } x \leq \alpha, \\ F_0(x) &= \frac{-k}{2x^2} + h && \text{for } x \geq \alpha. \end{aligned}\right\} \tag{61}$$

Since (56) holds for all $x > \alpha$, we easily conclude by continuity that it holds also for $x = \alpha$. Thus

$$\int_0^\alpha [\alpha - y - \alpha y]\,dF_0(y) + \int_\alpha^1 [\alpha - y + \alpha y]\,dF_0(y) = 0. \tag{62}$$

From the first equation of (61), however, we have

$$\int_0^\alpha [\alpha - y - \alpha y]\, dF_0(y) = 0,$$

and hence from (62)

$$\int_\alpha^1 [\alpha - y + \alpha y]\, dF_0(y) = 0,$$

or, by the second equation of (61),

$$\int_\alpha^1 [\alpha - y + \alpha y]\, d\left[\frac{-k}{2y^2} + h\right] = 0,$$

or

$$\int_\alpha^1 [\alpha - y + \alpha y]\left[\frac{k}{y^3}\right] dy = 0.$$

Since the function M has no saddle-point, however, we see that $k \neq 0$, and hence that

$$\int_\alpha^1 \left(\frac{\alpha}{y^3} - \frac{1}{y^2} + \frac{\alpha}{y^2}\right) dy = 0.$$

Carrying out the integration, we find that

$$\left[-\frac{\alpha}{2} + \frac{1}{1} - \frac{\alpha}{1}\right] - \left[-\frac{\alpha}{2\alpha^2} + \frac{1}{\alpha} - \frac{\alpha}{\alpha}\right] = 0,$$

and hence either $\alpha = 1/3$ or $\alpha = 1$. Since $\alpha = 1$, however, would imply that M had a saddle-point, we conclude that

$$\alpha = \frac{1}{3}. \tag{63}$$

Since F_0 is continuous, we must have by (61)

$$0 = F_0(\alpha) = F_0\left(\frac{1}{3}\right) = -\frac{9k}{2} + h,$$

so that

$$h = \frac{9k}{2}. \tag{64}$$

Since F_0 is a distribution function, we must have

$$1 = F_0(1) = -\frac{k}{2} + h,$$

so that

$$h = \frac{k}{2} + 1. \tag{65}$$

Solving equations (64) and (65) simultaneously, we find

$$\left.\begin{aligned} h &= \frac{9}{8}, \\ k &= \frac{1}{4}. \end{aligned}\right\} \tag{66}$$

Thus if there is any optimal strategy at all which satisfies the conditions (53), (54), and (55), then the following is a solution of the game

$$\left.\begin{aligned} v &= 0, & & \\ F_0(x) &= 0 & &\text{for } 0 \le x \le \frac{1}{3}, \\ F_0(x) &= \frac{9}{8} - \frac{1}{8x^2} & &\text{for } \frac{1}{3} \le x \le 1, \\ G_0(y) &= 0 & &\text{for } 0 \le y \le \frac{1}{3}, \\ G_0(y) &= \frac{9}{8} - \frac{1}{8y^2} & &\text{for } \frac{1}{3} \le y \le 1. \end{aligned}\right\} \tag{67}$$

We now wish to verify that (67) actually gives a solution; i.e., we want to show that if F and G are any distribution functions, then

$$E(F, G_0) \le E(F_0, G_0) \le E(F_0, G), \tag{68}$$

and that

$$E(F_0, G_0) = 0. \tag{69}$$

Now it is easy to verify that (69) is true; and it is easy, also, to show that (68) is true for all distribution functions which are continuous in the interval $(0, 1]$. But from the fact that $M(x, y)$ is discontinuous for $x = y \neq 0$, we see that the quantities $E(F, G_0)$ and $E(F_0, G)$ do not even exist if F and G have discontinuities in the half-open interval $(0, 1]$.

Thus, using our definition of a Stieltjes integral, inequalities (68) are meaningless when F and G have discontinuities in $(0, 1]$; hence we seem to be able to say merely that (68) is satisfied by all distribution functions F and G which are continuous in $(0, 1]$. This situation can be remedied, however, by a generalization of our definition of the Stieltjes integral—namely, by the introduction of the notion of a Lebesgue-Stieltjes integral, which is related to our notion (sometimes called the "Riemann-Stieltjes" integral) somewhat as the Lebesgue integral is related to the Riemann integral. If we set

$$E(F, G) = \int_0^1 \int_0^1 M(x, y)\, dF(x)\, dG(y),$$

where the integrals are interpreted as Lebesgue-Stieltjes integrals, then (68) becomes meaningful, and true, for all distribution functions F and G—even discontinuous ones.

We shall not introduce the integrals of Lebesgue-Stieltjes in this book, because to give any adequate theory of them would presuppose too much familiarity with other branches of mathematics. Even without introducing this notion, however, we can say that the given distribution functions constitute optimal strategies for the two players in the sense that, for every x and y in $[0, 1]$, we have $E(x, G_0) \leq E(F_0, G_0) \leq E(F_0, y)$.

REMARK 10.16. In Example 10.15 we succeeded in finding a solution of a continuous game with a discontinuous payoff function. Thus it appears that Theorem 10.4 is capable of being generalized so as to cover also certain discontinuous functions. Various results of this kind are available in the mathematical literature, but we do not propose to take them up here. It is perhaps worth while to notice, however, that we can obtain an immediate generalization of 10.4 by an intuitive argument. Suppose that we have a game Γ whose payoff function M is not continuous, and suppose that there exist functions Θ and Φ, which map $[0, 1]$ onto itself in a one-to-one way, and that the function

$$M'(x, y) = M(\Theta(x), \Phi(y))$$

is continuous. Since M' is continuous, we see that the game Γ', whose payoff function is M', has a solution. But since Γ results from Γ' by merely "relabeling" the x's and y's, it is seen that Γ also has a solution.

Thus suppose that we have a game whose (discontinuous) payoff function M is defined as follows:

$$M(x, y) = \begin{cases} (x - y)^2 & \text{if } 0 \leq x \leq \frac{1}{2} \text{ and } 0 \leq y \leq \frac{1}{2}, \\ \left(x + y - \frac{1}{2}\right)^2 & \text{if } 0 \leq x \leq \frac{1}{2} \text{ and } \frac{1}{2} < y, \\ \left(x + y - \frac{1}{2}\right)^2 & \text{if } \frac{1}{2} < x \text{ and } 0 \leq y \leq \frac{1}{2}, \\ (x - y)^2 & \text{if } \frac{1}{2} < x \text{ and } \frac{1}{2} < y. \end{cases}$$

Defining Θ by the conditions

$$\Theta(u) = \frac{1}{2} - u \qquad \text{if } 0 \leq u \leq \frac{1}{2},$$

$$\Theta(u) = u \qquad \text{if } \frac{1}{2} < u,$$

we see that Θ maps $[0, 1]$ onto itself in a one-to-one way and that, for all x and y,

$$\begin{aligned} M'(x, y) &= M(\Theta(x), \Theta(y)) \\ &= (x - y)^2. \end{aligned}$$

Since M' is continuous, we conclude that the game whose payoff function is M has a solution also.

It should be noted, however, that a game with a discontinuous payoff function can fail to have any value at all. An example of a game for which this seems intuitively plausible was given in Chap. 7 (Example 7.3).

The following theorems will be useful in later chapters, when we wish to find solutions of certain games.

THEOREM 10.17. Let M be the continuous payoff function of a continuous game, and suppose that there exists an optimal strategy of the form I_a for the first player. Then the value, v, of the game is given by the formula

$$v = \max_{0 \leq x \leq 1} \min_{0 \leq y \leq 1} M(x, y),$$

and the constant a can be taken to be any solution of the equation

$$\min_{0 \leq y \leq 1} M(a, y) = v.$$

Similarly, if there exists an optimal strategy of the form I_b for the second player, then

$$v = \min_{0 \leq y \leq 1} \max_{0 \leq x \leq 1} M(x, y),$$

and b can be taken to be any solution of the equation

$$\max_{0 \leq x \leq 1} M(x, b) = v.$$

PROOF. To prove the first part, we see by the hypothesis that there exists for the first player an optimal strategy that belongs to D_1 (the set of all step-functions with one step). Hence, making use also of Theorem 9.22, we have

$$\begin{aligned} v &= \max_{F \in \mathbf{D}} \min_{G \in \mathbf{D}} \int_0^1 \int_0^1 M(x, y)\, dF(x)\, dG(y) \\ &= \max_{I_a \in \mathbf{D}_1} \min_{G \in \mathbf{D}} \int_0^1 \int_0^1 M(x, y)\, dI_a(x)\, dG(y) \\ &= \max_{0 \le a \le 1} \min_{G \in \mathbf{D}} \int_0^1 M(a, y)\, dG(y) \\ &= \max_{0 \le a \le 1} \min_{0 \le y \le 1} M(a, y) \\ &= \max_{0 \le x \le 1} \min_{0 \le y \le 1} M(x, y), \end{aligned}$$

as was to be shown.

Moreover, by Theorem 10.9, if I_a is an optimal strategy for the first player, then

$$v = \min_{0 \le y \le 1} \int_0^1 M(x, y)\, dI_a(x),$$

or

$$v = \min_{0 \le y \le 1} M(a, y). \tag{70}$$

Thus if I_a is an optimal strategy for the first player, then a must satisfy Eq. (70). And if we suppose, on the other hand, that a is any number which satisfies (70), then we have, for all y,

$$v \le M(a, y) = \int_0^1 M(x, y)\, dI_a(x),$$

so that I_a is an optimal strategy by Theorem 10.6.

The proof of the second part of the theorem is similar.

REMARK 10.18. Theorem 10.17 has the obvious consequence that if M is the continuous payoff function of a continuous game, and if there exists an optimal strategy I_a for the first player and an optimal strategy I_b for the second player, then M has a saddle-point.

Theorem 10.17 can be readily generalized to the case of games for which there exist optimal strategies which are step-functions of any given finite order. We shall state this generalization, leaving the proof (which is very similar to the proof of Theorem 10.17) to be used as an exercise.

THEOREM 10.19. *Let M be the continuous payoff function of a continuous game, and suppose that there exists, for the first* player, an optimal strategy having the form

$$\alpha_1 I_{a_1}(x) + \cdots + \alpha_m I_{a_m}(x).$$

Then the value, v, of the game is given by the formula

$$v = \max_{0 \leq a_1 \leq \cdots \leq a_m \leq 1} \max_{\|\alpha_1 \cdots \alpha_m\| \in S_m} \min_{0 \leq y \leq 1} \sum_{i=1}^{m} \alpha_i M(a_i, y),$$

and the constants $\alpha_1, \cdots, \alpha_m$ and $a_1, \cdots, a_m$ can be taken to be any solutions of the equation

$$\min_{0 \leq y \leq 1} \sum_{i=1}^{m} \alpha_i M(a_i, y) = v.$$

Similarly, if there exists for the second player an optimal strategy having the form

$$\beta_1 I_{b_1}(y) + \cdots + \beta_n I_{b_n}(y),$$

then

$$v = \min_{0 \leq b_1 \leq \cdots \leq b_n \leq 1} \min_{\|\beta_1 \cdots \beta_n\| \in S_n} \max_{0 \leq x \leq 1} \sum_{i=1}^{n} \beta_i M(x, b_i),$$

and the constants $\beta_1, \cdots, \beta_n$ and $b_1, \cdots, b_n$ can be taken to be any solutions of the equation

$$\max_{0 \leq x \leq 1} \sum_{i=1}^{n} \beta_i M(x, b_i) = v.$$

For later reference, we include two theorems which will be used in Chap. 11. The proofs of these theorems will be left as exercises.

THEOREM 10.20. Any convex linear combination of optimal strategies is an optimal strategy.

THEOREM 10.21. If

$$F_1, F_2, \cdots, F_n, \cdots$$

is a sequence of optimal strategies which converges to a distribution function F at every point of continuity of F, then F is an optimal strategy.

HISTORICAL AND BIBLIOGRAPHICAL REMARKS

The first proof of Theorem 10.4 was given in Ville [1]. Proofs of generalizations of this theorem are given in Wald [6] and Karlin [1].

EXERCISES

1. Prove Lemma 10.2.

2. Show by a direct algebraic argument that if

$$a_{11}x_1 + a_{12}x_2,$$
$$a_{21}x_1 + a_{22}x_2$$

are two homogeneous linear forms such that, for every $\| x_1 \quad x_2 \|$ in $\mathbf{S}_2$, we have either

$$a_{11}x_1 + a_{12}x_2 \leq 0$$

or

$$a_{21}x_1 + a_{22}x_2 \leq 0,$$

then there exists an element $\| \overline{y}_1 \quad \overline{y}_2 \|$ of $\mathbf{S}_2$ such that, for all $\| x_1 \quad x_2 \|$ in $\mathbf{S}_2$,

$$(a_{11}x_1 + a_{12}x_2)\overline{y}_1 + (a_{21}x_1 + a_{22}x_2)\overline{y}_2 \leq 0.$$

3. Let Γ be a continuous game whose payoff function is

$$M(x, y) = \frac{1}{1 + \lambda(x - y)^2},$$

where $0 < \lambda \leq 4/3$. Show that the following is a solution of Γ:

$$F_0(x) = I_{1/2}(x),$$
$$G_0(y) = \frac{1}{2}I_0(y) + \frac{1}{2}I_1(y),$$
$$v = \frac{4}{4 + \lambda}.$$

4. Show that the game in Example 10.11 has no solution where the optimal strategy for the second player is a member of $\mathbf{D}_1$; i.e., where we have, for some a,

$$G_0(y) = I_a(y).$$

5. Show that if f is a continuous function such that, for all u,

$$f(u + 1) = f(u),$$

and if we set

$$M(x, y) = f(x - y),$$

then the continuous game whose payoff function is M has the following solution:

$$F_0(x) = x,$$
$$G_0(y) = y,$$
$$v = \int_0^1 f(u)\, du.$$

6. The payoff function of a continuous game is

$$M(x, y) = \frac{1}{1 + 2(x - y)^2}.$$

Show that there do not exist optimal strategies having the form

$$F_0(x) = I_a(x),$$
$$G_0(y) = \beta_1 I_{b_1}(y) + \beta_2 I_{b_2}(y).$$

7. Find optimal strategies for the game of Exercise 6, which have the form

$$F_0(x) = \alpha_1 I_{a_1}(x) + \alpha_2 I_{a_2}(x),$$
$$G_0(y) = \beta_1 I_{b_1}(y) + \beta_2 I_{b_2}(y).$$

8. Show that if

$$F_0(x) = x,$$
$$G_0(y) = y$$

are optimal strategies in the game whose (continuous) payoff function is M, then

$$F^*(x) = x^2,$$
$$G^*(y) = y^2$$

are optimal strategies in the game whose payoff function M^* is given by the equation

$$M^*(x, y) = M(x^2, y^2).$$

9. Generalize the result stated in Exercise 8.

10. Letting a be a fixed positive number, we set

$$M(x, y) = a(x - y) + xy \qquad \text{if } x < y,$$
$$M(x, y) = 0 \qquad \text{if } x = y,$$
$$M(x, y) = a(x - y) - xy \qquad \text{if } x > y.$$

Find a solution of the game whose payoff function is M.

11. Prove Theorem 10.19.

12. Show that there exists a solution for the game whose payoff function is defined as follows:

$$M(0, y) = 1 - y,$$
$$M(x, y) = |x - y| \qquad \text{for } 0 < x < 1,$$
$$M(1, y) = y.$$

Hint: Make use of the function Θ defined as follows:

$$\Theta(0) = 1,$$
$$\Theta(x) = x \qquad \text{for } 0 < x < 1,$$
$$\Theta(1) = 0.$$

13. The payoff function for a continuous game is

$$M(x, y) = e^{y-x} \qquad \text{if } y < x,$$
$$M(x, y) = e^{x-y} \qquad \text{if } y \geq x.$$

For the game above find optimal strategies having the form

$$F_0(x) = \alpha_1 F_1(x) + \alpha_2[I_0(x) + I_1(x)],$$
$$G_0(y) = \beta_1 G_1(y) + \beta_2[I_0(y) + I_1(y)],$$

where F_1 and G_1 are twice differentiable distribution functions with $dF_1(x)/dx \neq 0$, $dG_1(y)/dy \neq 0$, and α_1 and β_1 are positive.

14. For the game of Example 10.14 find optimal strategies having the form

$$F_0(x) = \alpha_1 I_{a_1}(x) + \alpha_2 I_{a_2}(x),$$
$$G_0(y) = \beta_1 I_{b_1}(y) + \beta_2 I_{b_2}(y).$$

15. Prove Theorem 10.20.

16. Prove Theorem 10.21.

CHAPTER 11

SEPARABLE GAMES

1. The Mapping Method. In this chapter we shall consider a certain rather wide class of games and shall explain a method of solving them. A function M of two variables is called *separable*, or sometimes *polynomial-like*, if there exist m continuous functions $r_1, \cdots, r_m$ and n continuous functions $s_1, \cdots, s_n$ and $m \cdot n$ constants $a_{11}, \cdots, a_{mn}$ such that, identically in x and y,

$$M(x, y) = \sum_{j=1}^{n} \sum_{i=1}^{m} a_{ij} r_i(x) s_j(y). \tag{1}$$

Thus the function M defined by the equation

$$M(x, y) \doteq x \sin y + x \cos y + 2x^2$$

is separable; here we can take

$$\begin{aligned} r_1(x) &= x, & s_1(y) &= \sin y, \\ r_2(x) &= x^2, & s_2(y) &= \cos y, \\ & & s_3(y) &= 1. \end{aligned}$$

It is clear that a given separable function can be represented in the form (1) in many ways. Thus, for the function given above, we can also take

$$\begin{aligned} r_1(x) &= x, & s_1(y) &= \sin y + \cos y, \\ r_2(x) &= 2x^2, & s_2(y) &= \frac{1}{2}. \end{aligned}$$

If a function M is separable, then by (1) we have, identically in x and y,

$$M(x, y) = \sum_{j=1}^{n} \sum_{i=1}^{m} a_{ij} r_i(x) s_j(y) = \sum_{j=1}^{n} \left[\sum_{i=1}^{m} a_{ij} r_i(x) \right] s_j(y);$$

hence, setting

$$t_j(x) = \sum_{i=1}^{m} a_{ij} r_i(x) \qquad (j = 1, \cdots, n),$$

we can write

$$M(x, y) = \sum_{j=1}^{n} t_j(x)s_j(y), \tag{2}$$

where the functions t_j and s_j are continuous.

A polynomial in two variables is, of course, a special case of a separable function. Thus the polynomial

$$xy + x^2 + xy^2 + 2x^2y + x^3y^3$$

can be represented in the form (2) by taking

$$\begin{aligned} r_1(x) &= x, & t_1(y) &= y + y^2, \\ r_2(x) &= x^2, & t_2(y) &= 1 + 2y, \\ r_3(x) &= x^3, & t_3(y) &= y^3. \end{aligned}$$

By a *separable game* we mean a continuous game whose payoff function is separable. Thus in a separable game the payoff function M satisfies (1), above, where r_i and s_j (for $1 \leq i \leq m$ and $1 \leq j \leq n$) are continuous functions over the closed unit interval.

If M is the payoff function of a separable game, and if P_1 uses a mixed strategy F and P_2 uses a mixed strategy G, then the expectation $E(F, G)$ of P_1 (making use of (1) and some simple properties of Stieltjes integrals) is given by:

$$\begin{aligned} E(F, G) &= \int_0^1 \int_0^1 M(x, y)\, dF(x)\, dG(y) \\ &= \int_0^1 \int_0^1 \left[\sum_{j=1}^{n} \sum_{i=1}^{m} a_{ij} r_i(x) s_j(y) \right] dF(x)\, dG(y) \\ &= \sum_{j=1}^{n} \sum_{i=1}^{m} a_{ij} \int_0^1 r_i(x)\, dF(x) \int_0^1 s_j(y)\, dG(y). \end{aligned} \tag{3}$$

Thus $E(F, G)$ depends on F and G only in so far as F and G affect the values of the components of the two vectors

$$\left\| \int_0^1 r_1(x)\, dF(x) \quad \cdots \quad \int_0^1 r_m(x)\, dF(x) \right\|$$

and

$$\left\| \int_0^1 s_1(y)\, dG(y) \quad \cdots \quad \int_0^1 s_n(y)\, dG(y) \right\|.$$

It is readily seen that

$$\left\| \int_0^1 r_1(x)\, dF^{(1)}(x) \quad \cdots \quad \int_0^1 r_m(x)\, dF^{(1)}(x) \right\|$$

and

$$\left\| \int_0^1 r_1(x)\, dF^{(2)}(x) \quad \cdots \quad \int_0^1 r_m(x)\, dF^{(2)}(x) \right\|$$

can be the same vector, even though $F^{(1)}$ and $F^{(2)}$ are not identical; however, if the two vectors are the same, regardless of whether $F^{(1)}$ and $F^{(2)}$ are identical, then

$$E(F^{(1)}, G) = E(F^{(2)}, G)$$

for every distribution function G, and it is immaterial whether P_1 uses strategy $F^{(1)}$ or strategy $F^{(2)}$. Therefore, when the two given vectors are the same, we shall call the strategies $F^{(1)}$ and $F^{(2)}$ *equivalent* (with respect to the given game, of course). Thus if we have a separable game whose payoff function satisfies (1), it is seen that for P_1 to choose a mixed strategy $F^{(1)}$ amounts to his choosing a point $\| u_1^{(1)} \quad \cdots \quad u_m^{(1)} \|$ from a certain subset U of m-dimensional Euclidean space. This set U, which we shall sometimes call the U-*space*, consists of all points $\| u_1 \quad \cdots \quad u_m \|$ such that, for some distribution function F,

$$\left.\begin{aligned} u_1 &= \int_0^1 r_1(x)\, dF(x), \\ &\vdots \\ u_m &= \int_0^1 r_m(x)\, dF(x). \end{aligned}\right\} \tag{4}$$

When $u = \| u_1 \quad \cdots \quad u_m \|$ and F satisfy (4), we say that u and F *correspond*. (A given point of the U-space will, in general, correspond to many different distribution functions.)

Similarly, for P_2 to choose a mixed strategy $G^{(1)}$ amounts to his choosing a point $\| w_1^{(1)} \quad \cdots \quad w_n^{(1)} \|$ from what we call the W-*space*, i.e., from the set W of all points $\| w_1 \quad \cdots \quad w_n \|$ of n-dimensional Euclidean space which are such that, for some distribution function G,

$$\begin{aligned} w_1 &= \int_0^1 s_1(y)\, dG(y), \\ &\vdots \\ w_n &= \int_0^1 s_n(y)\, dG(y). \end{aligned}$$

It should be noticed that the sets U and W are relative to the particular representation used for the payoff function M. Thus, as we have seen earlier, the same function M can satisfy both

$$M(x, y) = \sum_{j=1}^{2} \sum_{i=1}^{3} a_{ij} r_i(x) s_j(y)$$

and

$$M(x, y) = \sum_{i=1}^{2} r_i(x) t_i(y) .$$

With the first representation of M, U is a subset of Euclidean 3-space, while with the second representation, U is a subset of 2-space. However, we shall often omit reference to this dependence of U and W on the representation of M, since, in dealing with a given game, we shall ordinarily suppose that the representation is fixed once and for all.

So far we have considered the expectation function E as being defined only for distribution functions as arguments. For some of our later discussion, however, it will be convenient to be able to speak also of the expectation of P_1 when P_1 chooses a point u from the U-space and P_2 chooses point w from the W-space. We therefore extend the domain of the definition of E as follows: If u is any point of the U-space and w is any point of the W-space, and if F is any distribution function corresponding to u and G is any distribution function corresponding to w, then we set

$$E(u, w) = E(F, G) .$$

From the statement above, it is clear that $E(u, w)$ is thus well defined; i.e., if u corresponds to both F and F^1 and w corresponds to both G and G^1, then

$$E(F^1, G^1) = E(F, G) ,$$

and we could just as well write

$$E(u, w) = E(F^1, G^1) .$$

From (3) we see that if $u = \| u_1 \quad \cdots \quad u_m \|$ is any point of the U-space and $w = \| w_1 \quad \cdots \quad w_n \|$ is any point of the W-space, then

$$E(u, w) = \sum_{j=1}^{n} \sum_{i=1}^{m} a_{ij} u_i w_j ;$$

thus E is a bilinear form in the coordinates of u and w.

For our purposes it is now desirable to establish some geometrical properties of the U-space and the W-space and to characterize the relationship between each of these spaces and the set of all distribution functions. These results will be formulated in the next few theorems.

In Chap. 2 we defined the notion of a convex linear combination of a finite number of points of Euclidean n-space. By analogy with that notion, we say that a function F is a *convex linear combination* of functions $F^{(1)}, \cdots, F^{(r)}$, with weights $a_1, \cdots, a_r$, if $\| a_1 \quad \cdots \quad a_r \| \in \mathbf{S}_r$ and if, for every x in $[0, 1]$,

$$F(x) = a_1F^{(1)}(x) + a_2F^{(2)}(x) + \cdots + a_rF^{(r)}(x).$$

When F is such a convex linear combination of $F^{(1)}, \cdots, F^{(r)}$, we shall sometimes write simply

$$F = a_1F^{(1)} + a_2F^{(2)} + \cdots + a_rF^{(r)}.$$

THEOREM 11.1. Let $F^{(1)}, \cdots, F^{(p)}$ be distribution functions, let $u^{(1)}, \cdots, u^{(p)}$ be the corresponding points of the U-space of a separable game, and let $\| a_1 \quad \cdots \quad a_p \|$ be any member of $\mathbf{S}_p$. Then the point

$$u = a_1u^{(1)} + \cdots + a_pu^{(p)}$$

is a point of the U-space and corresponds to the distribution function

$$F = a_1F^{(1)} + \cdots + a_pF^{(p)};$$

and similarly for the W-space.

PROOF. Let the payoff function M of the separable game satisfy

$$M(x, y) = \sum_{j=1}^{n} \sum_{i=1}^{m} a_{ij}r_i(x)s_j(y).$$

Let $u^{(i)} = \| u_1^{(i)} \quad \cdots \quad u_m^{(i)} \|$ (for $i = 1, \cdots, p$). Since $u^{(i)}$ corresponds to $F^{(i)}$ (for $i = 1, \cdots, p$) we have

$$u_j^{(i)} = \int_0^1 r_j(x)\, dF^{(i)}(x) \qquad \text{for } i = 1, \cdots, p \text{ and } j = 1, \cdots, m. \tag{5}$$

Let $u = \| u_1 \quad \cdots \quad u_m \|$. Since

$$u = a_1u^{(1)} + \cdots + a_pu^{(p)},$$

we see that

$$u_j = a_1 u_j^{(1)} + \cdots + a_p u_j^{(p)} \qquad \text{for } j = 1, \cdots, m. \tag{6}$$

From (5) and (6) we conclude that, for $j = 1, \cdots, m$,

$$u_j = a_1 \int_0^1 r_j(x)\, dF^{(1)}(x) + \cdots + a_p \int_0^1 r_j(x)\, dF^{(p)}(x),$$

and hence, using the appropriate theorems about Stieltjes integrals, that, for $j = 1, \cdots, m$,

$$\begin{aligned} u_j &= \int_0^1 r_j(x)\, d[a_1F^{(1)}(x) + \cdots + a_pF^{(p)}(x)] \\ &= \int_0^1 r_j(x)\, dF(x). \end{aligned} \tag{7}$$

Since F is a distribution function by Theorem 8.3, we conclude that $u = \| u_1 \quad \cdots \quad u_m \|$ corresponds to the distribution function F, and hence, by definition, that u belongs to the U-space.

The proof for the W-space is similar.

We can classify the points of the U-space according to the kinds of distribution functions to which they correspond. An especially important subset of the U-space consists of those points which correspond to step-functions having just one step. We shall call this subset U*, and the analogous subset of the W-space will be called W*. We now have a theorem which, though easily proved, will turn out to be quite useful.

THEOREM 11.2. Let the payoff function M of a separable game be represented by the equation

$$M(x, y) = \sum_{j=1}^{n} \sum_{i=1}^{m} a_{ij} r_i(x) s_j(y),$$

where r_i and s_j are continuous functions of x and y, respectively. Then the set U* (with respect to the given representation) consists of all points $\| u_1 \quad \cdots \quad u_m \|$ such that, for some t in $[0, 1]$,

$$\begin{aligned} u_1 &= r_1(t), \\ &\vdots \\ u_m &= r_m(t). \end{aligned}$$

Similarly, the set W* consists of all points $\| w_1 \quad \cdots \quad w_n \|$

such that, for some t in $[0, 1]$,

$$w_1 = s_1(t),$$
$$\vdots$$
$$w_n = s_n(t).$$

PROOF. By the definition of U^*, a point $\| u_1 \quad \cdots \quad u_m \|$ belongs to U if, and only if, there is a step-function I_t with one step which corresponds to it, i.e., if, and only if, for some t in $[0, 1]$, we have

$$u_1 = \int_0^1 r_1(x)\, dI_t(x) = r_1(t),$$
$$u_2 = \int_0^1 r_2(x)\, dI_t(x) = r_2(t),$$
$$\vdots$$
$$u_m = \int_0^1 r_m(x)\, dI_t(x) = r_m(t).$$

The proof of the second part of the theorem is similar.

COROLLARY 11.3. For any separable game, the sets U^* and W^* are bounded, closed, connected sets.

PROOF. This follows from the fact that the functions r_i and s_j are continuous functions defined over a closed interval.

THEOREM 11.4. For any separable game, the U-space is the convex hull of the curve U^* and the W-space is the convex hull of the curve W^*. (Thus the U-space—and, similarly, the W-space—is a closed, bounded, convex set.)

PROOF. Let U' be the convex hull of U^*. We wish to show that U' coincides with the U-space, i.e., that $\mathsf{U}' = \mathsf{U}$. Since $\mathsf{U}^* \subseteq \mathsf{U}$, we see directly from Theorem 11.1 that $\mathsf{U}' \subseteq \mathsf{U}$. Hence it remains only to show that $\mathsf{U} \subseteq \mathsf{U}'$.

Suppose, then, if possible, that there is a point $z = \| z_1 \quad \cdots \quad z_m \|$ which belongs to U but not to U'. Since, by Corollary 11.3, U^* is bounded and closed, we see that U' also is bounded and closed; moreover, U' is convex. Since $z = \| z_1 \quad \cdots \quad z_m \|$ is not in U', we therefore conclude from Theorem 2.3 that there are constants $a_1, \cdots, a_m$, b, and δ, where $\delta > 0$, such that

$$a_1 z_1 + \cdots + a_m z_m + b > 0, \tag{8}$$

and, for any point $\| u_1 \quad \cdots \quad u_m \|$ of U',

$$a_1 u_1 + \cdots + a_m u_m + b < -\delta < 0. \tag{9}$$

From (8) and (9) we see that, for every point $\| u_1 \quad \cdots \quad u_m \|$ of U',

$$(a_1 z_1 + \cdots + a_m z_m + b) - (a_1 u_1 + \cdots + a_m u_m + b) > \delta,$$

and hence that

$$(a_1 z_1 + \cdots + a_m z_m) - (a_1 u_1 + \cdots + a_m u_m) > \delta. \tag{10}$$

Since (10) holds for every point $\| u_1 \quad \cdots \quad u_m \|$ in U', it holds also, in particular, for every point in U^*. Hence, making use of Theorem 11.2, we conclude that, for all t in the interval $[0, 1]$,

$$[a_1 z_1 + \cdots + a_m z_m] - [a_1 r_1(t) + \cdots + a_m r_m(t)] > \delta. \tag{11}$$

Now let F be a distribution function to which the point $\| z_1 \quad \cdots \quad z_m \|$ of U corresponds; i.e., such that

$$\left.\begin{aligned} z_1 &= \int_0^1 r_1(x)\, dF(x) = \int_0^1 r_1(t)\, dF(t), \\ &\vdots \\ z_m &= \int_0^1 r_m(x)\, dF(x) = \int_0^1 r_m(t)\, dF(t). \end{aligned}\right\} \tag{12}$$

From (11) we conclude that

$$\int_0^1 \{[a_1 z_1 + \cdots + a_m z_m] \\ - [a_1 r_1(t) + \cdots + a_m r_m(t)]\}\, dF(t) > \int_0^1 \delta\, dF(t),$$

and hence, since $\delta, a_1, \cdots, a_m, z_1, \cdots, z_m$ are all constants, that

$$[a_1 z_1 + \cdots + a_m z_m] \int_0^1 dF(t) - a_1 \int_0^1 r_1(t)\, dF(t) - \cdots \\ - a_m \int_0^1 r_m(t)\, dF(t) > \delta \int_0^1 dF(t),$$

or

$$[a_1 z_1 + \cdots + a_m z_m] - a_1 \int_0^1 r_1(t)\, dF(t) - \cdots$$

$$- a_m \int_0^1 r_m(t)\, dF(t) > \delta,$$

or, finally, making use of (12),

$$[a_1 z_1 + \cdots + a_m z_m] - [a_1 z_1 + \cdots + a_m z_m] > \delta,$$

or

$$0 > \delta,$$

contrary to the fact that δ is positive. Since we are thus led to a contradiction, we conclude that $\mathsf{U} \subseteq \mathsf{U}'$, as was to be shown.

Since $\mathsf{U}' = \mathsf{U}$, and since we have seen that U' is bounded, closed, and convex, we now conclude that U is bounded, closed, and convex, which completes the proof. (The proof for the W-space is similar.)

THEOREM 11.5. Let M be the payoff function of a separable game, and let

$$M(x, y) = \sum_{i=1}^{n} r_i(x) t_i(y).$$

Then every mixed strategy (for either player) is equivalent to a step-function with, at most, n steps; in particular, each player has an optimal strategy which is a step-function with, at most, n steps.

PROOF. Let F be any distribution function, and let $u = \| u_1 \quad \cdots \quad u_n \|$ be a point of the U-space which corresponds to F (the proof is similar if we take a distribution function G for P_2 and consider a point of the W-space which corresponds to G). Since the U-space, by Theorem 11.4, is the convex hull of the connected curve U^*, and since the U-space is a subset of Euclidean n-space, we see by Theorem 2.1 that there are points $u^{(1)}, \cdots, u^{(n)}$ of the curve U^*, and a member $\| a_1 \quad \cdots \quad a_n \|$ of S_n, such that

$$u = a_1 u^{(1)} + \cdots + a_n u^{(n)}.$$

Since $u^{(1)}, \cdots, u^{(n)}$ belong to U^*, there are step-functions $I_{t_1}, \cdots, I_{t_n}$, each having one step, such that $u^{(i)}$ (for $i = 1, \cdots, n$) corresponds to I_{t_i}. By

Theorem 11.1 we then see that the function I, given by the equation

$$I = a_1 I_{t_1} + \cdots + a_n I_{t_n},$$

corresponds to u. Clearly I is a step-function having, at most, n steps; and I and F are equivalent, since they both correspond to the same point of the U-space.

In particular, any optimal strategy F (and one exists by Theorem 10.4) is equivalent to a step-function I having, at most, n steps; and such an I is, of course, itself optimal.

We wish next to characterize those points of the U-space, and of the W-space, which correspond to optimal strategies. In order to do this, it is convenient to define a mapping of points of the U-space into sets of points of the W-space, and a mapping of points of the W-space into sets of points of the U-space. If u is any point of the U-space, then by the *image* of u we shall mean the set of points w of the W-space such that

$$E(u, w) = \min_{y \in \mathsf{W}} E(u, y) .$$

We shall denote the image of u by $\mathsf{W}(u)$. Similarly, if w is any point of the W-space, then by the image of w we shall mean the set of all points u of the U-space such that

$$E(u, w) = \max_{x \in \mathsf{U}} E(x, w) .$$

We shall denote the image of w by $\mathsf{U}(w)$.

If u is a point of the U-space and w is a point of the W-space such that $w \in \mathsf{W}(u)$ and $u \in \mathsf{U}(w)$, then we call u a *fixed-point* of the U-space and w a *fixed-point* of the W-space.

> THEOREM 11.6. If F is any distribution function, and if u is the corresponding point of the U-space, then F is an optimal strategy for P_1 if, and only if, u is a fixed-point. Similarly, a distribution function G is an optimal strategy for P_2 if, and only if, the corresponding point w of the W-space is a fixed-point.

PROOF. Let F_1 be a distribution function, and let u be the corresponding point of the U-space. To say that u is a fixed-point of U means that, for some point w of W, we have $w \in \mathsf{W}(u)$ and $u \in \mathsf{U}(w)$, i.e., that

$$\left.\begin{aligned} E(u, w) &= \min_{y \in \mathsf{W}} E(u, y) , \\ E(u, w) &= \max_{x \in \mathsf{U}} E(x, w) . \end{aligned}\right\} \qquad (13)$$

Letting G_1 be a distribution function to which w corresponds, we see by the definition of $E(u, w)$ that (13) is equivalent to

$$E(F_1, G_1) = \min_{F \in \mathsf{D}} E(F, G_1) = \max_{G \in \mathsf{D}} E(F_1, G),$$

which by Theorem 10.5 means that F_1 is an optimal strategy for P_1.

THEOREM 11.7. If u is any fixed-point of U and w is any fixed-point of W, then $u \in \mathsf{U}(w)$ and $w \in \mathsf{W}(u)$.

PROOF. Since u is a fixed-point of U, there exists a point w' of W such that $u \in \mathsf{U}(w')$ and $w' \in \mathsf{W}(u)$, i.e., such that

$$E(u, w') = \max_{x \in \mathsf{U}} E(x, w') \tag{14}$$

and

$$E(u, w') = \min_{y \in \mathsf{W}} E(u, y). \tag{15}$$

Similarly, since w is a fixed-point of W, there is a point u' of U such that

$$E(u', w) = \max_{x \in \mathsf{U}} E(x, w) \tag{16}$$

and

$$E(u', w) = \min_{y \in \mathsf{W}} E(u', y). \tag{17}$$

Then we have

$$\begin{aligned}
E(u, w) &\leq \max_{x \in \mathsf{U}} E(x, w), \text{ by the definition of a maximum} \\
&= E(u', w), \text{ by (16)} \\
&= \min_{y \in \mathsf{W}} E(u', y), \text{ by (17)} \\
&\leq E(u', w'), \text{ by the definition of a minimum} \\
&\leq \max_{x \in \mathsf{U}} E(x, w'), \text{ by the definition of a maximum} \\
&= E(u, w'), \text{ by (14)} \\
&= \min_{y \in \mathsf{W}} E(u, y), \text{ by (15)} \\
&\leq E(u, w), \text{ by the definition of a minimum.}
\end{aligned}$$

Since the first and last terms of this continued set of equalities and inequalities are equal, we therefore conclude that all the quantities involved are equal, and thus, in particular, that

$$E(u, w) = \max_{x \in \mathsf{U}} E(x, w)$$
$$= \min_{y \in \mathsf{W}} E(u, y),$$

which means that $u \in \mathsf{U}(w)$ and $w \in \mathsf{W}(u)$, as was to be shown.

2. An Illustrative Example. The theorems we have proved are sufficient to provide us with a general method of attack on the problem of finding solutions for separable games: (*a*) we plot the curves U^* and W^* (in m-dimensional space and n-dimensional space, respectively) and determine their convex hulls, which by Theorem 11.4 are U and W, respectively; (*b*) we find $\mathsf{W}(u)$ for every point u in U and $\mathsf{U}(w)$ for every point w in W (this is equivalent to finding the points in certain closed convex sets at which certain linear forms assume their minima or maxima); (*c*) using the results of (*b*), we find the fixed-points of U and of W; and (*d*) we express the fixed-points as convex linear combinations of points U^* and W^*, respectively, and make use of Theorem 11.1 to find distribution functions to which the fixed-points correspond—these distribution functions will be optimal strategies. We now illustrate this method by an example.

EXAMPLE 11.8. The payoff function M of a separable game satisfies the following equation (for any point $\| x \quad y \|$ of the closed unit square):

$$M(x, y) = \cos 2\pi x \cos 2\pi y + x + y.$$

Here we shall take

$$r_1(x) = x, \qquad s_1(y) = y,$$
$$r_2(x) = \cos 2\pi x, \qquad s_2(y) = \cos 2\pi y,$$
$$r_3(x) = 1, \qquad s_3(y) = 1,$$

so that M is represented as follows:

$$M(x, y) = r_2(x)s_2(y) + r_1(x)s_3(y) + r_3(x)s_1(y).$$

Clearly the U-space lies in the plane $u_3 = 1$ and the W-space in $w_3 = 1$. We shall, for simplicity, draw a 2-dimensional representation of each of these spaces.

The curve U^* is the curve whose parametric equations are

$$u_1 = t,$$
$$u_2 = \cos 2\pi t,$$
$$u_3 = 1,$$

The U-space is the convex hull of U*, which is that part of the plane $u_3 = 1$ bounded by the line-segments AB, AC, and BD and by the arc CQD of the cosine curve, as indicated in Fig. 1. (The W-space, which is the same as the U-space, is also indicated in Fig. 1.)

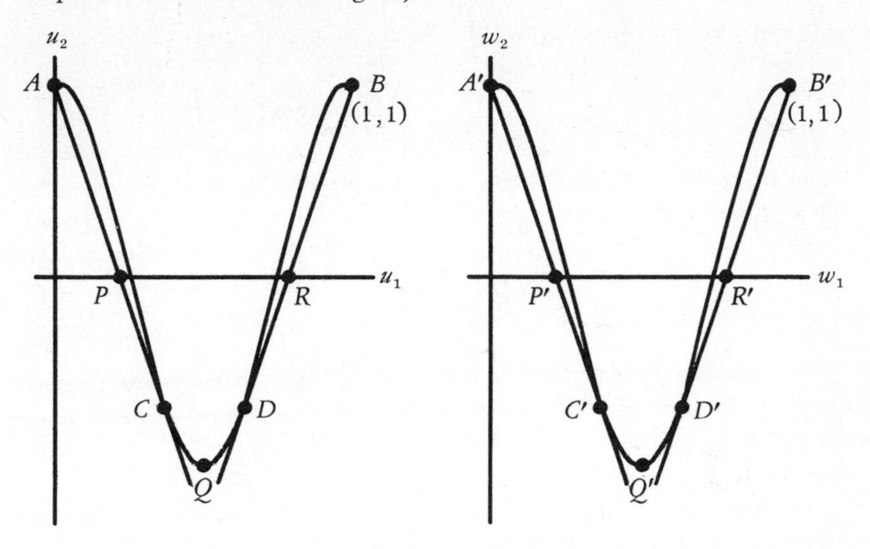

Fig. 1

Let us find the coordinates of C', since we shall need to use them later. We denote the point C' by $\| y_1 \quad \cos 2\pi y_1 \quad 1 \|$. The slope m of $A'C'$ is the same as the slope of the curve

$$w_2 = \cos 2\pi w_1$$

at $w_1 = y_1$. Hence we have

$$m = \frac{\cos 2\pi y_1 - 1}{y_1} = -2\pi \sin 2\pi y_1, \tag{18}$$

from which we obtain

$$2\pi y_1 = \frac{1 - \cos 2\pi y_1}{\sin 2\pi y_1} = \tan \pi y_1, \tag{19}$$

where $2\pi y_1$ is a second quadrant angle. Solving this equation, we find, to an accuracy of two places of decimals,

$$y_1 = .37,$$
$$m = -4.55.$$

Therefore the coordinates of C' are $\| .37 \quad -.68 \quad 1 \|$.

If F is any strategy for P_1 and G is any strategy for P_2, then the expectation $E(F, G)$ is given by

$$E(F, G) = \int_0^1 \cos 2\pi x \, dF(x) \int_0^1 \cos 2\pi y \, dG(y) + \int_0^1 x \, dF(x) + \int_0^1 y \, dG(y).$$

If we remember that a point $\| u_1 \quad u_2 \quad 1 \|$ belongs to the U-space if, and only if, for some F,

$$u_1 = \int_0^1 x \, dF(x),$$

$$u_2 = \int_0^1 \cos 2\pi x \, dF(x),$$

and that a point $\| w_1 \quad w_2 \quad 1 \|$ belongs to the W-space if, and only if, for some G,

$$w_1 = \int_0^1 y \, dG(y),$$

$$w_2 = \int_0^1 \cos 2\pi y \, dG(y),$$

we see that if P_1 chooses a point $u = \| u_1 \quad u_2 \quad 1 \|$ from U and if P_2 chooses a point $w = \| w_1 \quad w_2 \quad 1 \|$ from W, then the expectation $E(u, w)$ of P_1 is given by the formula:

$$E(u, w) = u_2 w_2 + u_1 + w_1. \tag{20}$$

Our problem is to find two points $\bar{u}$ and $\bar{w}$ which belong to U and W, respectively, and which have the properties that

$$\bar{u} \in \mathsf{U}(\bar{w}) \quad \text{and} \quad \bar{w} \in \mathsf{W}(\bar{u}),$$

i.e., such that

$$\max_{u \in \mathsf{U}} E(u, \bar{w}) = \min_{w \in \mathsf{W}} E(\bar{u}, w) = E(\bar{u}, \bar{w}).$$

For every point w of W, $\mathsf{U}(w)$ is a subset of the right-hand boundary BDQ of U. For suppose that $\mathsf{U}(w)$ contained a point $a = \| a_1 \quad a_2 \quad 1 \|$ not on BDQ. Let $a' = \| a_1 + d \quad a_2 \quad 1 \|$ be the horizontal projection of a on

BDQ, where d is the (positive) distance from a to a'. Then from (20) we see that

$$E(a', w) = E(a, w) + d,$$

which shows that $\max_{u \in U} E(u, w)$ does not occur at $u = a$, i.e., that a does not belong to $\mathsf{U}(w)$. Similarly, for every point u of U, the image $\mathsf{W}(u)$ is a subset of the left-hand boundary $A'C'Q'$ of W.

Next we shall find $\mathsf{W}(a)$ for an arbitrary point $a = \| a_1 \quad \cos 2\pi a_1 \quad 1 \|$ of the arc QD of the cosine curve which forms part of the boundary of U. A point b of W will belong to $\mathsf{U}(a)$ if, and only if,

$$\min_{w \in \mathsf{W}} E(a, w) = E(a, b).$$

From (20) we see that this means that we are to find the points $\| w_1 \quad w_2 \quad 1 \|$ on the left-hand boundary of W, which makes

$$w_1 + w_2 \cos 2\pi a_1 + a_1$$

a minimum. This problem is immediately seen to be equivalent to the following: Given the family of straight lines

$$w_1 + w_2 \cos 2\pi a_1 = k,$$

find the lines of this family which intersect the W-space and make k a minimum, and then find the points in which these particular straight lines intersect the W-space. The slope of each of the lines of this family is

$$\frac{-1}{\cos 2\pi a_1};$$

since $\| a_1 \quad \cos 2\pi a_1 \quad 1 \|$ is a point on the arc QD, $\cos 2\pi a_1$ is negative, and hence this slope is positive. Thus it is apparent that the minimum value of k will be obtained by taking the line

$$w_1 + w_2 \cos 2\pi a_1 = k$$

as far over to the left as possible, but so that it still intersects the W-space. This means that the line must pass through the point A', and, since its slope is positive, A' will be the only point of the W-space lying on the line. Thus $\mathsf{W}(a) = \| 0 \quad 1 \quad 1 \|$ when a lies on the arc QD.

Similarly, if $a = \| a_1 \quad a_2 \quad 1 \|$ is any point on the line-segment DR, then $\mathsf{W}(a)$ will consist of the points of W which minimize the expression

$$E(a, w) = w_1 + a_1 + w_2 a_2.$$

Hence again the problem is equivalent to finding the lines of the family

$$w_1 + a_2 w_2 = k \tag{21}$$

which intersect the W-space and make k a minimum. And since a_2 is negative and the slope of these lines is $-(1/a_2)$, we find, by the same argument we used before, that $\mathsf{W}(a) = \| 0 \quad 1 \quad 1 \|$.

Next we take $a = \| a_1 \quad a_2 \quad 1 \|$ along the line-segment RB and find $\mathsf{W}(a)$. Remembering that m is the slope of the line $A'C'$, we consider three cases, according as:

Case 1. $$\left| \frac{1}{a_2} \right| > | m |;$$

Case 2. $$\left| \frac{1}{a_2} \right| = | m |;$$

Case 3. $$\left| \frac{1}{a_2} \right| < | m |.$$

CASE 1. In this case we again find that $\mathsf{W}(a) = \| 0 \quad 1 \quad 1 \|$, by an argument essentially the same as that used above. The only difference is that now the lines of (21) have a negative slope; but since the absolute value of this slope is greater than the absolute value of the slope m of $A'C'$, it follows, as before, that when the line is taken as far to the left as possible, but still intersects the W-space, then it will have only the point A' in common with the W-space.

CASE 2. In this case the lines of (21) all have the same slope as the line $A'C'$; and, consequently, k is minimized when (21) coincides with $A'C'$. It is clear, then, that in this case $\mathsf{W}(a)$ consists of all points of the line-segment $A'C'$.

CASE 3. In this case the lines of (21) have a negative slope which is smaller in absolute value than the slope of the line $A'C'$. Hence the point $w = \| w_1 \quad w_2 \quad 1 \|$ in the W-space which will minimize k must be the point of tangency, along the arc $C'Q'$, of that line of the family (21) which lies as far down as possible but still intersects the W-space. This point of tangency is unique once a_2 is chosen subject to the defining condition of Case 3. Hence, in this case, $\mathsf{W}(a)$ is a single point somewhere along the arc $C'D'$, which is determined, of course, by the choice of a.

Thus we have shown how to find the image $\mathsf{W}(a)$ of all points a along the right-hand boundary QDB of the U-space.

By a completely analogous analysis it is possible to find the image $\mathsf{U}(b)$ of any point b along the left-hand boundary $A'C'Q'$ of the W-space. We state without proof the following results of such an analysis.

Let $b = \| b_1 \quad b_2 \quad 1 \|$ be a point of the left-hand boundary $A'C'Q'$ of the W-space. If $b_2 \geq 0$, then $\mathsf{U}(b)$ is the point $\| 1 \quad 1 \quad 1 \|$ of U. If $b_2 < 0$, then we have the following conditions:

(i) If the slope $-(1/b_2)$ of the lines

$$u_1 + b_2 u_2 = k \tag{22}$$

is less than the slope $-m$ of DB, then $\mathsf{U}(b)$ is a single point somewhere along the arc QD of the boundary of the U-space.

(ii) If the slope of the lines of (21) is greater than $-m$, then $\mathsf{U}(b)$ is the point B of U.

(iii) If $-(1/b_2) = -m$, then $\mathsf{U}(b)$ is the entire line-segment DB.

We are now ready to use these results to determine which points are fixed-points.

First we notice that A' is not a fixed-point; for $\mathsf{U}(A') = B$, while $\mathsf{W}(B)$ is a point of the arc $C'Q'$. Consequently, by Theorem 11.7, no point a of U for which $\mathsf{W}(a) = A'$ can be a fixed-point.

Next we observe that no point of the arc $C'Q'$ is a fixed-point, for if b is such a point, $\mathsf{U}(b)$ consists of a single point a on the arc QD, and $\mathsf{W}(a)$, for any point a on the arc QD, is the single point A'. Consequently, by Theorem 11.7 no point a of U for which $\mathsf{U}(a)$ is a point of the arc $C'Q'$ can be a fixed-point. This means that no point $a = \| a_1 \quad a_2 \quad 1 \|$ of the segment DB for which

$$\left| \frac{1}{a_2} \right| < | m |$$

can be a fixed-point of U.

Hence we are forced to conclude that the only fixed-points of U are those points which fall under Case 2, above, i.e., those points a for which

$$\left| \frac{1}{a_2} \right| = | m |. \tag{23}$$

In a similar way, we conclude that the only fixed-points of W are those points b for which

$$\left| \frac{1}{b_2} \right| = | m |. \tag{24}$$

We wish now to compute the coordinates of the points of U (but there will turn out to be only one) for which (23) is true. It is clear that, since the slope of $A'C'$ is m, the slope of AC is also m; hence, from the symmetry of the U-space with respect to the line $u_1 = ½$, we see that the slope of DB

is $-m$. Thus the equation of DB is

$$u_2 - 1 = -m(u_1 - 1).$$

Consequently,

$$-\frac{1}{a_2} = \frac{-1}{1 - m(a_1 - 1)};$$

thus we have

$$\frac{-1}{1 - m(a_1 - 1)} = m,$$

from which we find that

$$a_1 = \frac{1 + m + m^2}{m^2},$$

$$a_2 = -\frac{1}{m}.$$

Hence the only fixed-point of the **U**-space is the point

$$\bar{u} = \left\| \frac{1 + m + m^2}{m^2} \quad -\frac{1}{m} \quad 1 \right\|.$$

Similarly, the only fixed-point of the **W**-space is the point

$$\bar{w} = \left\| \frac{1 - m}{m^2} \quad \frac{1}{m} \quad 1 \right\|.$$

The value of the game (to P_1) is now seen to be

$$E(\bar{u}, \bar{w}) = \bar{u}_1 + \bar{w}_1 + \bar{u}_2\bar{w}_2 = \frac{1 + m^2}{m^2}.$$

We wish, finally, to find distribution functions F and G which correspond to the fixed-points

$$\bar{u} = \left\| \frac{1 + m + m^2}{m^2} \quad -\frac{1}{m} \quad 1 \right\|$$

and

$$\bar{w} = \left\| \frac{1 - m}{m^2} \quad \frac{1}{m} \quad 1 \right\|.$$

The point

$$\bar{w} = \left\| \frac{1 - m}{m^2} \quad \frac{1}{m} \quad 1 \right\|$$

can be expressed as a convex linear combination of $A' = \| 0 \quad 1 \quad 1 \|$ and $C' = \| y_1 \quad \cos 2\pi y_1 \quad 1 \|$ by setting

$$w_1 = \frac{1 - m}{m^2} = (1 - t) \cdot 0 + ty_1,$$

from which we obtain

$$t = \frac{1 - m}{y_1 m^2}. \tag{25}$$

The distribution function I_0 corresponds to $A' = \| 0 \quad 1 \quad 1 \|$, and the distribution function I_{y_1} corresponds to $\| y_1 \quad \cos 2\pi y_1 \quad 1 \|$. Hence by Theorem 11.1 the function

$$G = (1 - t)I_0 + tI_{y_1}, \tag{26}$$

where y_1 and t are determined by (18), (19), and (25), corresponds to $C' = \| y_1 \quad \cos 2\pi y_1 \quad 1 \|$, and hence is an optimal strategy for P_2. (It is possible, indeed, to show that the distribution function G given by (26) is the *only* optimal strategy for P_2, but we shall omit the proof of this.)

Substituting in (25) for y_1 and m, we find that, to an accuracy of two places of decimals,

$$t = .73.$$

Hence the value of the game is approximately equal to 1.05, and an approximation to the best strategy for P_2 is

$$.27I_0 + .73I_{.37}.$$

Making use of the fixed-point $\bar{u}$, we can find the (unique) optimal strategy for P_1 in an analogous way.

3. Fixed-points. In the discussion of the above example we tacitly admitted the possibility that the U-space might contain a fixed-point on the line BD and also a fixed-point on the arc QD. It turned out, however, that such was not the case. We are now going to prove a theorem which shows that if there had been two fixed-points on BDQ, one not in BD, then it would have followed that there was a fixed-point in the interior of U.

THEOREM 11.9. For any separable game, the set of fixed-points of the U-space is closed and convex; the same is true of the set of fixed-points of the W-space.

PROOF. We shall prove the theorem for the U-space; the proof for the W-space is similar.

Let $x^{(1)}, x^{(2)}, \cdots$ be a sequence of fixed-points of the U-space which converges to the point y; we wish (in order to show that the set of fixed-points is closed) to prove that y is a fixed-point. Let the distribution function H_i (for $i = 1, 2, \cdots$) correspond to the point $x^{(i)}$. The sequence $H_1, H_2, \cdots$ of distribution functions contains (by Theorem 8.2) a subsequence $H_{i_1}, H_{i_2}, \cdots$ which converges to a distribution function F at every point of continuity of F. To simplify the notation, we set

$$F_j = H_{i_j} \qquad \text{for } j = 1, 2, \cdots,$$

and

$$y^{(j)} = x^{(i_j)} \qquad \text{for } j = 1, 2, \cdots.$$

Then $y^{(1)}, y^{(2)}, \cdots$ is a sequence of fixed-points of U which converges to the point y; the distribution function F_i (for $i = 1, 2, \cdots$) corresponds to the point $y^{(i)}$; and the sequence $F_1, F_2, \cdots$ converges to F at every point of continuity of F. By Theorem 9.23 we then conclude that the distribution function F corresponds to the point y. By Theorem 11.6 we see that each of the functions F_i, since it corresponds to the fixed-point $y^{(i)}$, is an optimal strategy for P_1; and by Theorem 10.21 it then follows that F is an optimal strategy for P_1. Since y corresponds to F, we conclude, again by Theorem 11.6, that y is a fixed-point, as was to be shown.

To see that the set of fixed-points of U is convex, let $x^{(1)}, \cdots, x^{(r)}$ be any fixed-points, let $F_1, \cdots, F_r$ be corresponding distribution functions, and let $\| a_1 \quad \cdots \quad a_r \|$ be any member of $\mathbf{S}_r$. Then by Theorem 11.6 each F_i (for $i = 1, \cdots, r$) is an optimal strategy for P_1. Hence by Theorem 10.21 the function F, defined by

$$F = a_1F_1 + \cdots + a_rF_r,$$

is an optimal strategy for P_1. Moreover, by Theorem 11.1 the point

$$x = a_1x^{(1)} + \cdots + a_rx^{(r)}$$

is a point of U, and it corresponds to F; thus, again by Theorem 11.6, x is a fixed-point, as was to be shown.

The methods which we have used up to this point are applicable to any representation of a separable game. We are now going to develop a few theorems which apply only to separable games represented in a certain special way, but which, when they do apply, will frequently yield the solution to the game more easily than the methods used heretofore.

We consider a separable game whose payoff function M is represented in the form

$$M(x, y) = \sum_{j=1}^{n} \sum_{i=1}^{n} a_{ij} r_i(x) s_j(y) + \sum_{i=1}^{n} b_i r_i(x) + \sum_{j=1}^{n} c_j s_j(y) + d, \qquad (27)$$

where the following conditions are satisfied: (i) The functions $r_1, \cdots, r_n, s_1, \cdots, s_n$ are continuous over $[0, 1]$, and (ii) the determinant

$$\begin{vmatrix} a_{11} & \cdots & a_{1n} \\ \vdots & & \vdots \\ a_{n1} & \cdots & a_{nn} \end{vmatrix}$$

is different from 0. Any representation of a separable function in the form (27) which satisfies these two conditions we shall call a *canonical representation* (or a *canonical form*) of M.

We note that form (27) is neither more nor less general than form (1), so long as we do not impose condition (ii). The advantage, for our purposes, of form (27) over form (1) is that the use of (27) often enables us to avoid assigning functional symbols r_i or s_j to constants, and thus it reduces the dimension of the spaces in which U and W are embedded.

REMARK 11.10. Since, as we have seen at the beginning of the chapter, every separable function can be written in the form

$$\sum_{j=1}^{n} t_j(x) s_j(y),$$

where the determinant has value 1, it is clear that every separable function has a canonical form.

EXAMPLE 11.11. Let M be defined by the equation

$$M(x, y) = xy - xe^y + 2x \cos y + 2e^x y + 3e^x e^y + e^x \cos y + 5 \cos x e^y - 3 \cos x \cos y.$$

If we set here

$$\begin{aligned} r_1(x) &= x, & s_1(y) &= y, \\ r_2(x) &= e^x, & s_2(y) &= e^y, \\ r_3(x) &= \cos x, & s_3(y) &= \cos y, \end{aligned}$$

then

$$\begin{vmatrix} a_{11} & a_{12} & a_{13} \\ a_{21} & a_{22} & a_{23} \\ a_{31} & a_{32} & a_{33} \end{vmatrix} = \begin{vmatrix} 1 & -1 & 2 \\ 2 & 3 & 1 \\ 0 & 5 & -3 \end{vmatrix} = 0.$$

But if we set

$$r_1'(x) = x - 2\cos x, \qquad s_1'(y) = y + \frac{7}{5}\cos y,$$

$$r_2'(x) = e^x + \cos x, \qquad s_2'(y) = e^y - \frac{3}{5}\cos y,$$

then we have

$$M(x, y) = r_1'(x)s_1'(y) - r_1'(x)s_2'(y) + 2r_2'(x)s_1'(y) + 3r_2'(x)s_2'(y),$$

and

$$\begin{vmatrix} a_{11} & a_{12} \\ a_{21} & a_{22} \end{vmatrix} = \begin{vmatrix} 1 & -1 \\ 2 & 3 \end{vmatrix} = 5 \neq 0,$$

so that this representation is canonical.

Now if

$$M(x, y) = \sum_{j=1}^{n} \sum_{i=1}^{n} a_{ij} r_i(x) s_j(y) + \sum_{i=1}^{n} b_i r_i(x) + \sum_{i=1}^{n} c_j s_j(y) + d \tag{27}$$

is the canonical representation of the payoff function of a separable game, then we can represent the expectation of P_1 as follows:

$$E(u, w) = \sum_{i=1}^{n} \sum_{j=1}^{n} a_{ij} u_i w_j + \sum_{i=1}^{n} b_i u_i + \sum_{j=1}^{n} c_j w_j + d$$

$$= \sum_{j=1}^{n} \left(\sum_{i=1}^{n} a_{ij} u_i + c_j \right) w_j + \sum_{i=1}^{n} b_i u_i + d$$

$$= \sum_{i=1}^{n} \left(\sum_{j=1}^{n} a_{ij} w_j + b_i \right) u_i + \sum_{j=1}^{n} c_j w_j + d.$$

Since

$$\begin{vmatrix} a_{11} & \cdots & a_{1n} \\ \vdots & & \vdots \\ a_{n1} & \cdots & a_{nn} \end{vmatrix} \neq 0,$$

the system of equations

$$\left.\begin{aligned} \sum_{i=1}^{n} a_{i1}u_i + c_1 &= 0, \\ &\vdots \\ \sum_{i=1}^{n} a_{in}u_i + c_n &= 0 \end{aligned}\right\} \tag{28}$$

has a unique solution, and so has the system of equations

$$\left.\begin{aligned} \sum_{i=1}^{n} a_{ij}w_j + b_1 &= 0, \\ &\vdots \\ \sum_{j=1}^{n} a_{nj}w_j + b_n &= 0. \end{aligned}\right\} \tag{29}$$

We shall call the solution of (28) the *first critical point* of the game (with respect to the given representation), and we shall call the solution of (29) the *second* critical point of the game (with respect to the given representation).

It can happen, of course, that the first critical point does not lie in the U-space; we know only that it lies somewhere in n-space, and U need not include all the n-space. Similarly, the second critical point does not necessarily lie in the W-space.

THEOREM 11.12. Let the payoff function M of a separable game be given in canonical form by (27), let $p = \| p_1 \quad \cdots \quad p_n \|$ and $q = \| q_1 \quad \cdots \quad q_n \|$ be the first and second critical points, and suppose that $p \in \mathsf{U}$ and $q \in \mathsf{W}$. Then p and q are fixed-points, and the value of the game is given by

$$w = \sum_{i=1}^{n} b_i p_i + d = \sum_{j=1}^{n} c_j q_j + d.$$

PROOF. By the definition of critical points we have

$$\sum_{i=1}^{n} a_{ij}p_i + c_j = 0 \qquad \text{for } j = 1, \cdots, n, \tag{30}$$

and

$$\sum_{j=1}^{n} a_{ij}q_j + b_i = 0 \qquad \text{for } i = 1, \cdots, n. \tag{31}$$

We have also, for any point $w = \| w_1 \quad \cdots \quad w_n \|$ of W,

$$E(p, w) = \sum_{j=1}^{n} \left(\sum_{i=1}^{n} a_{ij}p_i + c_j \right) w_j + \sum_{i=1}^{n} b_i p_i + d,$$

or, in view of (30),

$$E(p, w) = \sum_{i=1}^{n} b_i p_i + d.$$

Since this is independent of w, it follows that

$$\mathsf{W}(p) = \mathsf{W}.$$

Similarly,

$$\mathsf{U}(q) = \mathsf{U}.$$

Hence $p \in \mathsf{U}(q)$ and $q \in \mathsf{W}(p)$, so that p and q are fixed-points, and

$$E(p, q) = \sum_{i=1}^{n} b_i p_i + d = \sum_{j=1}^{n} c_j q_j + d$$

is the value of the game.

REMARK 11.13. It is important to find a canonical representation of M which involves as few functions r_i and s_j as possible, since the following theorem becomes vacuous if we use more functions r_i and s_j than necessary. If M is expressed by (27), then U and W are subsets of n-dimensional Euclidean space; and if the functions r_i, for example, are linearly dependent, then U will lie in a hyperplane of n-dimensional Euclidean space, and hence its interior, $I(\mathsf{U})$, will be empty, so that the hypothesis of the second part of the theorem cannot be satisfied.

THEOREM 11.14. *Let the payoff function M of a separable game be given in canonical form by (27), and suppose that $I(\mathsf{W})$*

(the interior of W) contains a fixed-point; then the first critical point belongs to U and is the only fixed-point of U. Similarly, if $I(\mathsf{U})$ contains a fixed-point, then the second critical point belongs to W and is the only fixed-point of W.

PROOF. We prove only the first part of the theorem; the proof of the second part is similar.

Let $t = \| t_1 \quad \cdots \quad t_n \|$ be a fixed-point of W which lies in $I(\mathsf{W})$. Let $z = \| z_1 \quad \cdots \quad z_n \|$ be any fixed-point of U and let $p = \| p_1 \quad \cdots \quad p_n \|$ be the first critical point; we wish to show that $z = p$. Suppose, then, if possible, that $z \neq p$. Since p is the unique solution of (28), we have, for some $k \leq n$,

$$\sum_{i=1}^{n} a_{ik} z_i + c_k = g \neq 0. \tag{32}$$

Let h be a real number of opposite algebraic sign to g, and small enough to ensure that the point

$$\bar{t} = \| t_1 \quad \cdots \quad t_{k-1} \quad t_k + h \quad t_{k+1} \quad \cdots \quad t_n \|$$

lies in W.

Now

$$E(z, t) = \sum_{j=1}^{n} \left(\sum_{i=1}^{n} a_{ij} z_i + c_j \right) t_j + \sum_{i=1}^{n} b_i z_i + d,$$

and

$$\begin{aligned} E(z, \bar{t}) &= E(z, t) + h \cdot \left(\sum_{i=1}^{n} a_{ik} z_i + c_k \right) \\ &= E(z, t) + h \cdot g < E(z, t). \end{aligned} \tag{33}$$

But this means that $t \notin \mathsf{W}(z)$, and by Theorem 11.7 this contradicts our assumption that both z and t are fixed-points.

The following corollaries are immediate consequences of Theorems 11.12 and 11.14.

COROLLARY 11.16. Let the payoff function of a separable game be given in canonical form, and suppose that the first critical point does not belong to the U-space; then every fixed-point of W is in $B(\mathsf{W})$ (the boundary of W). Similarly, if the second critical point does not belong to W, then every fixed-point of U is in $B(\mathsf{U})$.

COROLLARY 11.16. Let the payoff function of a separable game be given in canonical form, let p and q, respectively, be the first critical point and the second critical point, and suppose that $p \in I(\mathsf{U})$ and $q \in I(\mathsf{W})$; then p is the only fixed-point of U and q is the only fixed-point of W.

REMARK 11.17. We note here that the payoff function M of Example 11.8 can be put in canonical form by setting

$$\begin{aligned} r_1(x) &= \cos 2\pi x, & s_1(y) &= \cos 2\pi y, \\ r_2(x) &= x, & s_2(y) &= y, \\ r_3(x) &= -x, & s_3(y) &= y, \end{aligned}$$

so that we have

$$M(x, y) = r_1(x)s_1(y) + r_2(x)s_2(y) + r_3(x)s_3(y) - r_3(x) + s_3(y).$$

Here we find

$$\begin{vmatrix} a_{11} & a_{12} & a_{13} \\ a_{21} & a_{22} & a_{23} \\ a_{31} & a_{32} & a_{33} \end{vmatrix} = \begin{vmatrix} 1 & 0 & 0 \\ 0 & 1 & 0 \\ 0 & 0 & 1 \end{vmatrix} = 1 \neq 0.$$

But with this choice of r_i and s_j we find that the first critical point does not belong to U and that the second critical point does not belong to W; hence the propositions 11.12, 11.14, 11.15, and 11.16 tell us only that the fixed-points must belong to the boundaries of U and W. But $B(\mathsf{U}) = \mathsf{U}$, since $r_2(x)$ and $r_3(x)$ are linearly dependent; and, similarly, $B(\mathsf{W}) = \mathsf{W}$. Hence the propositions 11.12, 11.14, 11.15, and 11.16 actually give us no information at all.

4. Further Examples.

EXAMPLE 11.18. The payoff function M of a separable game satisfies the following equation (for any point $\| x \quad y \|$ of the closed unit square):

$$M(x, y) = 3 \cos 7x \cos 8y + 5 \cos 7x \sin 8y + 2 \sin 7x \cos 8y + \sin 7x \sin 8y.$$

Here we shall take

$$\begin{aligned} r_1(x) &= \cos 7x, & s_1(y) &= \cos 8y, \\ r_2(x) &= \sin 7x, & s_2(y) &= \sin 8y, \end{aligned}$$

so that M is represented as follows:

$$M(x, y) = 3r_1(x)s_1(y) + 5r_1(x)s_2(y) + 2r_2(x)s_1(y) + r_2(x)s_2(y).$$

Thus both the U-space and the W-space are 2-dimensional.

In order to find U*, we plot the curve whose parametric equations are

$$u_1 = \cos 7t,$$
$$u_2 = \sin 7t.$$

This curve is immediately seen to be the circle whose radius is 1 and whose center is at the origin. (Since we plot the curve for $0 \leq t \leq 1$ and since $7 > 2\pi$, part of the circle is traced twice, but this is, at present, a matter of indifference to us.) Therefore U (the convex hull of U*) is this circle, together with its interior.

Similarly, since $8 > 2\pi$, the curve W* is the same as U*, and W is the same as U.

It is clear that the given representation is canonical, since

$$\begin{vmatrix} 3 & 5 \\ 2 & 1 \end{vmatrix} = -7 \neq 0.$$

In this example, (28) becomes

$$3u_1 + 2u_2 = 0,$$
$$5u_1 + u_2 = 0,$$

and (29) becomes

$$3w_1 + 5w_2 = 0,$$
$$2w_1 + w_2 = 0.$$

The solutions of these systems are clearly $p = \| 0 \quad 0 \|$ and $q = \| 0 \quad 0 \|$. Thus the first critical point belongs to $I(\mathsf{U})$, and the second critical point belongs to $I(\mathsf{W})$. Hence, by Corollary 11.16, p is the only fixed-point of U and q is the only fixed-point of W.

Thus the problem of finding solutions of our game reduces to the problem of finding distribution functions to which the point $\| 0 \quad 0 \|$ corresponds. To find a particular optimal strategy for P_1, we notice that the point $\| 0 \quad 0 \|$ is a convex linear combination of the two points $\| 1 \quad 0 \|$ and $\| -1 \quad 0 \|$; for we have

$$\| 0 \quad 0 \| = \frac{1}{2} \| 1 \quad 0 \| + \frac{1}{2} \| -1 \quad 0 \|. \tag{34}$$

Since $\| 1 \quad 0 \|$ is a point of U^*, there is a step-function I_t with one step which corresponds to this point; thus we have

$$1 = \int_0^1 r_1(x)\, dI_t(x) = r_1(t) = \cos 7t,$$

$$0 = \int_0^1 r_2(x)\, dI_t(x) = r_2(t) = \sin 7t.$$

These equations, together with the inequality

$$0 \leq t \leq 1,$$

imply that either $t = 0$ or $t = 2\pi/7$; thus the point $\| 1 \quad 0 \|$ corresponds to either of the two step-functions I_0 or $I_{2\pi/7}$. In a similar way we see that the point $\| -1 \quad 0 \|$ corresponds to the step-function $I_{\pi/7}$. From these results, together with (34), we now conclude by means of Theorem 11.1 that the distribution function

$$\frac{1}{2} I_0 + \frac{1}{2} I_{\pi/7}$$

is an optimal strategy for P_1, and the same is, of course, also true of the distribution function

$$\frac{1}{2} I_{2\pi/7} + \frac{1}{2} I_{\pi/7}.$$

Applying the same argument more generally, we find that the most general step-function with two steps to which the point $\| 0 \quad 0 \|$ of U corresponds is

$$\frac{1}{2} I_{t_1} + \frac{1}{2} I_{t_2},$$

where $0 \leq t_1 < t_2 \leq 1$ and $t_2 - t_1 = \pi/7$. Thus we have an infinite family of optimal strategies for P_1.

Similarly, the most general optimal strategy for P_2, which involves only two steps, can be written

$$\frac{1}{2} I_{t_1} + \frac{1}{2} I_{t_2},$$

where $0 \leq t_1 < t_2 \leq 1$ and $t_2 - t_1 = \pi/8$.

There are, however, also optimal strategies with more than two steps. If $u^{(1)}, \cdots, u^{(p)}$ are points of U^* such that $\| 0 \quad 0 \|$ lies in the convex hull of the set $\{u^{(1)}, \cdots, u^{(p)}\}$, then we can express $\| 0 \quad 0 \|$ as a convex linear combination of $u^{(1)}, \cdots, u^{(p)}$, and we can find a step-function with p steps corresponding to $\| 0 \quad 0 \|$. Thus, for example, suppose we consider the three points $\| 1 \quad 0 \|$, $\| 0 \quad 1 \|$, and $\| -\frac{1}{2} \quad -\sqrt{3}/2 \|$. To express $\| 0 \quad 0 \|$ as a convex linear combination of these three points, we must find non-negative numbers a_1, a_2, and a_3 satisfying

$$a_1 + a_2 + a_3 = 1,$$

$$a_1(1) + a_2(0) + a_3\left(-\frac{1}{2}\right) = 0,$$

$$a_1(0) + a_2(1) + a_3\left(-\frac{\sqrt{3}}{2}\right) = 0.$$

Solving the three equations simultaneously, we obtain

$$\left.\begin{aligned} a_1 &= \frac{3 - \sqrt{3}}{6}, \\ a_2 &= \frac{\sqrt{3} - 1}{2}, \\ a_3 &= \frac{3 - \sqrt{3}}{3}. \end{aligned}\right\} \tag{35}$$

Since it is easily seen that I_0 corresponds to $\| 1 \quad 0 \|$, that $I_{\pi/14}$ corresponds to $\| 0 \quad 1 \|$, and that $I_{4\pi/21}$ corresponds to $\| -\frac{1}{2} \quad -\sqrt{3}/2 \|$, we therefore conclude that the distribution function

$$a_1 I_0 + a_2 I_{\pi/14} + a_3 I_{4\pi/21},$$

where a_1, a_2, and a_3 are given by (27), is an optimal strategy for P_1.

There are, of course, also optimal strategies which are not step-functions. Thus the distribution function F, such that

$$F(x) = \frac{7x}{2\pi} \qquad \text{for } 0 \leq x \leq \frac{2\pi}{7},$$

$$F(x) = 1 \qquad \text{for } \frac{2\pi}{7} < x \leq 1,$$

is a continuous optimal strategy for P_1.

It is clear from Theorem 11.12 that the value of this game to P_1 is 0.

EXAMPLE 11.19. The payoff function of a separable game satisfies (for any point $\|x \quad y\|$ of the closed unit square)

$$\begin{aligned} M(x, y) = {} & 3 \cos 4x \cos 5y + 5 \cos 4x \sin 5y \\ & + \sin 4x \cos 5y + \sin 4x \sin 5y \\ & + 4 \cos 4x + \sin 4x + \cos 5y + 2 \sin 5y + 3. \end{aligned}$$

Here we shall take

$$r_1(x) = \cos 4x, \qquad s_1(y) = \cos 5y,$$
$$r_2(x) = \sin 4x, \qquad s_2(y) = \sin 5y,$$

so that M can be represented as follows:

$$M(x, y) = 3r_1(x)s_1(y) + 5r_1(x)s_2(y) + r_2(x)s_1(y) + r_2(x)s_2(y) + 4r_1(x) + r_2(x) + s_1(y) + 2s_2(y) + 3.$$

We note that the determinant

$$\begin{vmatrix} 3 & 5 \\ 1 & 1 \end{vmatrix} \neq 0,$$

and hence M is represented in canonical form.

Remembering that x is allowed to vary only between 0 and 1, we see that the curve **U*** is the part of the circle

$$u_1^2 + u_2^2 = 1,$$

indicated in Fig. 2 by the arc ABC (so that $\angle AOC$ is 4 radians), and that the **U**-space is the shaded region. Similarly, the **W**-space is the shaded region in Fig. 3.

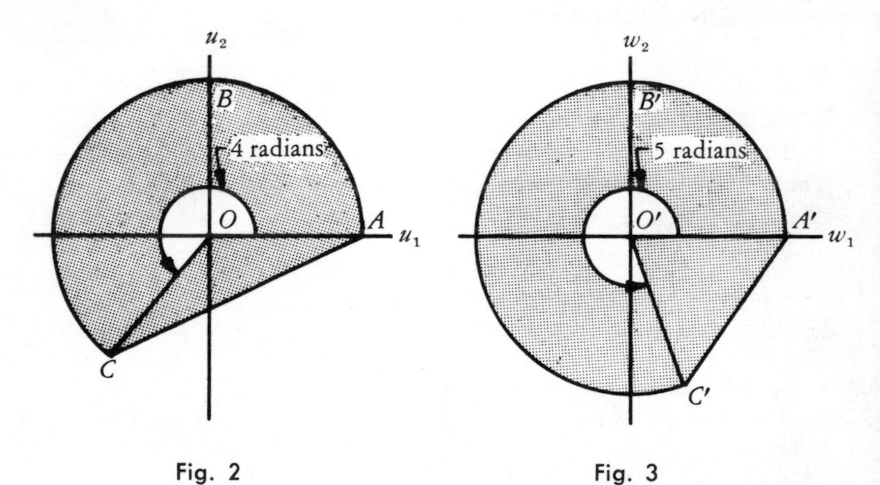

Fig. 2 Fig. 3

The first critical point is obtained by solving the equations

$$3u_1 + u_2 + 1 = 0,$$
$$5u_1 + u_2 + 2 = 0,$$

thus obtaining $\| u_1 \quad u_2 \| = \| -\frac{1}{2} \quad \frac{1}{2} \|$, which is an interior point of $\mathbf{U}$. Similarly, the second critical point is $\| w_1 \quad w_2 \| = \| -\frac{1}{2} \quad -\frac{1}{2} \|$, which is an interior point of $\mathbf{W}$. Thus it follows from Corollary 11.16 that $\| -\frac{1}{2} \quad \frac{1}{2} \|$ is the only fixed-point of $\mathbf{U}$ and that $\| -\frac{1}{2} \quad -\frac{1}{2} \|$ is the only fixed-point of $\mathbf{W}$.

To find solutions of this game, therefore, it is necessary only to find distribution functions which correspond to $\| -\frac{1}{2} \quad \frac{1}{2} \|$ and $\| -\frac{1}{2} \quad -\frac{1}{2} \|$; this can be done in a manner similar to that used in Example 11.18.

EXAMPLE 11.20. The payoff function M of a separable game satisfies the following equation (for any point $\| x \quad y \|$ of the closed unit square):

$$\begin{aligned} M(x, y) = {} & 3[1 + \cos 2\pi x] \cdot [1 + \cos 2\pi y] \\ & + 5[1 + \cos 2\pi x][1 + \sin 2\pi y] \\ & + 2[1 + \sin 2\pi x] \cdot [1 + \cos 2\pi y] \\ & + [1 + \sin 2\pi x][1 + \sin 2\pi y]. \end{aligned}$$

Setting

$$\begin{aligned} r_1(x) &= 1 + \cos 2\pi x, & s_1(y) &= 1 + \cos 2\pi y, \\ r_2(x) &= 1 + \sin 2\pi x, & s_2(y) &= 1 + \sin 2\pi y, \end{aligned}$$

we have

$$M(x, y) = 3r_1(x)s_1(y) + 5r_1(x)s_2(y) + 2r_2(x)s_1(y) + r_2(x)s_2(y).$$

It is now readily verified that the $\mathbf{U}$-space consists of points on, or within, the circle

$$(u_1 - 1)^2 + (u_2 - 1)^2 = 1,$$

and that the $\mathbf{W}$-space consists of points on, or within, the circle

$$(w_1 - 1)^2 + (w_2 - 1)^2 = 1.$$

The first and second critical points are $\| 0 \quad 0 \|$. Since the origin is not in the $\mathbf{U}$-space, and also not in the $\mathbf{W}$-space, we see from Corollary 11.15 that the only fixed-points of $\mathbf{U}$ are on the boundary—and similarly for $\mathbf{W}$. Moreover, the boundary of $\mathbf{U}$ is simply the circle

$$(u_1 - 1)^2 + (u_2 - 1)^2 = 1.$$

If $\mathbf{U}$ contained two distinct fixed-points (both of which would have to be on this circle), then by Theorem 11.9 all points on the chord joining these two fixed-points would also be fixed-points; this would contradict the fact that every fixed-point of $\mathbf{U}$ lies on the boundary of $\mathbf{U}$. Hence $\mathbf{U}$ contains just one fixed-point. Similarly, the $\mathbf{W}$-space contains just one fixed-point.

Since, for this example, U^* coincides with $B(\mathsf{U})$, we conclude, finally, that there is an optimal pure strategy for the first player and, similarly, that there is an optimal pure strategy for the second player. Thus we conclude that the game has a saddle-point.

The following example shows, on the other hand, that when $B(\mathsf{U})$ and $B(\mathsf{W})$ contain straight line-segments, then there may be infinitely many fixed-points—even though the first critical point does not belong to U and the second critical point does not belong to W.

EXAMPLE 11.21. The payoff function of a separable game satisfies (for any point $\| x \quad y \|$ of the closed unit square) the following equation:

$$M(x, y) = 5[2 + \cos\pi(1 + x)][\sin\pi y] + 2[\sin\pi(1 + x)][-2 + \cos\pi y] + [\sin\pi(1 + x)][\sin\pi y].$$

Setting

$$r_1(x) = 2 + \cos\pi(1 + x), \qquad s_1(y) = -2 + \cos\pi y,$$
$$r_2(x) = \sin\pi(1 + x), \qquad s_2(y) = \sin\pi y,$$

we have

$$M(x, y) = 5r_1(x)s_2(y) + 2r_2(x)s_1(y) + r_2(x)s_2(y),$$

and hence

$$E(u, v) = 5u_1v_2 + 2u_2v_1 + u_2v_2.$$

The first critical point is $\| 0 \quad 0 \|$, and so is the second. We now easily verify that the U-space and the W-space are the regions indicated by the shading in Figs. 4 and 5, respectively. Moreover, every point of the segment

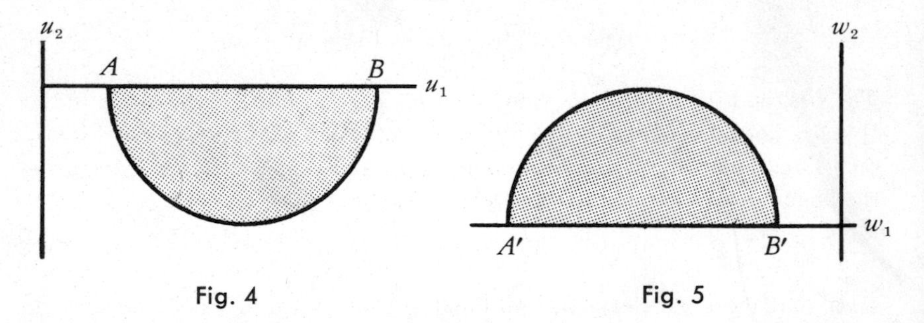

Fig. 4 Fig. 5

AB is a fixed-point of U, and every point of the segment $A'B'$ is a fixed-point of W.

In Example 11.20 we saw how one could make use of Corollary 11.15

in order to treat certain games where neither the **U**-space nor the **W**-space contains its critical point. The following example illustrates how this corollary can be used also in connection with games where one critical point lies in the space corresponding to it, but the other does not.

EXAMPLE 11.22. The payoff function M of a separable game satisfies the following equation (for any point $\| x \quad y \|$ of the closed unit square):

$$M(x, y) = 3[\cos 2\pi x][2 + \cos 2\pi y] + 5[\cos 2\pi x][2 + \sin 2\pi y] + 2[\sin 2\pi x][2 + \cos 2\pi y] + [\sin 2\pi x][2 + \sin 2\pi y].$$

Setting

$$r_1(x) = \cos 2\pi x, \qquad s_1(y) = 2 + \cos 2\pi y,$$
$$r_2(x) = \sin 2\pi x, \qquad s_2(y) = 2 + \sin 2\pi y,$$

we readily verify that the **U**-space consists of the circle

$$u_1^2 + u_2^2 = 1,$$

together with its interior, and that the **W**-space consists of the circle

$$(w_1 - 2)^2 + (w_2 - 2)^2 = 1,$$

together with its interior. In each case the critical point is $\| 0 \quad 0 \|$. This point is not in **W**, but it lies in the interior of **U**.

Since the second critical point does not belong to **W**, we see by Corollary 11.15 that the only fixed-points of **U** are in $B(\mathbf{U})$. By the same argument, which was used for this purpose in Example 11.20, we prove that there is just one fixed-point, say u^*, in **U**. Now if there were a fixed-point of **W** in $I(\mathbf{W})$, then by Theorem 11.14 the first critical point $\| 0 \quad 0 \|$ would have to be a fixed-point of **U**, which is contrary to the fact that the only fixed-point of **U** lies on $B(\mathbf{U})$. Hence the only fixed-points of **W** must lie on $B(\mathbf{W})$; and, since **W** is a circle plus its interior, we conclude, again by the argument used in Example 11.20, that there is just one fixed-point in **W**. Thus, as in Example 11.20, we see that the game has a saddle-point.

5. Rectangular Game Solved as a Separable Game. In this section we show, by an example, that a rectangular game can be solved by the methods used for separable games. Although the example involves a 3×3 payoff matrix, the extension to arbitrary payoff matrices will be obvious.

Consider the following payoff matrix:

$$A = \begin{Vmatrix} -3 & 2 & 0 \\ 0 & 1 & 2 \\ 1 & 2 & 1 \end{Vmatrix}.$$

A mixed strategy for P_1 is a vector $u = \| u_1 \quad u_2 \quad 1 - u_1 - u_2 \|$, where $0 \leq u_i \leq 1$ (for $i = 1, 2$) and $\sum_{i=1}^{2} u_i \leq 1$. A mixed strategy for P_2 is a vector $w = \| w_1 \quad w_2 \quad 1 - w_1 - w_2 \|$, where $0 \leq w_i \leq 1$ (for $i = 1, 2$) and $\sum_{i=1}^{2} w_i \leq 1$. The expectation $E(u, w)$ is given by

$$
\begin{aligned}
E(u, w) &= (-3u_1 - 2u_2)w_1 + (u_1 - 2u_2 + 1)w_2 + (-u_1 + u_2 + 1) \\
&= (-3w_1 + w_2 - 1)u_1 + (-2w_1 - 2w_2 + 1)u_2 + (w_2 + 1).
\end{aligned}
$$

First, we note that the payoff (expectation) is a bilinear form in the coordinates of u and w. From the above inequalities, we also see that the U-space is an isosceles right triangle with legs 1 unit long and that the W-space is the same as the U-space. Secondly, it follows that the U-space and W-space are closed, bounded, convex sets. Thus it is as though we had the following separable game: P_1 picks a point $u \in \mathsf{U}$ and P_2 picks a point $w \in \mathsf{W}$ and the payoff is the bilinear form $E(u, w)$.

To solve the game, we examine U and W for fixed-points. In determining the image $\mathsf{W}(u)$ of each point $u \in \mathsf{U}$, it is convenient to set

$$
p_1 = -3u_1 - 2u_2, \qquad p_2 = u_1 - 2u_2 + 1.
$$

Table 1 presents the images on W of each point u in U.

Table 1

IMAGES OF POINTS OF U

For Region	Defined by	The Image $\mathsf{W}(u)$ of u is
U_1	$p_1 > 0, \quad p_2 > 0$	$\| 0 \quad 0 \|$
U_2	$p_1 < 0, \quad p_1 < p_2$	$\| 1 \quad 0 \|$
U_3	$p_2 < 0, \quad p_2 < p_1$	$\| 0 \quad 1 \|$
U_4	$p_1 = p_2 < 0$	$\| \alpha \quad 1 - \alpha \|$, where $0 \leq \alpha \leq 1$
U_5	$p_1 = 0 < p_2$	$\| \alpha \quad 0 \|$, where $0 \leq \alpha \leq 1$
U_6	$p_2 = 0 < p_1$	$\| 0 \quad \alpha \|$, where $0 < \alpha < 1$
U_7	$p_1 = p_2 = 0$	$\| \alpha \quad \beta \|$, where $0 \leq \alpha, \beta \leq 1, \alpha + \beta \leq 1$

Let us now set

$$
q_1 = -3w_1 + w_2 - 1, \qquad q_2 = -2w_1 - 2w_2 + 1;
$$

then Table 2 presents the images on U of each point w in W.

Table 2

IMAGES OF POINTS OF W

For Region	Defined by	The Image $\mathsf{U}(w)$ of w is	
W_1	$q_1 < 0, \quad q_2 < 0$	$\Vert 0 \quad 0 \Vert$	
W_2	$q_2 > 0, \quad q_1 < q_2$	$\Vert 0 \quad 1 \Vert$	
W_3	$q_1 > 0, \quad q_2 < q_1$	$\Vert 1 \quad 0 \Vert$	
W_4	$q_1 = q_2 > 0$	$\Vert a \quad 1 - a \Vert$,	where $0 \leq a \leq 1$
W_5	$q_1 = 0 > q_2$	$\Vert a \quad 0 \Vert$,	where $0 \leq a \leq 1$
W_6	$q_2 = 0 > q_1$	$\Vert 0 \quad a \Vert$,	where $0 \leq a \leq 1$
W_7	$q_1 = q_2 = 0$	$\Vert a \quad b \Vert$,	where $0 \leq a, b \leq 1, a + b \leq 1$

Using these tables we can determine which points of U and of W are fixed-points and therefore which are the optimal strategies for P_1 and for P_2. For example, we notice that no point of U_1 is a fixed-point; for $\mathsf{W}(u) = \Vert 0 \quad 0 \Vert$, while $\mathsf{U}(\Vert 0 \quad 0 \Vert) = \Vert 0 \quad 1 \Vert$, which belongs to U_2 and not to U_1. In a like manner we can show that no point of U_2, U_3, or U_4 is a fixed-point. Now consider the region U_5. For this region we have $\mathsf{W}(u) = \Vert \alpha \quad 0 \Vert$. Letting $w = \Vert \alpha \quad 0 \Vert$, we have

$$\mathsf{U}(w) = \begin{cases} \Vert 0 \quad 0 \Vert & \text{if } \alpha \geq \dfrac{1}{2}, \\ \Vert 0 \quad a \Vert & \text{if } \alpha = \dfrac{1}{2}, \\ \Vert 0 \quad 1 \Vert & \text{if } \alpha < \dfrac{1}{2}. \end{cases}$$

Since $u = \Vert 0 \quad 0 \Vert \in \mathsf{U}_5$, it follows that $\Vert 0 \quad 0 \Vert$ is a fixed-point of U and that $\Vert \alpha \quad 0 \Vert$, where $\alpha \geq 1/2$, is a fixed-point of W. In other words, we have $\Vert 0 \quad 0 \Vert \in \mathsf{U}(\Vert \alpha \quad 0 \Vert)$ and $\Vert \alpha \quad 0 \Vert \in \mathsf{W}(\Vert 0 \quad 0 \Vert)$, where $\alpha \geq 1/2$. It is easily verified that these are the only fixed-points. Therefore, the optimal strategies of the game are given by

$$\Vert 0 \quad 0 \quad 1 \Vert \qquad (\text{for } P_1),$$

$$\Vert w_1 \quad 0 \quad 1 - w_1 \Vert, \quad \text{where } \frac{1}{2} \leq w_1 \leq 1 \qquad (\text{for } P_2).$$

The value of the game is 1.

6. Constrained Game Solved as a Separable Game. Suppose that the mixed strategies of a rectangular game must satisfy some further linear inequalities in addition to those defining a mixed strategy; we then have a *game with constraints.* Such games arise frequently in mathematical statistics (see Chap. 13) when we wish to consider statistical decision problems as games played between the statistician and nature—for the statistician often has enough experience with the past behavior of nature in a given domain that he is able to put at least upper and lower bounds to the frequencies with which things will happen.

The additional linear inequalities, in the case of a game with restraints, alter the U-space and the W-space, but the spaces still remain closed, bounded, and convex. The payoff, of course, remains unchanged and bilinear. Therefore, the methods of separable games are applicable. We can solve the game by examining for fixed-points.

For example, consider the game discussed in the preceding section. Suppose that we impose the following additional linear constraints on the mixed strategies:

$$u_1 \geq \frac{1}{3}, \qquad u_2 \leq \frac{1}{2}, \qquad w_2 - w_1 + \frac{2}{3} \leq 0.$$

The expectation remains the same as before, namely,

$$E(u, w) = (-3u_1 - 2u_2)w_1 + (u_1 - 2u_2 + 1)w_2 + (-u_1 + u_2 + 1).$$

The additional inequalities imply that the U-space is the quadrilateral *ABCD* (shaded area) as shown in Fig. 6 and that the W-space is the triangle *LMN* (shaded area) as shown in Fig. 7.

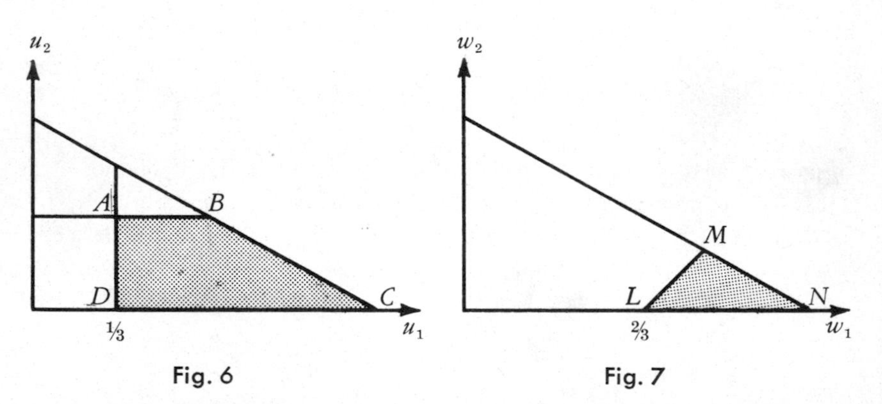

Fig. 6 Fig. 7

We solve this game by examining for fixed-points in the U-space and the W-space. From the geometry of the U-space, it is obvious that, for the entire

U-space, we have

$$p_1 < 0 < p_2.$$

Therefore the image $\mathsf{W}(u)$ of every point u in U is $\| 1 \quad 0 \|$. Now $\mathsf{U}(\| 1 \quad 0 \|) = \| \tfrac{1}{3} \quad 0 \|$. Therefore $\| \tfrac{1}{3} \quad 0 \|$ and $\| 1 \quad 0 \|$ are the fixed-points of U and of W. The optimal strategies are

$$\left\| \frac{1}{3} \quad 0 \quad \frac{2}{3} \right\| \qquad (\text{for } P_1),$$

$$\| 1 \quad 0 \quad 0 \| \qquad (\text{for } P_2).$$

The value of the game is $-\tfrac{1}{3}$.

HISTORICAL AND BIBLIOGRAPHICAL REMARKS

The work of this chapter is based largely on results to be found in Dresher, Karlin, and Shapley [1] and Dresher and Karlin [1], together with some private communications from Dresher. See also Dresher [1].

EXERCISES

1. Show that every separable game whose payoff function can be written in the form

$$M(x, y) = r(x)s(y) + ar(x) + bs(y) + c$$

has a saddle-point.

2. Complete the discussion of Example 11.19 by finding some optimal strategies for the two players.

3. Under what conditions on A, B, C, and D will the separable game whose payoff function is

$$M(x, y) = A \cos 7x \cos 8y + B \cos 7x \sin 8y + C \sin 7x \cos 8y + D \sin 7x \sin 8y$$

have exactly the same optimal strategies as has the game of Example 11.18?

4. Solve the separable game whose payoff function is

$$M(x, y) = 3 \cos \frac{\pi}{2} x \cos \frac{\pi}{2} y + 5 \cos \frac{\pi}{2} x \sin \frac{\pi}{2} y + 2 \sin \frac{\pi}{2} x \cos \frac{\pi}{2} y + \sin \frac{\pi}{2} x \sin \frac{\pi}{2} y.$$

5. Solve the separable game whose payoff function is

$$M(x, y) = (x - y)^2.$$

6. Solve the separable game whose payoff function is

$$M(x, y) = \left(x - \frac{1}{2}\right)\left(y - \frac{1}{2}\right) + 2\left(x - \frac{1}{2}\right)\left(y^3 - \frac{1}{4}\right)$$
$$+ 5\left(x^3 - \frac{1}{4}\right)\left(y - \frac{1}{2}\right) + 10\left(x^3 - \frac{1}{4}\right)\left(y^3 - \frac{1}{4}\right).$$

7. Solve the separable game whose payoff function is

$$M(x, y) = (x - 2)\left(y - \frac{1}{2}\right) + 2(x - 2)\left(y^3 - \frac{1}{4}\right)$$
$$+ 5(x^3 - 2)\left(y - \frac{1}{2}\right) + 10(x^3 - 2)\left(y^3 - \frac{1}{4}\right).$$

8. Solve the separable game whose payoff function is

$$M(x, y) = \cos 2\pi x \cos 2\pi y + x + 2y.$$

9. Solve the separable game whose payoff function is

$$M(x, y) = \cos 4\pi x \cos 4\pi y + x + y.$$

10. Prove the following theorem:
Let the payoff function of a separable game have a canonical representation

$$M(x, y) = \sum_{i=1}^{n} \sum_{j=1}^{n} a_{ij} r_i(x) s_j(y)$$
$$+ \sum_{i=1}^{n} b_i r_i(x) + \sum_{i=1}^{n} c_i s_j(y) + d,$$

and let u be any point of the **U**-space which is different from the first critical point; then

$$\mathbf{W}(u) \subseteq B(\mathbf{W}).$$

Similarly, if w is any point of the **W**-space which is different from the second critical point, then

$$\mathbf{U}(w) \subseteq B(\mathbf{U}).$$

11. Prove the following theorem:
If every fixed-point of the $\mathbf{U}$-space of a separable game belongs to $B(\mathbf{U})$, and if $B(\mathbf{U})$ contains no straight line-segments, then there is just one fixed-point in the $\mathbf{U}$-space. Similarly, if every fixed-point of the $\mathbf{W}$-space belongs to $B(\mathbf{W})$, and if $B(\mathbf{W})$ contains no straight line-segments, then there is just one fixed-point in the $\mathbf{W}$-space.

12. Let the payoff function of a separable game be

$$M(x, y) = Ae^x e^y + Be^x e^{y^2} + Ce^{x^2} e^y + De^{x^2} e^{y^2} + Ee^x + Fe^{x^2} + Ge^y + He^{y^2} + K,$$

where

$$\begin{vmatrix} A & B \\ C & D \end{vmatrix} \neq 0.$$

Find solutions of the game for various values of A, B, C, D, E, F, G, H, and K.

13. Find the solutions of the game whose payoff is

$$\begin{Vmatrix} 1 & 2 \\ 3 & 0 \\ 2 & 1 \end{Vmatrix}$$

and whose mixed strategies are subject to the following constraints:

$$\frac{1}{5} \leq y_1 \leq \frac{4}{5}, \qquad \frac{1}{3} \leq x_1 + x_2 \leq \frac{1}{2}.$$

14. Solve the game whose payoff is

$$\begin{Vmatrix} 3 & 39 & 30 \\ 33 & 9 & 0 \\ 28 & 4 & 25 \end{Vmatrix}$$

and whose mixed strategies are subject to the following constraints:

$$\frac{1}{10} \leq x_1 \leq \frac{4}{5}, \qquad \frac{1}{20} \leq x_2 \leq \frac{1}{2}, \qquad x_3 \geq \frac{6}{5}\left(1 - \frac{20}{9} x_1\right),$$

$$\frac{1}{10} \leq y_2 \leq 2y_1, \qquad y_3 \geq \frac{1}{6}.$$

CHAPTER 12

GAMES WITH CONVEX PAYOFF FUNCTIONS

1. Convex Functions. Besides games with separable payoff functions, another class of games for which it is relatively easy to find solutions is the class of games in which the payoff function is continuous and convex in one variable. In this section we shall present some of the known results for this case.

A function f of a real variable is called *convex* in an interval (a, b) if, for every member $\| \lambda_1 \quad \lambda_2 \|$ of $\mathbf{S}_2$ and for every pair of distinct numbers x_1 and x_2 of (a, b), we have

$$f(\lambda_1 x_1 + \lambda_2 x_2) \leq \lambda_1 f(x_1) + \lambda_2 f(x_2) .$$

If the equality never holds for $\lambda_1 \neq 0 \neq \lambda_2$, we call f *strictly convex.*

To understand the geometrical significance of the notion of convexity, consider the diagram in Fig. 1.

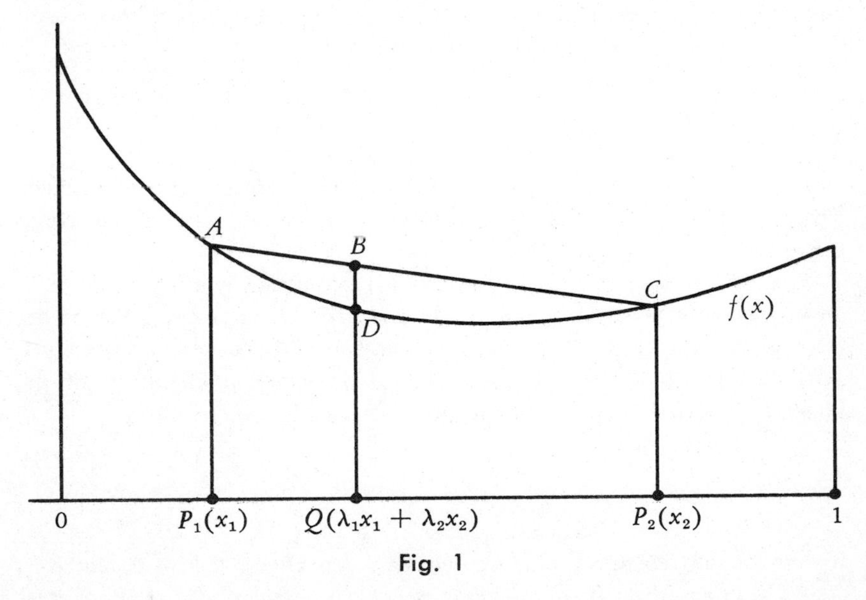

Fig. 1

Since the abscissa of P_1 is x_1, the ordinate to the curve at P_1 is $f(x_1)$; i.e.,

$$P_1 A = f(x_1) .$$

Similarly,

$$P_2C = f(x_2),$$

and

$$QD = f(\lambda_1 x_1 + \lambda_2 x_2).$$

Since the abscissa of Q is $\lambda_1 x_1 + \lambda_2 x_2$, we see that Q divides the segment P_1P_2 in the ratio λ_1/λ_2; moreover, since λ_1 and λ_2 are both positive, Q lies between P_1 and P_2. By an argument involving similar triangles, we then see that B divides AC in the ratio λ_1/λ_2, and hence that

$$QB = \lambda_1 f(x_1) + \lambda_2 f(x_2).$$

Thus the inequality defining a convex function means that

$$QD \leq QB;$$

i.e., between any two points of the graph of the function, the graph never lies above the segment connecting the points. And a function is strictly convex if the graph of the function always lies *below* the line-segment in question.

It is readily seen that the only functions which are convex, without being strictly convex, are functions whose graphs consist in part or in whole of straight line-segments. Thus the following functions are convex in the interval $(-\infty, +\infty)$, i.e., in every finite interval:

$$x^4, \qquad 3 - 7x + x^2, \qquad e^x, \qquad e^{-x}, \qquad |x|.$$

All of these functions but the last are also strictly convex.

A function can, of course, be convex in some intervals but not in others. The function $\sin x$, for example, is convex in the interval $(-\pi, 0)$ but is not convex in the interval $(0, \pi)$.

From differential calculus we know that a function is strictly convex in an interval if it possesses a positive second derivative at every point of the interval. On the other hand, a function can be strictly convex in an interval, even though it does not possess a second derivative at all points of the interval. Thus, for instance, the function f defined by the conditions

$$f(x) = 2 + (x + 4)^2 \qquad \text{for } x \geq 0,$$
$$f(x) = 2 + (x - 4)^2 \qquad \text{for } x < 0$$

has no second derivative (indeed, no first derivative) at $x = 0$; but it is easily verified that this function is strictly convex (in the interval $(-\infty, +\infty)$).

If one variable, in a function of two variables, is held fixed, we obtain a function of one variable, and it may happen that this function is convex (or strictly convex). Thus the function

$$f(x, y) = 5x^2 - 10y^2$$

is a strictly convex function of x for each y, since

$$\frac{\partial^2 f}{\partial x^2} = 10 > 0;$$

it is not a convex function of y for any x, however, since

$$\frac{\partial^2 f}{\partial y^2} = -20 < 0.$$

The function

$$f(x, y) = x^2 + y^2 - 3xy + 1$$

is a strictly convex function of x for each y and a strictly convex function of y for each x.

It is also possible to generalize the notion of convexity to functions of more than one variable. If f is a function of n variables, then we say that f is *convex* within an n-dimensional interval if, for every member $\| \lambda_1 \quad \lambda_2 \|$ of $\mathbf{S}_2$, and for every pair of distinct points $\| x_1 \quad \cdots \quad x_n \|$ and $\| \overline{x}_1 \quad \cdots \quad \overline{x}_n \|$ of the interval,

$$f(\lambda_1 x_1 + \lambda_2 \overline{x}_1, \cdots, \lambda_1 x_n + \lambda_2 \overline{x}_n) \leq \lambda_1 f(x_1, \cdots, x_n) + \lambda_2 f(\overline{x}_1, \cdots, \overline{x}_n).$$

As in the case of a function of one variable, we call f *strictly convex* if the equality never holds when $\lambda_1 \neq 0 \neq \lambda_2$.

This last notion of convexity does not reduce to the notion of convexity in each variable separately. Thus $f(x, y)$ can be a convex function of x for each y and a convex function of y for each x without being convex (in the two variables simultaneously). For instance, the function

$$f(x_1, x_2) = x_1^2 + x_2^2 - 3x_1 x_2 + 1$$

is, as was mentioned above, a convex function of x_1 for each x_2 and a convex function of x_2 for each x_1. But it is not a convex function in x_1 and x_2 simultaneously, as can be seen by taking $\lambda_1 = \lambda_2 = ½$ and $\| x_1 \quad x_2 \| = \| -1 \quad -1 \|$ and $\| y_1 \quad y_2 \| = \| 1 \quad 1 \|$.

A notion which is, in a sense, dual to that of convexity is the notion of concavity. A function f is called *concave* if $-f$ is convex. We can also speak of concavity in several variables, strict concavity, and so on. The properties of concavity follow easily from the corresponding properties of convexity.

LEMMA 12.1. Let f be a function which is continuous and strictly convex in a closed interval. Then there is precisely one point of the interval at which f assumes its minimum value.

PROOF. Since f is continuous, it assumes its minimum value at at least one point. But if f assumed its minimum at two distinct points, a and b, we should have, by the strict convexity,

$$f\left(\frac{a+b}{2}\right) < \frac{1}{2} f(a) + \frac{1}{2} f(b) = \frac{1}{2} f(a) + \frac{1}{2} f(a) = f(a),$$

contrary to the hypothesis that the minimum is assumed at a.

2. A Unique Strategy for One Player. We turn now to the consideration of continuous games whose payoff functions are convex for the minimizing player.

THEOREM 12.2. Let M be the payoff function of a continuous game, and suppose that M is continuous in both variables and that $M(x, y)$ is strictly convex in y for every x. Then there is a unique optimal strategy for the second player, which is a step-function of first order; i.e., there is a number c in the closed interval $[0, 1]$ such that the (unique) optimal strategy for the second player is the step-function I_c. The value v of the game is given by the formula

$$v = \min_{0 \leq y \leq 1} \max_{0 \leq x \leq 1} M(x, y),$$

and the constant c is the unique solution of the equation

$$\max_{0 \leq x \leq 1} M(x, c) = v.$$

PROOF. Since M is continuous, it follows from Theorem 10.4 that there are optimal mixed strategies for the two players. Let v be the value of the game and let F^* be any fixed optimal mixed strategy for the first player. We set

$$\phi(y) = \int_0^1 M(x, y)\, dF^*(x). \tag{1}$$

From the continuity of M it is easily shown that ϕ is a continuous function of y. For, since M is continuous, for every ε there is a δ such that, for $|y_1 - y_2| < \delta$, we have

$$|M(x, y_1) - M(x, y_2)| < \varepsilon.$$

Then

$$
\begin{aligned}
|\phi(y_1) - \phi(y_2)| &= \left| \int_0^1 [M(x, y_1) - M(x, y_2)]\, dF^*(x) \right| \\
&\leq \int_0^1 |M(x, y_1) - M(x, y_2)|\, dF^*(x) \\
&\leq \int_0^1 \varepsilon\, dF^*(x) = \varepsilon \int_0^1 dF^*(x) = \varepsilon .
\end{aligned}
$$

Moreover, from the fact that $M(x, y)$ is strictly convex in y for each x, it follows that $\phi(y)$ is a strictly convex function of y. For by Theorem 9.15

$$
\begin{aligned}
\phi[\lambda_1 y_1 + \lambda_2 y_2] &= \int_0^1 M(x, \lambda_1 y_1 + \lambda_2 y_2)\, dF^*(x) \\
&< \int_0^1 [\lambda_1 M(x, y_1) + \lambda_2 M(x, y_2)]\, dF^*(x) \\
&= \lambda_1 \int_0^1 M(x, y_1)\, dF^*(x) + \lambda_2 \int_0^1 M(x, y_2)\, dF^*(x) \\
&= \lambda_1 \phi(y_1) + \lambda_2 \phi(y_2) .
\end{aligned}
$$

Hence the hypothesis of Lemma 12.1 is satisfied by ϕ, and we conclude that ϕ assumes its minimum at precisely one point, c, of $[0, 1]$. Thus

$$\phi(c) = \min_{0 \leq y \leq 1} \phi(y),$$

and

$$\phi(c) < \phi(y) \qquad \text{if } y \neq c .$$

We wish now to show that the only optimal strategy for the second player is the distribution function I_c. Since we know that there is certainly at least one optimal strategy for this player, it suffices to show that every optimal strategy for him is identical with I_c.

Let G^*, then, be any optimal strategy for the second player. We are to show that $G^* = I_c$, that is to say, we are to show that $G^*(y) = 0$, for $y < c$, and that $G^*(c) = 1$. To do this, it clearly suffices to show that, for every positive number ε, we have

$$G^*(c + \varepsilon) - G^*(c - \varepsilon) = 1 .$$

But since ϕ is continuous and assumes its minimum only at c, there is a positive δ such that, for $0 \leq y \leq c - \varepsilon$ and for $c + \varepsilon \leq y \leq 1$, we have $\phi(y) \geq \phi(c) + \delta$, and such that, for $c - \varepsilon \leq y \leq c + \varepsilon$, we have

$\phi(y) \geq \phi(c)$. Hence we conclude that

$$\begin{aligned}
\int_0^1 \phi(y)\, dG^*(y) &= \int_0^{c-\varepsilon} \phi(y)\, dG^*(y) + \int_{c-\varepsilon}^{c+\varepsilon} \phi(y)\, dG^*(y) \\
&\quad + \int_{c+\varepsilon}^1 \phi(y)\, dG^*(y) \\
&\geq \int_0^{c-\varepsilon} [\phi(c) + \delta]\, dG^*(y) + \int_{c-\varepsilon}^{c+\varepsilon} \phi(c)\, dG^*(y) \\
&\quad + \int_{c+\varepsilon}^1 [\phi(c) + \delta]\, dG^*(y) \\
&= \int_0^1 \phi(c)\, dG^*(y) + \int_0^{c-\varepsilon} \delta\, dG^*(y) + \int_{c+\varepsilon}^1 \delta\, dG^*(y) \\
&= \phi(c) + \delta[G^*(c - \varepsilon) - G^*(0) \\
&\quad + G^*(1) - G^*(c + \varepsilon)] \\
&= \phi(c) + \delta[1 + G^*(c - \varepsilon) - G^*(c + \varepsilon)].
\end{aligned}$$

Moreover, using Theorem 9.22,

$$\begin{aligned}
v &= \int_0^1 \int_0^1 M(x, y)\, dF^*(x)\, dG^*(y) = \int_0^1 \phi(y)\, dG^*(y) \\
&= \min_{G \in D} \int_0^1 \phi(y)\, dG(y) = \min_{0 \leq y \leq 1} \phi(y) = \phi(c),
\end{aligned}$$

and hence

$$\phi(c) \geq \phi(c) + \delta[1 + G^*(c - \varepsilon) - G^*(c + \varepsilon)].$$

Since $\delta > 0$, we therefore conclude that

$$1 + G^*(c - \varepsilon) - G^*(c + \varepsilon) \leq 0,$$

and hence that

$$G^*(c + \varepsilon) - G^*(c - \varepsilon) \geq 1.$$

But clearly

$$G^*(c + \varepsilon) - G^*(c - \varepsilon) \leq 1,$$

and hence

$$G^*(c + \varepsilon) - G^*(c - \varepsilon) = 1,$$

as was to be shown.

The truth of the last sentence of our theorem follows directly from Theorem 10.17.

By an entirely analogous argument we obtain the following theorem:

THEOREM 12.3. Let M be the payoff function of a continuous game, and suppose that M is continuous in both variables and that $M(x, y)$ is strictly concave in x for each y. Then there is a unique optimal strategy for the first player, which is a step-function of first order; i.e., there is a number c in the closed interval $[0, 1]$ such that the (unique) optimal strategy for the first player is the step-function I_c. The value v of the game is given by the formula

$$v = \max_{0 \leq x \leq 1} \min_{0 \leq y \leq 1} M(x, y),$$

and the constant c is the unique solution of the equation

$$\min_{0 \leq y \leq 1} M(c, y) = v.$$

EXAMPLE 12.4. Let

$$M(x, y) = \sin \frac{\pi(x + y)}{2}.$$

Since

$$\frac{\partial^2 M}{\partial x^2} = -\left(\frac{\pi}{2}\right)^2 \sin \frac{\pi(x + y)}{2} \leq 0 \qquad \text{for } 0 \leq x \leq 1 \text{ and } 0 \leq y \leq 1,$$

$M(x, y)$ is a concave function of x for each y. Hence by Theorem 12.3

$$v = \max_{0 \leq x \leq 1} \min_{0 \leq y \leq 1} \sin \frac{\pi(x + y)}{2}.$$

By considering the graph of $\sin[\pi(x + y)/2]$ for various fixed values of x, we easily verify that if $0 \leq x \leq ½$, then

$$\min_{0 \leq y \leq 1} \sin \frac{\pi(x + y)}{2} = \sin \frac{\pi x}{2}, \tag{2}$$

and if $½ < x \leq 1$, then

$$\min_{0 \leq y \leq 1} \sin \frac{\pi(x + y)}{2} = \sin \frac{\pi(x + 1)}{2}. \tag{3}$$

Thus

$$v = \max\left[\max_{0\le x\le 1/2}\ \min_{0\le y\le 1} \sin\frac{\pi(x+y)}{2},\ \max_{1/2\le x\le 1}\ \min_{0\le y\le 1} \sin\frac{\pi(x+y)}{2}\right]$$

$$= \max\left[\max_{0\le x\le 1/2} \sin\frac{\pi x}{2},\ \max_{1/2\le x\le 1} \sin\frac{\pi(x+1)}{2}\right]$$

$$= \max\left(\frac{\sqrt{2}}{2}, \frac{\sqrt{2}}{2}\right) = \frac{\sqrt{2}}{2}.$$

Moreover,

$$\min_{0\le y\le 1} \sin\frac{\pi\left(\frac{1}{2}+y\right)}{2} = \frac{\sqrt{2}}{2}.$$

Hence the value of this game is $\sqrt{2}/2$, and the unique optimal strategy for the first player is the distribution function $I_{1/2}$.

3. Strategies for the Other Player. In the above example we were able to determine an optimal strategy for the first player but not for the second. The following two theorems enable us to determine optimal strategies for both players.

In these theorems we use the notations $M^{(1)}(x, y)$ and $M^{(2)}(x, y)$ to mean the partial derivatives of $M(x, y)$ with respect to x and y, respectively. Thus

$$M^{(1)}(x, y) = \lim_{z\to 0} \frac{M(x+z, y) - M(x, y)}{z} \tag{4}$$

and

$$M^{(2)}(x, y) = \lim_{z\to 0} \frac{M(x, y+z) - M(x, y)}{z}. \tag{5}$$

Since we are confining ourselves to functions defined over the closed unit square ($0 \le x \le 1$ and $0 \le y \le 1$), $M^{(1)}(0, y)$, $M^{(1)}(1, y)$, $M^{(2)}(x, 0)$, and $M^{(2)}(x, 1)$ would be meaningless according to these definitions. But we shall take $M^{(1)}(0, y)$ to mean the limit in (4) when z is restricted to positive values, and we shall take $M^{(1)}(1, y)$ to mean this limit when z is restricted to negative values, and similarly for $M^{(2)}(x, 0)$ and $M^{(2)}(x, 1)$.

THEOREM 12.5. Let M be the payoff function of a continuous game, and suppose that M is continuous in both variables, that $M^{(2)}(x, y)$ exists for each x and y in the unit square, and that $M(x, y)$ is a strictly convex function of y for each x. Let I_{y_0} be the unique optimal strategy for the second player, and let v be the value of the game. If $y_0 = 0$ or $y_0 = 1$, then there is an optimal strategy I_{x_0} for the first player; the constant x_0 can be taken to be any number satisfying the conditions

$$0 \le x_0 \le 1,$$
$$M(x_0, y_0) = v,$$
$$M^{(2)}(x_0, y_0) \begin{cases} \ge 0 & \text{if } y_0 = 0, \\ \le 0 & \text{if } y_0 = 1. \end{cases}$$

If $0 < y_0 < 1$, then there is an optimal strategy for the first player, which has the form

$$\alpha I_{x_1}(x) + (1 - \alpha) I_{x_2}(x);$$

and the constants α, x_1, and x_2 can be taken to be any numbers satisfying the conditions

$$0 \le x_1 \le 1, \qquad 0 \le x_2 \le 1, \qquad 0 \le \alpha \le 1,$$
$$M(x_1, y_0) = v, \qquad M(x_2, y_0) = v,$$
$$M^{(2)}(x_1, y_0) \ge 0, \qquad M^{(2)}(x_2, y_0) \le 0,$$
$$\alpha M^{(2)}(x_1, y_0) + (1 - \alpha) M^{(2)}(x_2, y_0) = 0.$$

PROOF. Suppose first that $y_0 = 0$. Since I_0 is an optimal strategy for the second player, we see by Theorem 10.6 that, for all x,

$$\int_0^1 M(x, y)\, dI_0(y) \le v,$$

and thus that, for all x,

$$M(x, 0) \le v; \tag{6}$$

moreover, since by Theorem 12.2 we have

$$v = \max_{0 \le x \le 1} M(x, 0),$$

we see that there exists an x_0 in the closed interval $[0, 1]$ such that

$$M(x_0, 0) = v. \tag{7}$$

Now suppose, if possible, that every number x_0 which satisfies (7) is such that

$$M^{(2)}(x_0, 0) < 0.$$

Then, for every x in $[0, 1]$ there exists a positive ε such that

$$M(x, y) < v \qquad \text{for } 0 < y < \varepsilon. \tag{8}$$

For each x, we define $\varepsilon(x)$ to be the least upper bound of all numbers ε satisfying (8). From the continuity of M it is seen that $\varepsilon(x)$ is a continuous function of x in the closed interval $[0, 1]$; moreover, $\varepsilon(x)$ is always positive and hence has a positive minimum. Let $\varepsilon_0 > 0$ be the minimum of $\varepsilon(x)$. Now if we choose y_1 so that $0 < y_1 < \varepsilon_0$, we have

$$\max_{0 \leq x \leq 1} M(x, y_1) < v,$$

and hence, applying Theorem 12.2,

$$v = \min_{0 \leq y \leq 1} \max_{0 \leq x \leq 1} M(x, y) \leq \max_{0 \leq x \leq 1} M(x, y_1) < v.$$

Since this is absurd, we are forced to conclude that there exists a number x_0 which satisfies (7) and, in addition,

$$M^{(2)}(x_0, 0) \geq 0. \tag{9}$$

Now let x_0 be any number in $[0, 1]$ which satisfies conditions (7) and (9); we wish to show that I_{x_0} is an optimal strategy for the first player. But, since $M(x_0, y)$ is a convex function of y, we see that (7) and (9) imply that v is the minimum of $M(x_0, y)$, so that, for every y,

$$v \leq M(x_0, y) = \int_0^1 M(x, y)\, dI_{x_0}(x),$$

and hence by Theorem 10.6 an optimal strategy for the first player is I_{x_0}, as was to be shown. This completes the proof for the case $y_0 = 0$.

The proof for the case $y_0 = 1$ is similar.

Now suppose that $0 < y_0 < 1$. As in the proof for the case $y_0 = 0$, we see that, for all x,

$$M(x, y_0) \leq v, \tag{10}$$

and that, for some x,

$$M(x, y_0) = v. \tag{11}$$

If every x satisfying (11) were such that

$$M^{(2)}(x, y_0) < 0, \tag{12}$$

then we would be led to the same absurdity as in the argument for the case

$y_0 = 0$. Thus there is a number x which satisfies (11) and, also,

$$M^{(2)}(x, y_0) \geq 0;$$

i.e., there is a number x_1 such that

$$\left.\begin{aligned} 0 \leq x_1 \leq 1, \\ M(x_1, y_0) = v, \\ M^{(2)}(x_1, y_0) \geq 0. \end{aligned}\right\} \tag{13}$$

In a similar way we see that there is a number x_2 such that

$$\left.\begin{aligned} 0 \leq x_2 \leq 1, \\ M(x_2, y_0) = v, \\ M^{(2)}(x_2, y_0) \leq 0. \end{aligned}\right\} \tag{14}$$

Now consider the function

$$f(\xi) = \xi M^{(2)}(x_1, y_0) + (1 - \xi)M^{(2)}(x_2, y_0).$$

We notice that

$$f(0) = M^{(2)}(x_2, y_0) \leq 0$$

and that

$$f(1) = M^{(2)}(x_1, y_0) \geq 0.$$

Since f is a continuous function of ξ, we conclude that there exists an α satisfying

$$\left.\begin{aligned} 0 \leq \alpha \leq 1, \\ f(\alpha) = \alpha M^{(2)}(x_1, y_0) + (1 - \alpha)M^{(2)}(x_2, y_0) = 0. \end{aligned}\right\} \tag{15}$$

To complete the proof of our theorem we need only show that if x_1, x_2, and α are any numbers satisfying conditions (13), (14), and (15), then the distribution function

$$\alpha I_{x_1}(x) + (1 - \alpha)I_{x_2}(x)$$

is an optimal strategy for the first player. From the fact that $M(x, y)$ is a convex function of y for each x, we easily conclude that the function

$$g(y) = \alpha M(x_1, y) + (1 - \alpha)M(x_2, y)$$

is a convex function of y. Moreover, from the equation in (15) we see that the derivative of $g(y)$ vanishes at $y = y_0$. Hence $g(y)$ assumes its minimum at y_0. Thus, since by (13) and (14) we have

$$g(y_0) = \alpha M(x_1, y_0) + (1 - \alpha)M(x_2, y_0) = \alpha v + (1 - \alpha)v = v,$$

we see that v is the minimum value of $g(y)$, i.e., that, for all y,

$$v \leq \alpha M(x_1, y) + (1 - \alpha)M(x_2, y)$$

or

$$v \leq \int_0^1 M(x, y)\, d[\alpha I_{x_1}(x) + (1 - \alpha)I_{x_2}(x)].$$

Thus our theorem follows by Theorem 10.6.

The following dual theorem can be proved in an analogous way.

THEOREM 12.6. Let M be the payoff function of a continuous game, and suppose that M is continuous in both variables, that $M^{(1)}(x, y)$ exists for each x and y in the unit square, and that $M(x, y)$ is a strictly concave function of x for each y. Let I_{x_0} be the unique optimal strategy for the first player, and let v be the value of the game. If $x_0 = 0$ or $x_0 = 1$, then there is an optimal strategy I_{y_0} for the second player; the constant y_0 can be taken to be any number satisfying the conditions

$$0 \leq y_0 \leq 1,$$

$$M(x_0, y_0) = v,$$

$$M^{(1)}(x_0, y_0)\begin{cases} \leq 0 & \text{if } x_0 = 0, \\ \geq 0 & \text{if } x_0 = 1. \end{cases}$$

If $0 < x_0 < 1$, then there is an optimal strategy for the second player, which has the form

$$\alpha I_{y_1}(y) + (1 - \alpha)I_{y_2}(y);$$

and the constants α, y_1, and y_2 can be taken to be any numbers satisfying the conditions

$$0 \leq y_1 \leq 1, \qquad 0 \leq y_2 \leq 1, \qquad 0 \leq \alpha \leq 1,$$

$$M(x_0, y_1) = v, \qquad M(x_0, y_2) = v,$$

$$M^{(1)}(x_0, y_1) \geq 0, \qquad M^{(1)}(x_0, y_2) \leq 0,$$

$$\alpha M^{(1)}(x_0, y_1) + (1 - \alpha)M^{(1)}(x_0, y_2) = 0.$$

4. Remarks and an Example.

REMARK 12.7. It should be noticed that the optimal strategies whose existence was asserted in the two preceding theorems are not necessarily unique. Thus, for example, let

$$M(x, y) = \left(y + \frac{1}{2}\right)^2.$$

Then M satisfies the hypothesis of Theorem 12.5, so there exists an optimal strategy for the first player, which has the form

$$\alpha I_{x_1}(x) + (1 - \alpha)I_{x_2}(x);$$

but this is not the only optimal strategy available to the first player. Indeed, since $M(x, y)$ is independent of x, the payoff is unaffected by what the first player does; so that *every* strategy is optimal for the first player.

REMARK 12.8. By means of Theorem 12.6 it is an easy matter to find an optimal strategy for the second player in the game described in Example 12.4. Since for this game we found that $x_0 = \frac{1}{2}$, we conclude that there is an optimal strategy for the second player, which has the form

$$\alpha I_{y_1}(y) + (1 - \alpha)I_{y_2}(y),$$

where α, y_1, and y_2 satisfy the conditions

$$0 \leq y_1 \leq 1, \qquad 0 \leq y_2 \leq 1, \qquad 0 \leq \alpha \leq 1,$$

$$\sin\left[\frac{\pi}{2}\left(\frac{1}{2} + y_1\right)\right] = \frac{\sqrt{2}}{2}, \qquad \sin\left[\frac{\pi}{2}\left(\frac{1}{2} + y_2\right)\right] = \frac{\sqrt{2}}{2},$$

$$\frac{\pi}{2}\cos\left[\frac{\pi}{2}\left(\frac{1}{2} + y_1\right)\right] \geq 0, \qquad \frac{\pi}{2}\cos\left[\frac{\pi}{2}\left(\frac{1}{2} + y_2\right)\right] \leq 0,$$

$$\alpha \cdot \frac{\pi}{2}\cos\left[\frac{\pi}{2}\left(\frac{1}{2} + y_1\right)\right] + (1 - \alpha) \cdot \frac{\pi}{2}\cos\left[\frac{\pi}{2}\left(\frac{1}{2} + y_2\right)\right] = 0.$$

Since the only values of y in the closed interval $[0, 1]$ which satisfy the equation

$$\sin\left[\frac{\pi}{2}\left(\frac{1}{2} + y\right)\right] = \frac{\sqrt{2}}{2}$$

are 0 and 1, and since

$$\frac{\pi}{2}\cos\left[\frac{\pi}{2}\left(\frac{1}{2} + 0\right)\right] = \frac{\pi}{2} \cdot \frac{\sqrt{2}}{2} > 0$$

and

$$\frac{\pi}{2}\cos\left[\frac{\pi}{2}\left(\frac{1}{2}+1\right)\right]=\frac{\pi}{2}\cdot\left(-\frac{\sqrt{2}}{2}\right)<0,$$

we conclude that $y_1 = 0$ and $y_2 = 1$. From the equation

$$\alpha\cdot\frac{\pi}{2}\cos\left[\frac{\pi}{2}\left(\frac{1}{2}+0\right)\right]+(1-\alpha)\frac{\pi}{2}\cos\left[\frac{\pi}{2}\left(\frac{1}{2}+1\right)\right]=0,$$

we then conclude that $\alpha = ½$. Hence an optimal strategy for the second player is the distribution function

$$\frac{1}{2}I_0(y)+\frac{1}{2}I_1(y).$$

Thus an optimal way for the second player to play is to choose 0 and 1 with equal frequencies.

EXAMPLE 12.9. We shall now find a solution, by the methods of this chapter, of the game whose payoff function is

$$M(x, y) = 16y^6 - 3xy + x^2.$$

Since

$$\frac{\partial^2 M}{\partial y^2} = 16\cdot 6\cdot 5y^4 \geq 0,$$

the function M is convex in y for each x. Thus by Theorem 12.2

$$v = \min_{0\leq y\leq 1}\ \max_{0\leq x\leq 1}\ (16y^6 - 3xy + x^2) = \frac{16}{729}.$$

Moreover, we have

$$\max_{0\leq x\leq 1}\left[16\left(\frac{1}{3}\right)^6 - 3\left(\frac{1}{3}\right)x + x^2\right]=\frac{16}{729},$$

so by Theorem 12.2 we conclude that the (unique) optimal strategy for the second player is the distribution function

$$I_{1/3}(y).$$

Since our function M has derivatives everywhere, the hypothesis of Theorem 12.5 is satisfied. Since $y_0 = 1/3$ and $v = 16/729$, we wish to find solutions of the equation

$$M\left(x, \frac{1}{3}\right) = \frac{16}{729},$$

i.e., of the equation

$$16\left(\frac{1}{3}\right)^6 - 3\left(\frac{1}{3}\right)x + x^2 = \frac{16}{729}.$$

It is immediately seen that the solutions are $x = 0$ and $x = 1$.

Since

$$M^{(2)}(x, y) = \frac{\partial M(x, y)}{\partial y} = 96y^5 - 3x,$$

we have

$$M^{(2)}\left(0, \frac{1}{3}\right) = \frac{32}{81}$$

and

$$M^{(2)}\left(1, \frac{1}{3}\right) = -\frac{211}{81}.$$

Hence, using the notation of Theorem 12.5, we can take $x_1 = 0$ and $x_2 = 1$.

Now, solving the equation

$$\alpha M^{(2)}\left(0, \frac{1}{3}\right) + (1 - \alpha)M^{(2)}\left(1, \frac{1}{3}\right) = 0,$$

we find that $\alpha = 211/243$.

Hence an optimal strategy for the first player is the distribution function

$$\frac{211}{243} I_0(x) + \frac{32}{243} I_1(x).$$

REMARK 12.10. We have formulated our theorems about convex functions in a rather weak and special form, simply to avoid complications in the proofs. The results we have established in this chapter can be strengthened and generalized in several ways.

First, the condition in Theorems 12.5 and 12.6 that certain derivatives exist can be dropped. In this case, however, it is necessary to make use of the fact that a convex, or concave, function possesses left-hand and right-hand derivatives at each point of its interval of definition, with the possible exception of end points. The conditions on $M^{(1)}(x_0, y_1)$, $M^{(1)}(x_0, y_2)$, $M^{(2)}(x_1, y_0)$, and

$M^{(2)}(x_2, y_0)$ in the conclusions of these theorems are then replaced by appropriate conditions on the right-hand and left-hand derivatives at the points in question.

Secondly, the condition that the payoff function be strictly convex, or strictly concave, may be relaxed to the condition that it be merely convex, or concave, respectively. In this case, however, the optimal strategy for the second player in Theorem 12.2 (or for the first player in Theorem 12.3) is, in general, no longer unique.

Finally, the results can be extended to the case in which the players, instead of choosing simply numbers from the closed unit interval, choose points from the n-dimensional unit cube. In this case it is necessary to use the more general notion of convexity (convexity in several variables).

BIBLIOGRAPHICAL REMARK

The theorems which have been proved in this chapter are special cases of more general theorems whose proofs can be found in Bohnenblust, Karlin, and Shapley [2].

EXERCISES

1. Give an example of a function which is strictly concave in the interval $(-\infty, +\infty)$ and which fails to have a derivative at the points $x = 0$, $x = 1$, and $x = 2$.

2. Give an example of a function f of two variables such that $f(x, y)$ is concave in x for every y and is concave in y for every x, while $f(x, y)$ is not concave in x and y simultaneously.

3. Show that if a function f is continuous and strictly concave in a closed interval, then there is exactly one point of the interval at which f assumes its maximum value.

4. Show that the sum of two convex functions is convex. Show that the product of two convex functions is not necessarily convex.

5. The payoff function of a continuous game is

$$M(x, y) = \sin(2x + y).$$

Find the value of the game and optimal strategies for the two players.

6. Solve Exercise 5 of Chap. 11 by the methods of the present chapter.

7. The payoff function of a continuous game is

$$M(x, y) = 80y^8 - 5xy + x^2.$$

Find the value of the game and optimal strategies for the two players.

8. The payoff function of a continuous game is

$$M(x, y) = Ay^{10} - 4xy + x^6,$$

where A is a number satisfying the inequalities

$$0 < A < \frac{2^{19}}{5}.$$

Find the value of the game and optimal strategies for the two players (some of the answers may, of course, involve the parameter A).

9. Prove the following theorem:

Let $M(x, y)$ and $M_1(x, y), M_2(x, y), \cdots$ be continuous over the unit square, and suppose that

$$| M_n(x, y) - M(x, y) | < \frac{1}{n}$$

for each n and for all points $\| x \quad y \|$ in the unit square. For each n, let v_n be the value of the continuous game whose payoff function is $M_n(x, y)$, and let $F_n(x)$ and $G_n(y)$ be optimal strategies in this game for the first and second player, respectively. Suppose, moreover, that

$$\lim_{n \to \infty} v_n = v,$$

$$\lim_{n \to \infty} F_n(x) = F(x),$$

$$\lim_{n \to \infty} G_n(y) = G(y),$$

where F and G are distribution functions. Then v is the value of the continuous game whose payoff function is $M(x, y)$, and $F(x)$ and $G(y)$ are optimal strategies in this game for the first and second player, respectively.

10. Show that if $M(x, y)$ is a convex function of y for each x and if n is any positive integer, then the function

$$M_n(x, y) = M(x, y) + \frac{y^2 - y}{n}$$

is a strictly convex function of y for each x.

11. Show (by making use of Exercises 9 and 10) how we can strengthen Theorems 12.2 and 12.5 by replacing the words "strictly convex" by "convex."

12. Show that a function which is convex in an open interval is continuous, and give an example of a function which is convex in a closed interval, but which has a discontinuity.

13. Show that a convex function has both a right-hand derivative and a left-hand derivative at each point (though they are not necessarily equal), except possibly at the end points of the interval over which it is defined.

14. Formulate and prove (by making use of Exercise 13) a generalization of Theorem 12.5, where the existence of $M^{(2)}(x, y)$ is not assumed.

15. A certain game is played as follows: The first player chooses a point $\| x_1 \quad x_2 \|$ in the closed unit square and the second player, not being informed about the first player's choice, chooses a point $\| y_1 \quad y_2 \|$ in the closed unit square. The payoff (to the first player) is then

$$M(x_1, x_2, y_1, y_2),$$

where $M(x_1, x_2, y_1, y_2)$ is strictly concave in $\| x_1 \quad x_2 \|$ for each $\| y_1 \quad y_2 \|$ and is strictly convex in $\| y_1 \quad y_2 \|$ for each $\| x_1 \quad x_2 \|$.

Show that there is a unique pure strategy for each player.

CHAPTER 13
APPLICATIONS TO STATISTICAL INFERENCE

In Chap. 1 we pointed out that when a man is interested in maximizing something, it makes a great difference whether he must contend only against the forces of nature or must take into account also the behavior of some other rational being—one who perhaps wishes to make small the very quantity the first man wants to make large. Both types of situation can be regarded as games: the first type gives a one-person game and the second type gives an n-person game with $n > 1$. Nature, we pointed out, cannot properly be conceived as trying to outwit us and as possessing the direct antagonism to us which we would find, for example, in playing a zero-sum two-person game. Thus the non-zero-sum one-person game (the zero-sum one-person game is, of course, completely trivial) can be regarded as a pure maximization problem in the classical sense, where there is no question of countering the moves of another rational creature.

Despite this great difference between the two situations, however, even in the case of a (non-zero-sum) game played against nature, it can happen that the player will be interested in determining what is the worst nature can do to him; i.e., he may wish to calculate what is the very minimum he can guarantee himself, even if nature turns out to be completely unfavorable.

Situations of this sort arise particularly in connection with statistics, for the statistician is often concerned with such problems as the following: to maximize the accuracy of the determination of a quantity for a given cost; or to minimize the cost of determining something to a given accuracy; or to maximize the profit of a manufacturer by devising a suitable method of testing his output (this application of statistics is called *quality control*). The relation of the theory of games to statistics has indeed turned out to be so intimate that, in recent years, mathematical statisticians have devoted much attention to this subject. We shall not attempt to formulate the general theorems which have been established in this connection, however, but shall confine ourselves to a discussion of some specific examples of the application of the theory of games to statistical problems.

The examples we discuss may appear simple almost to the point of triviality. This is because we have tried to avoid any previous familiarity with statistical theory and to keep the matrices small enough so that they can be solved without the use of computing machines. The same principles, however, are involved in these simple examples as would be involved in more realistic problems.

One of the most common kinds of problem confronting the statistician is that of making some estimate about a large class of things on the basis of the examination of a sample. Thus a political pollster, for instance, may wish to make a prediction about the outcome of an approaching election from interviews with citizens. The statistician can ordinarily increase the reliability of his estimate by increasing the size of his sample; but extra expense is involved in making more tests. Thus the statistician is presented with the problem of how large a sample it is best for him to examine. The following example gives a highly simplified and idealized model of such a situation.

EXAMPLE 13.1. A certain urn is known to contain two balls, each of which is either black or white. A statistician, S, wishes to make a guess as to how many (if any) of the balls are black. If he guesses right, he is to be paid an amount α; if his answer differs from the correct answer by 1 (e.g., if he guesses 1 when there are actually 2, or guesses 2 when there is actually 1, etc.), he is to be paid the amount β; if his answer differs from the correct answer by 2 (so that he guesses 0 when there are actually 2, or guesses 2 when there is actually 0), he is to be paid the amount γ. We suppose that $\alpha \geq \beta \geq \gamma$; but we make no assumption as to whether these three quantities are positive or negative. It costs S the amount δ to examine one of the balls. With the case as described, it would seem natural to suppose that δ is negligibly small, but we do not impose this restriction. To make the situation more plausible in this regard, the reader may wish to think of the balls not as being white or black, but as being two shades of grey; if the two shades are almost identical, elaborate physical tests may be required to determine whether a given ball is of the first shade or the second.

There are eight possible ways for S to proceed (i.e., eight pure strategies) in order to arrive at his guess as to how many of the balls are black:

I. Make no test and guess that both balls are black.

II. Make no test and guess that one ball is black and one is white.

III. Make no test and guess that both balls are white.

IV. Test one ball and guess that the other ball is of the same color as the one tested.

V. Test one ball and, regardless of what color it turns out to be, guess that the other ball is black.

VI. Test one ball and, regardless of what color it turns out to be, guess that the other ball is white.

VII. Test one ball and guess that the other ball is of the opposite color to the one tested.

VIII. Test both balls and announce the correct number of black balls (which will, of course, now be known).

(We have not listed procedures which, though logically possible, are stupid —for instance, we do not consider the possibility that S tests one ball and then, even if the ball tested is white, guesses that both are black.)

Moreover, there are just three possibilities open to nature: It can happen that neither ball is black, that just one is black, or that both are black; we indicate these strategies by the numerals 0, 1, and 2.

Let us examine the payoff to S for various combinations of these strategies.

If S uses strategy I and nature uses strategy 0, then S guesses that 2 balls are black, whereas in actuality neither is black. Thus S is off by 2, and the payoff to him is γ.

Similar simple arguments enable us to take care of all the cases in which S uses strategies I, II, III, or VIII and of the cases in which nature uses strategy 0 or strategy 2.

To see how the payoff is calculated in the other cases, suppose, for instance, that S uses strategy V and that nature uses strategy 1. Then, on the single test which S makes, the probability is ½ that the ball tested will be black and ½ that it will be white. If it is black, then, since S is using V, he will guess that both are black, so that he will be in error by 1; thus the payoff in this case, taking into account the cost of making the test, will be $\beta - \delta$. If, on the other hand, the ball tested happens to be white, then S will guess that just one ball is white, which is correct, and hence he will get $\alpha - \delta$. Thus the expectation of S is

$$\frac{1}{2}(\beta - \delta) + \frac{1}{2}(\alpha - \delta) = \frac{1}{2}(\alpha + \beta) - \delta .$$

By continuing in this way we arrive at payoff Matrix 1.

Now, of course, if S knew the probability with which nature played her various strategies, then he would be presented with a simple maximization problem—he would need merely to pick the row which, for the given frequencies of columns, would give him his maximum expectation. But we are supposing that S has no such knowledge of the way nature behaves. In this case, however, he can at least calculate a minimum which he can expect to receive under the most unfavorable possible choice of probabilities by nature. This problem is solved by treating Matrix 1 as the payoff matrix for a zero-sum two-person game. If S has no reason for expecting nature to do one thing rather than another, he may very well feel that the wisest (certainly the most conservative) thing he can do is to choose his strategy as if he were playing such a game against nature.

MATRIX 1

	0	1	2
I	γ	β	α
II	β	α	β
III	α	β	γ
IV	$\alpha - \delta$	$\beta - \delta$	$\alpha - \delta$
V	$\beta - \delta$	$\frac{1}{2}(\alpha + \beta) - \delta$	$\alpha - \delta$
VI	$\alpha - \delta$	$\frac{1}{2}(\alpha + \beta) - \delta$	$\beta - \delta$
VII	$\beta - \delta$	$\alpha - \delta$	$\beta - \delta$
VIII	$\alpha - 2\delta$	$\alpha - 2\delta$	$\alpha - 2\delta$

The value of this game to S and optimal strategies for him in playing it depend on the relative values of α, β, γ, and δ.

Thus if we take $\alpha = 100$, $\beta = 0$, $\gamma = -100$, and $\delta = 1$, then we obtain Matrix 2.

MATRIX 2

	0	1	2
I	-100	0	100
II	0	100	0
III	100	0	-100
IV	99	-1	99
V	-1	49	99
VI	99	49	-1
VII	-1	99	-1
VIII	98	98	98

This matrix has no saddle-point. It is easily verified that the value of the game to S is 98, that an optimal strategy for S is the vector $\|0\ \ 0\ \ 0\ \ 0\ \ 0\ \ 0\ \ 0\ \ 1\|$, and that an optimal strategy for nature is $\|\frac{1}{3}\ \ \frac{1}{3}\ \ \frac{1}{3}\|$. Thus in this case the best thing for the statistician to do is to test both balls. This is not surprising in view of the fact that the cost of testing is so small in proportion to the other quantities involved.

On the other hand, if the cost of the test is very high, the best thing for the statistician to do may be to make no test at all. Thus suppose that $\alpha = 100$, $\beta = 0$, $\gamma = -100$, and $\delta = 200$. Then we obtain Matrix 3.

MATRIX 3

	0	1	2
I	−100	0	100
II	0	100	0
III	100	0	−100
IV	−100	−200	−100
V	−200	−150	−100
VI	−100	−150	−200
VII	−200	−100	−200
VIII	−300	−300	−300

It is now easily verified that the value of the game to S is 0, that an optimal strategy for S is $\| 0 \quad 1 \quad 0 \quad 0 \quad 0 \quad 0 \quad 0 \quad 0 \|$, and that an optimal strategy for nature is $\| \frac{1}{2} \quad 0 \quad \frac{1}{2} \|$. Thus the best thing for the statistician to do is always to guess (without making any test at all) that one ball is black and one is white.

Finally, if δ assumes an intermediate value, it may turn out that the best thing for S to do is to use a mixed strategy. Thus, for example, we obtain Matrix 4 by taking $\alpha = 100$, $\beta = 0$, $\gamma = -100$, and $\delta = 50$.

MATRIX 4

	0	1	2
I	−100	0	100
II	0	100	0
III	100	0	−100
IV	50	−50	50
V	−50	0	50
VI	50	0	−50
VII	−50	50	−50
VIII	0	0	0

It is easily verified that the value of this game to S is 25, that an optimal strategy for S is $\| 0 \quad \frac{1}{2} \quad 0 \quad \frac{1}{2} \quad 0 \quad 0 \quad 0 \quad 0 \|$, and that an optimal strategy for nature is $\| \frac{3}{8} \quad \frac{1}{4} \quad \frac{3}{8} \|$. (Since the minimum of each row is less than 25, S has no optimal pure strategy.) Thus an optimal procedure for S is the following: he tosses an (unbiased) coin; if the coin shows heads, he guesses (without any test) that one ball is white and one is black; if the coin shows tails, he tests one ball and guesses that both balls are of the same color as the one tested.

REMARK 13.2. Another way to solve problems such as the above (but one for which there seems to be little rational justification) is to make what might be called an "argument from ignorance." This consists in saying that, since we are quite ignorant of the probabilities with which the balls are distributed, each of the following alternatives is equally likely: (1) that both balls are black, (2) that the first is black and the second is white, (3) that the first is white and the second is black, and (4) that both are white. Since the case that just one ball is black is a combination of (2) and (3), this amounts to the assumption that nature uses the mixed strategy $\| \frac{1}{4} \quad \frac{1}{2} \quad \frac{1}{4} \|$. Using this assumption, we see from Matrix 1 that if S plays I, his expectation will be $\frac{1}{4}\gamma + \frac{1}{2}\beta + \frac{1}{4}\alpha = \frac{1}{4}(\alpha + 2\beta + \gamma)$; similarly, the expectations of S, for the various strategies open to him, are as follows:

I. $\frac{1}{4}(\alpha + 2\beta + \gamma)$,

II. $\frac{1}{4}(2\alpha + 2\beta)$,

III. $\frac{1}{4}(\alpha + 2\beta + \gamma)$,

IV. $\frac{1}{4}(2\alpha + 2\beta - 4\delta)$,

V. $\frac{1}{4}(2\alpha + 2\beta - 4\delta)$,

VI. $\frac{1}{4}(2\alpha + 2\beta - 4\delta)$,

VII. $\frac{1}{4}(2\alpha + 2\beta - 4\delta)$,

VIII. $\frac{1}{4}(4\alpha - 8\delta)$.

Under the assumptions that $\alpha > \beta > \gamma$ and $\delta > 0$, we see that the quantity corresponding to II is the largest of the first seven quantities. Thus S need consider only strategies II and VIII, i.e.,

$$\frac{1}{4}(2\alpha + 2\beta) \qquad \text{and} \qquad \frac{1}{4}(4\alpha - 8\delta).$$

Hence if $2\alpha + 2\beta < 4\alpha - 8\delta$, i.e., if $\delta < \frac{1}{4}(\alpha - \beta)$, then S should test both balls; if $\delta \geq \frac{1}{4}(\alpha - \beta)$, he should make no test at all, but should simply guess that one ball is black and one is white. This method leads to the same answer as does the game-theoretic approach for Matrices 2 and 3, but to a different answer in the case of Matrix 4. In the latter case, the argument from ignorance prescribes that S should always play strategy II. In this case, if nature does indeed use the mixed strategy $\| \frac{1}{4} \quad \frac{1}{2} \quad \frac{1}{4} \|$, then S will obtain 50, which is greater than the 25 he can ensure himself by playing the mixed strategy $\| 0 \quad \frac{1}{2} \quad 0 \quad \frac{1}{2} \quad 0 \quad 0 \quad 0 \quad 0 \|$; but it should be noticed that S cannot guarantee that he will get even 25 by playing strategy II exclusively, since if he does so, he can expect only 0 in the event nature happens to play strategy $\| 1 \quad 0 \quad 0 \|$. Thus it appears that in this case the argument from ignorance does not give so safe a strategy for S as does the strategy obtained by a game-theoretic analysis.

REMARK 13.3. Although we have treated Example 13.1 as an ordinary two-person zero-sum game, it should be remembered that nature is not in actuality a conscious, rational creature. If S plays his optimal strategy for this game, he is really merely behaving in such a way as to set an absolute lower bound to his expectation; he may feel that it is reasonable for him to act in this way, but he does not thereby commit himself to the animistic view that nature is a malevolent intelligence. His position is somewhat like that of a man who wishes to arrange his investments in such a way that he will not be bankrupt by either inflation or deflation. Such a man does not necessarily believe that the market will always move in the way most unfavorable for him personally; but if he is not in a position to predict the movement of prices with any accuracy, he may want to be adequately prepared for any contingency.

Thus if we let A_1 be the set of mixed strategies available to S, and A_2 the set of mixed strategies available to nature, then S wants to calculate the value of

$$v_1 = \max_{\xi \in \mathsf{A}_1} \min_{\eta \in \mathsf{A}_2} E(\xi, \eta),$$

where E is the expectation function; and perhaps, if he is rather conservative, he may even want to behave in such a way as to be sure of getting v_1. He is not especially interested in the quantity

$$v_2 = \min_{\eta \in \mathsf{A}_2} \max_{\xi \in \mathsf{A}_1} E(\xi, \eta),$$

nor in the fact that $v_2 = v_1$.

The remark finds practical application in the case in which S is not completely ignorant of the possible mixed strategies which nature may use. It can happen, for instance, that although S does not know exactly what mixed strategy nature uses, he may know enough to restrict it to some subset of the logically possible mixed strategies; thus he may know that any mixed strategy $\| \eta_1 \quad \eta_2 \quad \eta_3 \|$ used by nature will be such that $0 \leq \eta_1 \leq 1/10$ and $1/4 \leq \eta_2 \leq 1/3$, for instance, or that $\eta_1^2 + \eta_2^2 + \eta_3^2 = 3/4$. In this case we are confronted with a game with constraints, where the max-min theorem may no longer be true (the theorem has not been shown to be true, in general, unless the set of strategies available to nature constitutes a convex subset of Euclidean space); this will be of no concern to the statistician, however, who is interested only in the question of the existence and value of

$$\max_{\xi \in \mathbf{A}_1} \min_{\eta \in \mathbf{A}_2} E(\xi, \eta)$$

(where, of course, $\mathbf{A}_2$ now denotes the set of strategies to which nature is limited by the knowledge available to S).

We turn now to an example which illustrates the application of game theory to quality control.

EXAMPLE 13.4. A certain very costly object is to be manufactured, which consists of three similar but connected parts such that the whole object will be satisfactory only if each of the three parts is satisfactory. For the sake of definiteness we can think of the object, for instance, as a wheel with three spokes; for the wheel to be satisfactory, each spoke must, let us say, have a certain tensile strength. (In order to understand why the wheel is expensive, we can think of it as being rather large and as being cut, perhaps, out of a single piece of quartz.)

The consumer, A, of this wheel (the government, or it may be an astronomical laboratory) is not itself prepared to manufacture wheels; therefore the following contract is made with a manufacturer, M: A agrees to pay M a certain amount to make the wheel in accordance with certain gross specifications (as to material, dimensions, etc.); after the wheel is completed subject to these specifications, M can either junk it (its salvage value will be taken to be 0) or can turn it over to A, who will test it in operation; if it is found satisfactory, A pays M an additional amount α; if it is unsatisfactory, M pays A a penalty β (α and β are, of course, both positive).

Since A has already paid M to produce the wheel, however, and because A does not wish to leave open the possibility that M might manufacture the wheel merely for the sake of this initial payment, A imposes the additional condition that M is not to junk the wheel unless a certain test shows it to be defective (though, if M wishes to do so, he may turn it over to A without making any test). This test is one which can be made on each of the three spokes, and it costs M the amount γ to test each spoke. The test is adequate in the following sense: the wheel will be found satisfactory by A if, and only if, each spoke of it would pass the test, if the test were performed on it.

The problem of whether to test some, or all, of the spokes before accepting the wheel (i.e., before turning it over to A) now confronts M. There are four possible courses of action (pure strategies) open to him:

I. Accept the wheel without any test at all.

II. Choose one of the three spokes at random and test it. If this spoke is satisfactory, accept the wheel. If it is unsatisfactory, reject the wheel.

III. Test a spoke chosen at random. If this spoke is defective, reject the wheel. If it is satisfactory, choose one of the remaining two spokes at random and test it. If this spoke is defective, reject the wheel. If it is satisfactory, accept the wheel.

IV. Test a spoke chosen at random. If this spoke is defective, reject the wheel. If it is satisfactory, choose one of the remaining two spokes at random and test it. If this spoke is defective, reject the wheel. If it is satisfactory, test the third spoke and accept or reject the wheel according as this last spoke does or does not pass the test.

Moreover, there are just four possibilities open to nature. It can happen that none, one, two, or three of the spokes are defective. We indicate these strategies for nature by the numerals 0, 1, 2, and 3.

Let us examine what will be the profit to M for various combinations of these strategies.

If M plays strategy I and nature plays strategy 0, then M makes no test, and none of the spokes are defective. Thus A will find the wheel satisfactory, and will pay α to M. The payoff to M in this case is α.

If M plays strategy II and nature plays strategy 0, then M will make just one test, and the wheel will be satisfactory to A. Thus M will be paid α by A, but will have to spend δ for the test. Hence the payoff to M is $\alpha - \delta$.

Similarly, if M plays III and nature plays 0, then the payoff to M will be $\alpha - 2\delta$. And if M plays IV and nature plays 0, then the payoff to M will be $\alpha - 3\delta$.

If M plays I and nature plays 1, then M turns the wheel over to A, who finds it defective. Thus M must pay A the penalty β; hence the payoff is $-\beta$. (In this case, since M makes no test, there is no cost to him for testing.)

It is seen that the payoff to M is also $-\beta$ in case M plays I and nature plays either 2 or 3.

If nature plays 3, then all the spokes are defective. Thus, if M makes any test at all, he will discover on the first test that the wheel is defective and hence will reject it. Thus the payoff to M is merely the cost of testing the one spoke, namely $-\gamma$. This holds if nature plays 3 and M plays either II, III, or IV.

If M plays II and nature plays 1, then the probability that M discovers the defective spoke is ⅓, and the probability that he does not is ⅔. If he discovers the defective spoke, the payoff to him is $-\gamma$. If he does not discover it, then he has to pay the penalty β, besides having to pay for the test; thus in this case the payoff is $-\beta - \gamma$. Hence the expectation of M is

$$\frac{1}{3}(-\gamma) + \frac{2}{3}(-\beta - \gamma) = -\frac{2}{3}\beta - \gamma.$$

In a similar fashion, we see that in case M plays II and nature plays 2, then the expected payoff to M is

$$\frac{2}{3}(-\gamma) + \frac{1}{3}(-\beta - \gamma) = -\frac{1}{3}\beta - \gamma.$$

If M plays III and nature plays 1, then the probability that M discovers the defective spoke on his first test is ⅓; the probability that he discovers it on the second test is therefore ⅔ · ½ = ⅓; and hence the probability that the defective spoke goes undetected is 1 − (⅓ + ⅓) = ⅓. If the defective spoke is discovered on the first test, the payoff to M is $-\gamma$; if it is discovered on the second test, the payoff is -2γ; and if it remains undiscovered, the payoff is $-\beta - 2\gamma$. Hence the expected payoff to M is

$$\frac{1}{3}(-\gamma) + \frac{1}{3}(-2\gamma) + \frac{1}{3}(-\beta - 2\gamma) = -\frac{1}{3}\beta - \frac{5}{3}\gamma.$$

By continuing in this way we arrive at Matrix 5.

MATRIX 5

	0	1	2	3
I	α	$-\beta$	$-\beta$	$-\beta$
II	$\alpha - \gamma$	$-\frac{2}{3}\beta - \gamma$	$-\frac{1}{3}\beta - \gamma$	$-\gamma$
III	$\alpha - 2\gamma$	$-\frac{1}{3}\beta - \frac{5}{3}\gamma$	$-\frac{4}{3}\gamma$	$-\gamma$
IV	$\alpha - 3\gamma$	-2γ	$-\frac{4}{3}\gamma$	$-\gamma$

The value of this game to M, and the optimal strategies for him in playing it, depend on the relative values of α, β, and γ.

Thus, for instance, if we take $\alpha = 100$, $\beta = 300$, and $\gamma = 3$ (so that the penalty for delivering a defective wheel is very large in comparison with the cost of testing), then we obtain Matrix 6.

MATRIX 6

	0	1	2	3
I	100	−300	−300	−300
II	97	−203	−103	−3
III	94	−105	−4	−3
IV	91	−6*	−4	−3

The place in this matrix marked with an asterisk is a saddle-point. Thus the worst nature can do to M is to make just one spoke of the wheel defective, and the best strategy for M is to use strategy IV (i.e., to test all spokes of the wheel). By playing strategy IV, M can be certain that his loss will not be greater than 6. (In making the original contract, therefore, M might reasonably have insisted that the initial payment from A be at least 6 units greater than his cost of production.)

On the other hand, if we have $\alpha = 100$, $\beta = 300$, and $\gamma = 303$ (so that it is costly to make the test), then we obtain Matrix 7.

MATRIX 7

	0	1	2	3
I	100	−300*	−300*	−300*
II	−203	−503	−403	−303
III	−506	−605	−404	−303
IV	−809	−606	−404	−303

Here the three elements marked with asterisks are all saddle-points. Thus the worst thing nature can do to M is to make one or more spokes (it does not matter how many) defective. The best thing for M to do is to play strategy I (i.e., to make no test at all). Thus with the increased cost of testing he cannot now be sure but that his loss will fall as low as 300.

By proper choice of α, β, and γ, it can even happen that the matrix has no saddle-point. Thus if we take $\alpha = 100$, $\beta = 900$, and $\gamma = 300$, we obtain

Matrix 8, which has no saddle-point.

MATRIX 8

	0	1	2	3
I	100	−900	−900	−900
II	−200	−900	−600	−300
III	−500	−800	−400	−300
IV	−800	−600	−400	−300

It is readily verified that an optimal strategy for nature is now $\| \frac{1}{4} \;\; \frac{3}{4} \;\; 0 \;\; 0 \|$, that an optimal strategy for M is $\| \frac{1}{6} \;\; 0 \;\; 0 \;\; \frac{5}{6} \|$, and that the value of the game is -650. Thus the most disagreeable thing nature can do is to make the probability ¼ that the wheel will be perfect, and ¾ that it will have just one defective spoke. The best way for M to act is to throw a die and then behave as follows: if the die shows a 6, to pass the wheel without any test; otherwise, to test all three spokes. Since the value is -650, M might be justified in demanding at least 650, in addition to his cost of production, as an initial payment.

BIBLIOGRAPHICAL REMARKS

For applications of the theory of games to statistics, the reader is referred to the following: Wald [5], [6], [7], and [8]; Arrow, Blackwell, and Girshick [1]; and Dvoretzky, Wald, and Wolfowitz [1].

EXERCISES

1. Find the value of the game and an optimal strategy for S if in Matrix 1 of Example 13.1 we take $\alpha = 100$, $\beta = -100$, $\gamma = -200$, and $\delta = 110$.

2. Show by means of Matrix 1 of Example 13.1 that any optimal mixed strategy for S in Example 13.1 will always assign frequency 0 to strategy VII. What is the intuitive significance of this fact?

3. Show that if we have $\beta \geq (\alpha + \gamma)/2$ and $\beta \geq \alpha - \delta$, then there exists an optimal strategy for nature in Matrix 1 of Example 13.1 which assigns frequency 0 to strategy 1.

4. Show that, under the supposition that $0 < \delta$ and $\gamma < \beta < \alpha$, Matrix 1 does not have a saddle-point.

5. In Example 13.1 we supposed that the payments made to S for guessing correctly or incorrectly are as follows:

	0	1	2
0	α	β	γ
1	β	α	β
2	γ	β	α

where the 0, 1, and 2 at the left represent the number of black balls that S guesses are in the urn, and the numbers at the top indicate the number of black balls that are actually in the urn; thus if S guesses that one ball is black while two are in fact black, then he gets paid β. It can happen, however, that the payments (or penalties) are different according to whether the guess is too high or too low. Thus suppose that the payments are as follows:

	0	1	2
0	α	β	γ
1	β'	α	β
2	γ'	β'	α

Find the payoff matrix (corresponding to Matrix 1) in this case. Find the value of the game and an optimal strategy for S in case we take $\alpha = 100$, $\beta = 1$, $\beta' = 0$, $\gamma = -100$, $\gamma' = -101$, and $\delta = 50$.

6. In many cases it happens that it is less than twice as expensive to make a test twice as it is to make it once. How would Matrix 1 be affected if we suppose that it costs δ to test one ball, but only $\delta + \delta'$ (where $0 < \delta' < \delta$) to test two?

7. Formulate a problem similar to Example 13.1 for the case in which the urn contains three balls, and write out the matrix corresponding to Matrix 1 (it will have 19 rows and 4 columns). Find optimal strategies for S for some typical values of the parameters.

8. Formulate a problem similar to Example 13.1 for the case in which the urn contains two balls, each of which may be either black, white, or red, and write out the matrix corresponding to Matrix 1 (it will have 34 rows and 6 columns). Find optimal strategies for S for some typical values of the parameters.

9. Find an optimal strategy for M in Example 13.4 if we have $\alpha = 100$, $\beta = 210$, and $\gamma = 102$.

10. Show that, under the assumption that $\beta > 0$ and $\gamma > 0$, there is an optimal strategy for nature in Example 13.4, which assigns the frequency 0 to the strategies 2 and 3.

11. For what values of the parameters will Matrix 5 have a saddle-point?

12. Manufacturer M is to produce an article for consumer A. The consumer agrees to pay M a certain amount to make the article in accordance with certain gross specifications. After it is turned over to A, if A finds it satisfactory, he is to pay M an additional amount α. If it is unsatisfactory, M is to pay A a penalty β. There are two independent tests which can be made on the article by M and which are such that if the article passes both tests, it will be found satisfactory by A. It costs M the amount γ to make one of these tests and the amount γ' to make the other. We suppose that $0 < \alpha$, $0 < \beta$, and $0 < \gamma < \gamma'$. Enumerate the possible strategies open to M and write down the matrix similar to Matrix 5 of Example 13.4. Find optimal strategies for M for some representative values of the parameters.

13. Suppose that, in the situation corresponding to Matrix 2, the statistician has past experience which tells him that the probability that neither of the balls is black is not greater than $1/20$. Find the value of the game and an optimal strategy for S.

CHAPTER 14

LINEAR PROGRAMMING

In this chapter we shall discuss special types of minimization, and maximization, problems, called *linear programming*, which often arise in economic theory and which are closely related to the theory of games. We begin with an example.

Suppose that a man has found that in order to be healthy it is necessary for him to take daily at least b_1 units of a chemical compound B_1 (a vitamin, perhaps, or a particular metallic salt) and at least b_2 units of a chemical compound B_2. Suppose, moreover, that it is not possible for him to buy either B_1 or B_2 in pure form but that he can buy a medicine C_1 at p_1 cents per ounce, each ounce of which contains a_{11} units of B_1 and a_{12} units of B_2, and that he can also buy a medicine C_2 at p_2 cents per ounce, each ounce of which contains a_{21} units of B_1 and a_{22} units of B_2. The man wishes to decide what proportions of C_1 and C_2 he should purchase daily in order to obtain the required amount of B_1 and B_2 at minimum cost. Now if the man buys x_1 ounces of C_1 and x_2 ounces of C_2, then the cost to him will be

$$p_1x_1 + p_2x_2,$$

and the amounts of B_1 and B_2 which he will obtain will be, respectively,

$$a_{11}x_1 + a_{21}x_2$$

and

$$a_{12}x_1 + a_{22}x_2;$$

hence he wishes to choose x_1 and x_2 non-negative and such as to satisfy

$$a_{11}x_1 + a_{21}x_2 \geq b_1,$$
$$a_{12}x_1 + a_{22}x_2 \geq b_2,$$

while at the same time making

$$p_1x_1 + p_2x_2$$

as small as possible.

For given numerical values of the parameters, this problem can be solved by simple methods of analytic geometry. Thus, for example, suppose that

$$a_{11} = a_{22} = 1, \qquad a_{12} = a_{21} = 5, \qquad b_1 = b_2 = 15,$$
$$p_1 = 1, \qquad p_2 = 3.$$

Then we want to pick out, from the pairs of numbers $\| x_1 \quad x_2 \|$ satisfying

$$x_1 \geq 0,$$
$$x_2 \geq 0,$$
$$x_1 + 5x_2 \geq 15,$$
$$5x_1 + x_2 \geq 15,$$

a pair $\| x_1 \quad x_2 \|$ which will make the linear form

$$x_1 + 3x_2$$

a minimum. We now plot the lines

$$x_1 + 5x_2 = 15$$

and

$$5x_1 + x_2 = 15$$

as in Fig. 1, and we conclude that the pairs $\| x_1 \quad x_2 \|$ which satisfy the given

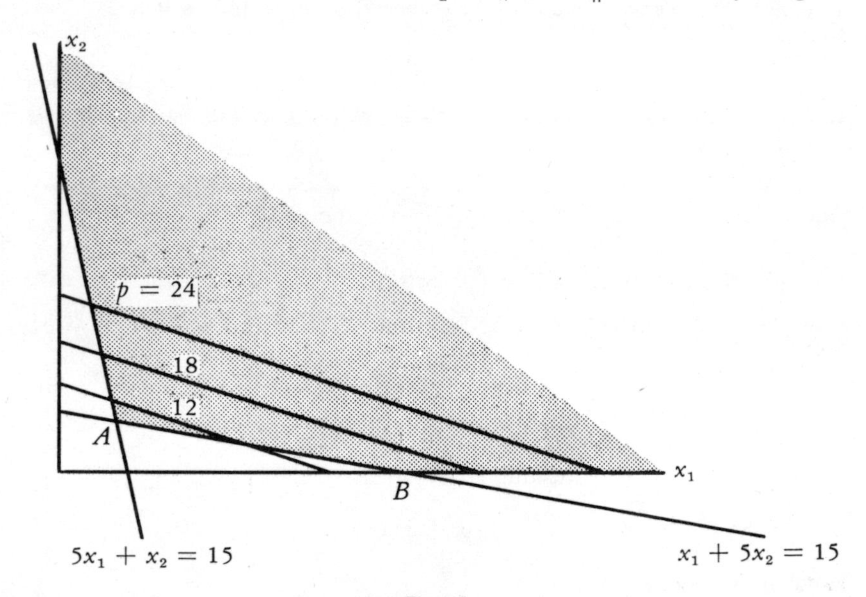

Fig. 1

inequalities are the coordinates of points in the shaded region of the figure. We also plot the family of parallel lines

$$x_1 + 3x_2 = p,$$

for various values of p, as indicated in the figure. It is now clear from Fig. 1 that the smallest value of p will be obtained by choosing $\| x_1 \quad x_2 \|$ in such a way that the line

$$x_1 + 3x_2 = p$$

passes through A, the point of intersection of the lines

$$x_1 + 5x_2 = 15,$$

$$5x_1 + x_2 = 15.$$

Solving these two equations simultaneously, we find

$$x_1 = x_2 = \frac{5}{2}.$$

Thus the required medication will be obtained most cheaply by taking 5/2 ounces of each of the two medicines C_1 and C_2; by doing this, the exact required amount of both B_1 and B_2 will be obtained at a cost of

$$\frac{5}{2} + 3 \cdot \frac{5}{2} = 10.$$

It should be remarked that for some values of p_1 and p_2 the solution of the above problem will not make the man obtain exactly b_1 and b_2 units of B_1 and B_2, respectively. Thus suppose that the values of a_{11}, a_{12}, a_{21}, a_{22}, b_1, and b_2 are as given above, but that $p_1 = 1$ and $p_2 = 20$. In this case it is found that the cheapest combination of medicines is obtained by taking $\| x_1 \quad x_2 \|$ to be the coordinates of the point B in Fig. 1, i.e., by taking $x_1 = 15$ and $x_2 = 0$. In this case the man obtains the required medication most cheaply by buying only C_1, and no C_2 at all; by doing this, he obtains at a cost of

$$15 + 20 \cdot 0 = 15$$

the amount

$$15 + 5 \cdot 0 = 15$$

of B_1, but the amount

$$5 \cdot 15 + 0 = 75 > 15$$

of B_2.

We also notice that the usual methods of finding maxima and minima by differential calculus are not very helpful in solving problems like those given above. This is because the minimum, or maximum, values of the functions are obtained by taking points on the boundary of the shaded region—not at points where derivatives vanish.

Finally, it should be remarked that the geometrical method we have used to solve the above problem would become extremely cumbrous for problems involving a large number of variables. Thus, for instance, if the man had to choose among six medicines, instead of between two, this method would involve the discussion of regions of 6-dimensional space. Since problems of this kind are very important (they arise, for instance, in connection with the proper routing of freight cars so as to transport a given quantity of goods at a minimum cost, or in a minimum time), it becomes important to have general methods of dealing with them.

Thus a special branch of mathematics has arisen which is called *linear programming.* By a linear-programming problem is meant a problem of finding the minimum, or the maximum, value assumed by a linear function, where the variables are subject to linear inequalities. The general linear-programming problem can be formulated as follows: To pick out, from all n-tuples $\| y_1 \quad \cdots \quad y_n \|$ satisfying

$$\begin{array}{ccccccc} a_{11}y_1 & + & \cdots & + & a_{1n}y_n & \geq & b_1, \\ \vdots & & & & \vdots & & \vdots \\ a_{m1}y_1 & + & \cdots & + & a_{mn}y_n & \geq & b_m, \end{array}$$

an n-tuple $\| y_1 \quad \cdots \quad y_n \|$ which will minimize, or maximize, the linear function

$$p_1y_1 + \cdots + p_ny_n.$$

We shall not give anything like a general theory of linear programming, since this would be outside the scope of this book, but shall confine ourselves to indicating some relations between this subject and the theory of games. We shall see that the problem of solving an arbitrary rectangular game can be regarded as a special linear-programming problem and, conversely, that many linear-programming problems can be reduced to problems in game theory.

Before proceeding to our discussion of the relation of linear programming to game theory, however, we want to make a few obvious remarks about the existence of solutions of linear-programming problems.

In the first place, it should be clear that a linear-programming problem does not necessarily have a solution. It may happen, for instance, that the given inequalities are inconsistent. Thus there exists no solution to the problem of minimizing

$$y_1 + 2y_2,$$

subject to the inequalities

$$y_1 + y_2 \geq 3,$$
$$-y_1 - y_2 \geq 4.$$

Even when the inequalities are consistent, moreover, the problem may fail to have a solution because of the unboundedness of the set of points whose coordinates satisfy the inequalities. Thus there exists no solution to the problem of minimizing

$$1 - y_1 - y_2,$$

subject to the inequalities

$$y_1 \geq 1,$$
$$y_2 \geq 2.$$

On the other hand, a linear-programming problem can have more than one solution. Thus *any* $\| y_1 \quad y_2 \|$ is a solution to the problem of minimizing the function

$$0y_1 + 0y_2,$$

subject to the inequalities

$$0y_1 \geq 0,$$
$$0y_2 \geq 0.$$

It can easily be shown that any convex linear combination of solutions of a linear-programming problem is also a solution; thus a linear-programming problem, if it has more than one solution, has infinitely many.

Finally, it should be pointed out that many problems which are not prima-facie problems in linear programming can easily be transformed into such problems. Thus the problem of minimizing

$$p_1y_1 + p_2y_2 + p_3y_3$$

subject to the equations

$$a_{11}y_1 + a_{12}y_2 = b_1,$$
$$a_{21}y_1 + a_{22}y_2 = b_2,$$

is clearly equivalent to the problem of minimizing

$$p_1y_1 + p_2y_2 + p_3y_3$$

subject to the inequalities

$$\begin{aligned} a_{11}y_1 + a_{12}y_2 &\geq b_1, \\ -a_{11}y_1 - a_{12}y_2 &\geq -b_1, \\ a_{21}y_1 + a_{22}y_2 &\geq b_2, \\ -a_{21}y_1 - a_{22}y_2 &\geq -b_2, \end{aligned}$$

which is itself a linear-programming problem.

We wish now to show that the problem of solving an arbitrary rectangular game can be reduced to the problem of solving a linear-programming problem. We shall discuss this problem in connection with a game with a 3×2 matrix. (This restriction is imposed only in order to simplify the notation; the discussion would be exactly analogous for the case of an $m \times n$ game.)

Let the matrix of the game be

$$\begin{Vmatrix} a_{11} & a_{12} \\ a_{21} & a_{22} \\ a_{31} & a_{32} \end{Vmatrix},$$

and let v be the value of the game.

The value v is the smallest number z such that, for some element $\| y_1 \quad y_2 \|$ of $\mathbf{S}_2$,

$$\left.\begin{aligned} a_{11}y_1 + a_{12}y_2 &\leq z, \\ a_{21}y_1 + a_{22}y_2 &\leq z, \\ a_{31}y_1 + a_{32}y_2 &\leq z, \end{aligned}\right\} \tag{1}$$

and by Theorem 2.10 a strategy $\| y_1 \quad y_2 \|$ is optimal for P_1 if, and only if, y_1 and y_2 satisfy (1) when z is replaced by v. Now it is clear that three numbers y_1, y_2, and z satisfy (1) if, and only if, there are non-negative numbers z_1, z_2, and z_3 such that

$$\begin{aligned} a_{11}y_1 + a_{12}y_2 + z_1 &= z, \\ a_{21}y_1 + a_{22}y_2 + z_2 &= z, \\ a_{31}y_1 + a_{32}y_2 + z_3 &= z, \end{aligned}$$

i.e., such that

$$z = a_{11}y_1 + a_{12}y_2 + z_1,$$

$$(a_{21} - a_{11})y_1 + (a_{22} - a_{12})y_2 + z_2 - z_1 = 0,$$
$$(a_{31} - a_{11})y_1 + (a_{32} - a_{12})y_2 + z_3 - z_1 = 0.$$

Thus if we consider the system

$$\left.\begin{aligned} z = a_{11}y_1 + a_{12}y_2 + z_1, \\ (a_{21} - a_{11})y_1 + (a_{22} - a_{12})y_2 + z_2 - z_1 = 0, \\ (a_{31} - a_{11})y_1 + (a_{32} - a_{12})y_2 + z_3 - z_1 = 0, \\ y_1 + y_2 = 1, \quad y_1 \geq 0, \quad y_2 \geq 0, \\ z_1 \geq 0, \quad z_2 \geq 0, \quad z_3 \geq 0, \end{aligned}\right\} \tag{2}$$

we see that v is the smallest number z such that, for some numbers y_1, y_2, z_1, z_2, and z_3, the system (2) is satisfied. Moreover, a strategy $\| y_1 \quad y_2 \|$ is optimal for P_2 if, and only if, y_1, y_2, z_1, z_2, and z_3 (for some z_1, z_2, and z_3) satisfy (2) when z is replaced by v. Thus the problem of finding the value of the game and optimal strategies for player P_2 is reduced to the problem of solving system (2) as a linear-programming problem. In a similar way we can take care of the problem of finding optimal strategies for player P_1; here, of course, the linear-programming problem becomes a problem of maximizing a certain quantity.

Thus we have seen that the problem of solving an arbitrary rectangular game is reducible to linear programming of a special form:

$$\left.\begin{aligned} z = p_1u_1 + \cdots + p_nu_n, \\ a_{11}u_1 + \cdots + a_{1n}u_n \geq b_1, \\ \vdots \qquad \vdots \qquad \vdots \\ a_{m1}u_1 + \cdots + a_{mn}u_n \geq b_m, \\ u_1 \geq 0, \\ \vdots \\ u_n \geq 0. \end{aligned}\right\} \tag{3}$$

(In the case of the game just discussed we have $n = 5$ and $u_1 = y_1$, $u_2 = y_2$, $u_3 = z_1$, $u_4 = z_2$, $u_5 = z_3$. The feature which makes this a *special* problem in linear programming is the inclusion of the last n special inequalities $u_1 \geq 0, \cdots, u_n \geq 0$.) It is true, conversely, that every linear-programming problem of this special form is equivalent to a problem about a rectangular game. We shall not carry out the proof of this reduction, however, but shall confine ourselves to writing down the matrix of the corresponding game. For the linear-programming problem given by system (3), supposing that we wish to minimize z, the matrix of the game is:

$$B = \begin{Vmatrix} 0 & \cdots & 0 & a_{11} & \cdots & a_{m1} & -p_1 \\ \vdots & & \vdots & \vdots & & \vdots & \vdots \\ 0 & \cdots & 0 & a_{1n} & \cdots & a_{mn} & -p_n \\ -a_{11} & \cdots & -a_{1n} & 0 & \cdots & 0 & b_1 \\ \vdots & & \vdots & \vdots & & \vdots & \vdots \\ -a_{m1} & \cdots & -a_{mn} & 0 & \cdots & 0 & b_m \\ p_1 & \cdots & p_n & -b_1 & \cdots & -b_m & 0 \end{Vmatrix}.$$

The situation can now be shown to be as follows: the linear-programming problem determined by system (3) has a solution if, and only if, there is an optimal strategy $\| x_1 \quad \cdots \quad x_{m+n+1} \|$, for the game whose matrix is B, such that $x_{m+n+1} \neq 0$; moreover, if $\| x_1 \quad \cdots \quad x_{m+n+1} \|$ is any optimal strategy for the game such that $x_{m+n+1} \neq 0$, and if we set

$$u_1 = \frac{x_1}{x_{m+n+1}},$$

$$\vdots$$

$$u_n = \frac{x_n}{x_{m+n+1}},$$

and

$$z = p_1 u_1 + \cdots + p_n u_n,$$

then $u_1, \cdots, u_n$ and z constitute a solution of the linear-programming problem.

BIBLOGRAPHICAL REMARKS

For a proof of the equivalence of the problem of solving a game with a linear-programming problem, see Dantzig [2] or Gale, Kuhn, and Tucker [3]. For other results in connection with linear programming, see Dantzig [1], [3], and [4] and Dantzig and Wood [1].

EXERCISES

1. Find the maximum value of the linear function

$$x + y,$$

for x and y satisfying the inequalities

$$y - x \leq 10,$$
$$2y + x \leq 36,$$
$$y + 2x \leq 45.$$

2. Find the maximum value of the linear function

$$2y - x,$$

for x and y satisfying the inequalities of Exercise 1.

3. Find the maximum value of the linear function

$$y + 2x,$$

for x and y satisfying the inequalities of Exercise 1. Find all pairs $\| x \quad y \|$ which give this maximum.

4. Let p, q, a, b, and c be numbers such that

$$\begin{vmatrix} p & q \\ a & b \end{vmatrix} \neq 0.$$

Show that the following linear-programming problem has no solution: To find the maximum value of the function

$$px + qy,$$

for x and y satisfying the inequality

$$ax + by \geq c.$$

5. Supposing that p, q, b_1, b_2, a_{11}, a_{12}, a_{21}, and a_{22} are numbers such that

$$\begin{vmatrix} a_{11} & a_{12} \\ a_{21} & a_{22} \end{vmatrix} > 0,$$

show that the following linear-programming problems are equivalent:

Problem A. To find the maximum of the function

$$px + qy,$$

where x and y are subject to the inequalities

$$a_{11}x + a_{12}y \geq b_1,$$
$$a_{21}x + a_{22}y \geq b_2;$$

Problem B. To find the maximum of the function

$$(a_{22}p - a_{21}q)u_1 + (a_{11}q - a_{12}p)u_2,$$

where u_1 and u_2 are subject to the inequalities

$$u_1 \geq b_1, \qquad u_2 \geq b_2;$$

Problem C. To find the maximum of the function

$$(a_{22}p - a_{21}q)w_1 + (a_{11}q - a_{12}p)w_2 + (a_{22}p - a_{21}q)b_1 + (a_{11}q - a_{12}p)b_2,$$

where w_1 and w_2 are subject to the inequalities

$$w_1 \geq 0, \qquad w_2 \geq 0.$$

6. Formulate and solve an exercise similar to Exercise 5, under the hypothesis that

$$\begin{vmatrix} a_{11} & a_{12} \\ a_{21} & a_{22} \end{vmatrix} < 0.$$

7. Discuss Problem *A* of Exercise 5 under the assumption that

$$\begin{vmatrix} a_{11} & a_{12} \\ a_{21} & a_{22} \end{vmatrix} = 0.$$

8. Let $u = \| u_1 \quad \cdots \quad u_n \|$ and $u' = \| u'_1 \quad \cdots \quad u'_n \|$ be two vectors which satisfy the inequalities of (3) and make z a maximum. Show that every convex linear combination of u and u' also satisfies these inequalities and makes z a maximum.

9. Let $\| x_1 \quad \cdots \quad x_{m+n+1} \|$ be an optimal strategy for the game whose matrix is

$$B = \left\| \begin{matrix} 0 & \cdots & 0 & a_{11} & \cdots & a_{m1} & -p_1 \\ \vdots & & \vdots & \vdots & & \vdots & \vdots \\ 0 & \cdots & 0 & a_{1n} & \cdots & a_{mn} & -p_n \\ -a_{11} & \cdots & -a_{1n} & 0 & \cdots & 0 & b_1 \\ \vdots & & \vdots & \vdots & & \vdots & \vdots \\ -a_{m1} & \cdots & -a_{mn} & 0 & \cdots & 0 & b_m \\ p_1 & \cdots & p_n & -b_1 & \cdots & -b_m & 0 \end{matrix} \right\|,$$

and suppose that $x_{m+n+1} \neq 0$. Let

$$u_1 = \frac{x_1}{x_{m+n+1}},$$

$$\vdots$$

$$u_{m+n} = \frac{x_{m+n}}{x_{m+n+1}}.$$

Show that $\| u_1 \quad \cdots \quad u_n \|$ is a solution of the following linear-programming problem: To find numbers $u_1, \cdots, u_n$ which will minimize the function

$$p_1u_1 + \cdots + p_nu_n,$$

subject to the inequalities

$$\begin{aligned} a_{11}u_1 + \cdots + a_{1n}u_n &\geq b_1, \\ \vdots \qquad\qquad \vdots \quad & \quad \vdots \\ a_{m1}u_1 + \cdots + a_{mn}u_n &\geq b_n, \\ u_1 &\geq 0, \\ &\vdots \\ u_n &\geq 0. \end{aligned}$$

Hint: Notice first that the matrix B is antisymmetric and hence that the value of the game is 0.

CHAPTER 15

ZERO-SUM n-PERSON GAMES

1. Characteristic Functions. So far we have confined our attention to two-person games, and in this domain we have been able to give intuitively acceptable definitions of the value of the game (to each player) and of optimal strategies. We now turn our attention to finite games with more than two players.

For this wider class of games there is not available, unfortunately, any theory which is intuitively as acceptable as is the theory for two-person games. Although a large part of von Neumann and Morgenstern's book [1] (roughly 400 out of 600 pages) is devoted to games with more than two players, mathematicians generally seem to have been dissatisfied with the theory there developed. Comparatively little research has been done in this branch of game theory during the last few years.

Despite the fact that the theory of n-person games, for $n > 2$, is not in an altogether satisfactory state, it is important for the student to become familiar with the theory in its present form; for there are certainly some elements of soundness in the present theory, even though it may be improved in the future. This section will accordingly be devoted to an exposition of the elements of the theory of zero-sum n-person games as it was given by von Neumann and Morgenstern; when possible, we shall, of course, present intuitive justifications for the definitions introduced, but we shall give the major definitions of von Neumann and Morgenstern in any case, even when the notions introduced appear strange and irrelevant.

We begin by remarking that when a finite n-person game involves partial information, more than one move by some of the players, chance moves, and the like, it is still possible, by introducing the notion of a strategy, to describe an equivalent "rectangular" game; i.e., a game in which each player makes just one choice from a finite set and in which each player chooses in ignorance of the choices of the other players. This has been illustrated in Example 5.10.

Thus we can confine our discussion to *rectangular* zero-sum n-person games. Such a game has just n moves; in the ith move (for $i = 1, 2, \cdots, n$) player P_i, not being informed about the outcome of any of the previous moves, chooses a number x_i from the finite set C_i. After the n moves are completed, player P_i (for $i = 1, \cdots, n$) receives the amount

$$M_i(x_1, x_2, \cdots, x_n).$$

Since the game is zero-sum, the functions $M_1, M_2, \cdots, M_n$ satisfy (identically in $x_1, \cdots, x_n$) the equation

$$\sum_{i=1}^{n} M_i(x_1, \cdots, x_n) = 0.$$

It is necessary now to make a few remarks about the nature of the values assumed by our payoff functions. It is clear that these values should be entities regarded as good or bad by the various players and that $+2$, for example, should be preferred to $+1$ or 0 or -1. But in order to give any reasonably adequate treatment of games with more than two players, it seems to be necessary to make the additional assumption that the values of the payoff functions be *objective* and *transferable*. Thus $+2$ must be understood to mean something like "getting two dollars" or "getting two pounds of spaghetti," and not to mean something like "receiving two units of gustatory gratification"; for spaghetti is something external to us, which can be passed about from one man to another, while gustatory gratification exists only in the mind and cannot readily be transferred. (Indeed, it is by no means obvious that any operational meaning could be assigned to such a statement as "Tony obtains just as much gustatory gratification from a plate of spaghetti as Chang does from a bowl of rice.")

Since the payoff is thus regarded as objective and transferable, there is nothing to keep the several players from making payments to each other ("on the side," so to speak) in compensation for certain kinds of cooperation. Thus it may happen, for example, that player P_1 will make a great deal more if he persuades P_2 to choose a certain number (e.g., $M_1(2, 2, z)$ may be large in comparison to $M_1(x, y, z)$ for $y \neq 2$), even though this choice will not result in any immediate benefit to P_2 (e.g., $M_2(x, y, z)$ may be independent of x, y, and z and may always assume a constant value v). In such a case it is clear that it will behoove P_1 to make a side-payment to P_2 in order to persuade him to make the choice advantageous to P_1. The theory of n-person games is largely concerned with the questions of what combinations of players ("coalitions") will be formed and what payments the players can be expected to make to each other as inducements to join the various coalitions.

To denote the players of a game we have heretofore used the symbols $P_1, P_2, \cdots, P_n$. It is slightly more convenient, however, henceforth to denote them merely by $1, 2, \cdots, n$. We shall denote the set of players by N, so that

$$\mathsf{N} = \{1, 2, \cdots, n\}.$$

Now suppose that the players of set N group themselves into two coalitions, T and $\mathsf{N} - \mathsf{T} = -\mathsf{T}$, so that the members of T cooperate with each other in choosing strategies, and similarly the members of $-\mathsf{T}$ cooperate with each other. Then we can consider T and $-\mathsf{T}$ as the two players of a two-person game. Thus suppose that, in the original game, player i (for $i = 1, \cdots, n$) chooses a number from the finite set C_i and suppose that $\mathsf{T} = \{i_1, \cdots, i_r\}$ and $-\mathsf{T} = \{j_1, \cdots, j_s\}$. Then, in the artificial two-person game with players T and $-\mathsf{T}$, the composite player T chooses an element from the Cartesian product C of the sets $\mathsf{C}_{i_1}, \cdots, \mathsf{C}_{i_r}$, and similarly the composite player $-\mathsf{T}$ chooses an element from the Cartesian product C' of the sets $\mathsf{C}_{j_1}, \cdots, \mathsf{C}_{j_s}$.

Let $\mathsf{C} = \{A_1, \cdots, A_u\}$ and $\mathsf{C}' = \{B_1, \cdots, B_v\}$. Then it is clear that there are functions $M_1, M_2, \cdots, M_n$ such that

$$M_i(A_j, B_k)$$

is the payoff to player i when T uses strategy A_j and $-\mathsf{T}$ uses strategy B_k; in fact, if $A_j = \| x_{j_1} \quad \cdots \quad x_{j_r} \|$ and $B_k = \| x_{k_1} \quad \cdots \quad x_{k_s} \|$, then

$$M_i(A_j, B_k) = M_i(x_1, \cdots, x_n),$$

where M_i is the ith payoff function of the original game. If T uses strategy A_j and $-\mathsf{T}$ uses strategy B_k, then the total payoff to T is

$$\sum_{i \in \mathsf{T}} M_i(A_j, B_k)$$

and the total payoff to $-\mathsf{T}$ is

$$\sum_{i \in -\mathsf{T}} M_i(A_j, B_k).$$

We denote the first of the above sums by

$$M_{\mathsf{T}}(A_j, B_k)$$

and the second by

$$M_{-\mathsf{T}}(A_j, B_k).$$

From the fact that the original game was zero-sum, it follows immediately that

$$M_{-\mathsf{T}}(A_j, B_k) = -M_{\mathsf{T}}(A_j, B_k).$$

Since the set C contains u elements, a mixed strategy for T is a member of S_u. Similarly, a mixed strategy for $-\mathsf{T}$ is a member of S_v.

Now if T uses the mixed strategy $\| \alpha_1 \quad \cdots \quad \alpha_u \|$ and $-\mathsf{T}$ uses the mixed strategy $\| \beta_1 \quad \cdots \quad \beta_v \|$, then the total expectation of T is

$$\sum_{k=1}^{v} \sum_{j=1}^{u} M_{\mathsf{T}}(A_j, B_k)\alpha_j\beta_k$$

and the total expectation of $-\mathsf{T}$ is

$$\sum_{k=1}^{v} \sum_{j=1}^{u} M_{-\mathsf{T}}(A_j, B_k)\alpha_j\beta_k .$$

Setting $\alpha = \| \alpha_1 \quad \cdots \quad \alpha_u \|$ and $\beta = \| \beta_1 \quad \cdots \quad \beta_v \|$, we can represent these expectations by

$$E_{\mathsf{T}}(\alpha, \beta) \qquad \text{and} \qquad E_{-\mathsf{T}}(\alpha, \beta),$$

respectively. From the fact that, for all j and k, $M_{-\mathsf{T}}(A_j, A_k) = M_{\mathsf{T}}(A_j, A_k)$, it follows immediately that

$$E_{-\mathsf{T}}(\alpha, \beta) = -E_{\mathsf{T}}(\alpha, \beta) . \tag{1}$$

From the theory of rectangular two-person games we now see that

$$\max_{\alpha \in \mathsf{S}_u} \min_{\beta \in \mathsf{S}_v} E_{\mathsf{T}}(\alpha, \beta) = \min_{\beta \in \mathsf{S}_v} \max_{\alpha \in \mathsf{S}_u} E_{\mathsf{T}}(\alpha, \beta).$$

We set

$$v(\mathsf{T}) = \max_{\alpha \in \mathsf{S}_u} \min_{\beta \in \mathsf{S}_v} E_{\mathsf{T}}(\alpha, \beta).$$

We have thus a function v, which is defined for every subset T of N, and whose value, for each T, represents the total amount the members of T can expect to obtain if they make a coalition; we call v the *characteristic function* of the game. In view of the fact that the payoff is transferable, it seems fairly plausible that all questions about coalitions and side-payments can be settled solely on the basis of the characteristic function; e.g., if a man's price for joining a certain coalition in a given game is π, then it will also be π in any other game with the same characteristic function. This question could, of course, be definitely settled if we could give an adequate theory of n-person games solely in terms of characteristic functions. Von Neumann and Morgenstern believe that they have succeeded in doing this.

We shall now establish some of the mathematical properties of characteristic functions.

Theorem 15.1. Let v be the characteristic function of an n-person zero-sum game, where N is the set of players. Then

(i) $v(\mathsf{N}) = 0$;
(ii) for any subset T of N, $v(-\mathsf{T}) = -v(\mathsf{T})$;
(iii) if R and T are mutually exclusive subsets of N, then

$$v(\mathsf{R} \cup \mathsf{T}) \geq v(\mathsf{R}) + v(\mathsf{T}).$$

Proof. To prove (i), we notice that if we divide the players of set N into the two coalitions N and $\mathsf{N} - \mathsf{N} = \Lambda$, then there are no strategies for the second coalition (the empty set of players), and a (pure) strategy for the first coalition is an ordered n-tuple $\| x_1 \quad \cdots \quad x_n \|$, where $x_i \in \mathsf{C}_i$ (for $i = 1, \cdots, n$). Hence each of the functions M_i is a function of only one variable, namely, an element A_j of C. Thus, for any strategy A_j chosen by N, the total payoff to N (making use of the fact that the original game was zero-sum) is

$$M_{\mathsf{N}}(A_j) = \sum_{i \in \mathsf{N}} M_i(A_j) = 0.$$

From this it immediately follows that

$$v(\mathsf{N}) = \max_{\alpha \in \mathsf{S}_u} E_{\mathsf{N}}(\alpha) = \max_{\alpha \in \mathsf{S}_u} \left[\sum_{j=1}^{u} M_{\mathsf{N}}(A_j)\alpha_j \right]$$

$$= \max_{\alpha \in \mathsf{S}_u} \left[\sum_{j=1}^{u} 0 \cdot \alpha_j \right] = 0,$$

as was to be shown.

To prove (ii), we have

$$v(-\mathsf{T}) = \max_{\alpha \in \mathsf{S}_v} \min_{\beta \in \mathsf{S}_u} E_{-\mathsf{T}}(\alpha, \beta) = \max_{\alpha \in \mathsf{S}_v} \min_{\beta \in \mathsf{S}_u} - E_{\mathsf{T}}(\alpha, \beta)$$

$$= - \min_{\alpha \in \mathsf{S}_v} \max_{\beta \in \mathsf{S}_u} E_{\mathsf{T}}(\alpha, \beta) = - \max_{\beta \in \mathsf{S}_u} \min_{\alpha \in \mathsf{S}_v} E_{\mathsf{T}}(\alpha, \beta) = -v(\mathsf{T}).$$

We shall prove (iii) only for the case of a three-person game; the proof for the general case is essentially the same but requires a more complicated notation. Suppose, then, that $\mathsf{N} = \{1, 2, 3\}$; we are going to show that

$$v(\{1, 2\}) \geq v(\{1\}) + v(\{2\}).$$

Let the (pure) strategies for players 1, 2, and 3 be, respectively,

$$\mathsf{C}_1 = \{1, 2, \cdots, n_1\},$$

$$\mathsf{C}_2 = \{1, 2, \cdots, n_2\},$$
$$\mathsf{C}_3 = \{1, 2, \cdots, n_3\}.$$

Then the pure strategies for a two-person coalition $\{i, j\}$, where $i < j$, is the set C_{ij} of ordered couples $\| x \quad y \|$, where $x \in \mathsf{C}_i$ and $y \in \mathsf{C}_j$. Thus C_i has n_i pure strategies and C_{ij} has $n_i \cdot n_j$ pure strategies.

A mixed strategy for player 1 is a vector $\| \alpha_1 \quad \cdots \quad \alpha_{n_1} \|$ of S_{n_1}, which assigns the frequency α_i (for $i = 1, \cdots, n_1$) to the pure strategy i; there are similar strategies for players 2 and 3. A mixed strategy for the coalition $\{1, 2\}$ is a vector

$$\| \alpha_{1,1} \quad \alpha_{1,2} \quad \cdots \quad \alpha_{1,n_2} \quad \alpha_{2,1} \quad \cdots \quad \alpha_{n_1,n_2} \|$$

of $\mathsf{S}_{n_1 \cdot n_2}$, which assigns the frequency $\alpha_{i,j}$ to the pure strategy $\| i \quad j \|$; there are similar strategies for the coalitions $\{1, 3\}$ and $\{2, 3\}$.

Let $\alpha^* = \| \alpha_1^* \quad \cdots \quad \alpha_{n_1}^* \|$ be a member of S_{n_1} such that

$$v(\{1\}) = \max_{\eta \in \mathsf{S}_{n_1}} \min_{\xi \in \mathsf{S}_{n_2 \cdot n_3}} E_{\{1\}}(\eta, \xi) = \min_{\xi \in \mathsf{S}_{n_2 \cdot n_3}} E_{\{1\}}(\alpha^*, \xi). \tag{2}$$

Let $\beta^* = \| \beta_1^* \quad \cdots \quad \beta_{n_2}^* \|$ be a member of S_{n_2} such that

$$v(\{2\}) = \max_{\eta \in \mathsf{S}_{n_2}} \min_{\xi \in \mathsf{S}_{n_1 \cdot n_3}} E_{\{2\}}(\eta, \xi) = \min_{\xi \in \mathsf{S}_{n_1 \cdot n_3}} E_{\{2\}}(\beta^*, \xi). \tag{3}$$

Since it is readily verified that the vector

$$\| \alpha_1^* \cdot \beta_1^* \quad \alpha_1^* \cdot \beta_2^* \quad \cdots \quad \alpha_{n_1}^* \cdot \beta_{n_2}^* \|$$

belongs to $\mathsf{S}_{n_1 \cdot n_2}$, we now have

$$\begin{aligned}
v(\{1, 2\}) &= \max_{\eta \in \mathsf{S}_{n_1 \cdot n_2}} \min_{\xi \in \mathsf{S}_{n_3}} E_{\{1,2\}}(\eta, \xi) \\
&= \max_{\eta \in \mathsf{S}_{n_1 \cdot n_2}} \min_{\xi \in \mathsf{S}_{n_3}} \sum_{i,j,k} M_{\{1,2\}}(i, j, k)\, \eta_{i,j} \xi_k \\
&\geq \min_{\xi \in \mathsf{S}_{n_3}} \sum_{i,j,k} M_{\{1,2\}}(i, j, k)\, \alpha_i^* \beta_j^* \xi_k \\
&= \min_{\xi \in \mathsf{S}_{n_3}} \sum_{i,j,k} [M_1(i, j, k) + M_2(i, j, k)]\, \alpha_i^* \beta_j^* \xi_k \\
&= \min_{\xi \in \mathsf{S}_{n_3}} \Big[\sum_{i,j,k} M_1(i, j, k)\, \alpha_i^* \beta_j^* \xi_k \\
&\qquad + \sum_{i,j,k} M_2(i, j, k)\, \alpha_i^* \beta_j^* \xi_k \Big].
\end{aligned} \tag{4}$$

Since the minimum of the sum of two functions is never less than the sum of their minima, we conclude from (4) that

$$v(\{1, 2\}) \geq \min_{\xi \in S_{n_3}} \left[\sum_{i,j,k} M_1(i, j, k) \alpha_i^* \beta_j^* \xi_k \right] + \min_{\xi \in S_{n_3}} [M_2(i, j, k) \alpha_i^* \beta_j^* \xi_k]. \tag{5}$$

Now let $\| \gamma_1^* \quad \cdots \quad \gamma_{n_3}^* \|$ be a member of S_{n_3} such that

$$\sum_{i,j,k} M_1(i, j, k) \alpha_i^* \beta_j^* \gamma_k^* = \min_{\xi \in S_{n_3}} \sum_{i,j,k} M_1(i, j, k) \alpha_i^* \beta_j^* \xi_k ; \tag{6}$$

and let $\| \delta_1^* \quad \cdots \quad \delta_{n_3}^* \|$ be a member of S_{n_3} such that

$$\sum_{i,j,k} M_2(i, j, k) \alpha_i^* \beta_j^* \delta_k^* = \min_{\xi \in S_{n_3}} \sum_{i,j,k} M_2(i, j, k) \alpha_i^* \beta_j^* \xi_k . \tag{7}$$

From (5), (6), and (7) we obtain

$$v(\{1, 2\}) \geq \sum_{i,j,k} M_1(i, j, k) \alpha_i^* \beta_j^* \gamma_k^* + \sum_{i,j,k} M_2(i, j, k) \alpha_i^* \beta_j^* \delta_k^* . \tag{8}$$

Since $\| \beta_1^* \quad \cdots \quad \beta_{n_2}^* \| \in S_{n_2}$ and $\| \gamma_1^* \quad \cdots \quad \gamma_{n_3}^* \| \in S_{n_3}$, we conclude that the vector

$$\| \beta_1^* \cdot \gamma_1^* \quad \beta_1^* \cdot \gamma_2 \quad \cdots \quad \beta_{n_2}^* \cdot \gamma_{n_3}^* \|$$

belongs to $S_{n_2 \cdot n_3}$. Hence we have

$$\sum_{i,j,k} M_1(i, j, k) \alpha_i^* \beta_j^* \gamma_k^* \geq \min_{\xi \in S_{n_2 \cdot n_3}} \sum_{i,j,k} M_1(i, j, k) \alpha_i^* \xi_{j,k} = \min_{\xi \in S_{n_2 \cdot n_3}} E_{\{1\}}(\alpha^*, \xi). \tag{9}$$

From (2) and (9) we conclude that

$$\sum_{i,j,k} M_1(i, j, k) \alpha_i^* \beta_j^* \gamma_k^* \geq v(\{1\}). \tag{10}$$

In a like manner we conclude that

$$\sum_{i,j,k} M_2(i, j, k) \alpha_i^* \beta_j^* \delta_k^* \geq v(\{2\}). \tag{11}$$

From (8), (9), and (11), finally, we see that

$$v(\{1, 2\}) \geq v(\{1\}) + v(\{2\}),$$

as was to be shown.

We shall shortly prove the converse of Theorem 15.1, namely, that every function v which satisfies the three conditions of Theorem 15.1 is the characteristic function of some zero-sum game. In order to do this it is convenient first to derive some simple mathematical consequences of the three conditions of Theorem 15.1.

LEMMA 15.2. Let N be a set, and let v be a real-valued function which is defined for every subset T of N and which satisfies the conditions of 15.1. Then

(i) $v(\Lambda) = 0$;

(ii) if $\mathsf{T}_1, \cdots, \mathsf{T}_r$ are mutually exclusive subsets of N, then

$$v(\mathsf{T}_1 \cup \cdots \cup \mathsf{T}_r) \geq v(\mathsf{T}_1) + \cdots + v(\mathsf{T}_r);$$

(iii) if $\mathsf{T}_1, \cdots, \mathsf{T}_r$ are mutually exclusive subsets of N, whose union is N, then

$$v(\mathsf{T}_1) + \cdots + v(\mathsf{T}_r) \leq 0.$$

PROOF. Condition (i) follows immediately from conditions (i) and (ii) of Theorem 15.1; for

$$v(\Lambda) = v(-\mathsf{N}) = -v(\mathsf{N}) = -0 = 0.$$

Condition (ii) follows from (iii) of 15.1 by an induction on r. Condition (iii), finally, follows from (ii) above and from (i) of 15.1; for

$$v(\mathsf{T}_1) + \cdots + v(\mathsf{T}_r) \leq v(\mathsf{T}_1 \cup \cdots \cup \mathsf{T}_r) = v(\mathsf{N}) = 0.$$

We now establish the previously mentioned converse of Theorem 15.1.

THEOREM 15.3. Let N be a finite set containing n persons, and let v be a real-valued function which is defined for every subset T of N and which satisfies the three conditions of 15.1. Then there exists an n-person zero-sum game of which v is the characteristic function.

PROOF. We define a game whose players are the members of N as follows: Each member x of N makes just one move, which consists in choosing a subset T_x of N such that $x \in \mathsf{T}_x$; each of these moves is made in ignorance of the moves of the other players.

In order to define the payoff functions, we first introduce the auxiliary notion of a distinguished subset of N. A subset T of N is called *distinguished* (with respect to a given play of the game) if either

(*a*) for every x in T, $\mathsf{T}_x = \mathsf{T}$,

or

(*b*) T is a set containing just one element x, and x belongs to no set satisfying (*a*).

It is easily seen that, for a given play of the game, the distinguished subsets of N are mutually exclusive and their union is N.

Now suppose that, for a given play of the game, the distinguished subsets of N are $\mathsf{T}_1, \cdots, \mathsf{T}_p$ and that T_j (for $j = 1, \cdots, p$) contains n_j elements. Then if player i (for $i = 1, \cdots, n$) belongs to T_j, the payoff to player i is defined to be

$$\frac{1}{n_j} v(\mathsf{T}_j) - \frac{1}{n} \sum_{r=1}^{p} v(\mathsf{T}_r),$$

where v is the given function.

We shall show first that the game defined in this way is zero-sum. Suppose, as before, that, for a given play, the distinguished subsets are $\mathsf{T}_1, \cdots, \mathsf{T}_p$, where T_j contains n_j elements. We notice that the payoff to any two members of the same set T_j is the same. Hence, for each j, the sum of the payments to the members of T_j is simply the product of n_j by the payoff to each member of T_j, i.e.,

$$n_j \left[\frac{1}{n_j} v(\mathsf{T}_j) - \frac{1}{n} \sum_{r=1}^{p} v(\mathsf{T}_r) \right].$$

The sum of the payments to all the members of N is therefore

$$\begin{aligned} &\sum_{j=1}^{p} n_j \left[\frac{1}{n_j} v(\mathsf{T}_j) - \frac{1}{n} \sum_{r=1}^{p} v(\mathsf{T}_r) \right] \\ &= \sum_{j=1}^{p} v(\mathsf{T}_j) - \frac{1}{n} \cdot \sum_{j=1}^{p} n_j \cdot \sum_{r=1}^{p} v(\mathsf{T}_r) \\ &= \sum_{j=1}^{p} v(\mathsf{T}_j) - \frac{1}{n} \cdot n \cdot \sum_{r=1}^{p} v(\mathsf{T}_r) = 0, \end{aligned}$$

as was to be shown.

To complete the proof of our theorem, it now remains to show that v is the characteristic function of the game defined above. Let $\bar{v}$ be the characteristic function of the game defined; we are to prove that, for every subset T of N, $\bar{v}(\mathsf{T}) = v(\mathsf{T})$.

By hypothesis, v satisfies the three conditions of Theorem 15.1; and $\bar{v}$ satisfies these conditions because it is the characteristic function of a zero-sum game. It follows that v and $\bar{v}$ also satisfy the three conditions of Lemma 15.2.

We shall first show that, for every subset T of N,

$$\bar{v}(\mathsf{T}) \geq v(\mathsf{T}).$$

For $\mathsf{T} = \Lambda$, this follows from (i) of 15.2. In case $\mathsf{T} \neq \Lambda$, the players in T can form a coalition and can agree that each member x of T will choose $\mathsf{T}_x = \mathsf{T}$. This will make T a distinguished set with respect to the play in question. Suppose, now, that the distinguished sets of such a play are $\mathsf{T}_1, \cdots, \mathsf{T}_p$, where $\mathsf{T}_1 = \mathsf{T}$. Then each player in T gets exactly

$$\frac{1}{n_1} v(\mathsf{T}) - \frac{1}{n} \sum_{r=1}^{p} v(\mathsf{T}_r),$$

and hence the total payment to the members of T is

$$n_1\left[\frac{1}{n_1} v(\mathsf{T}) - \frac{1}{n} \sum_{r=1}^{p} v(\mathsf{T}_r)\right] = v(\mathsf{T}) - \frac{n_1}{n} \sum_{r=1}^{p} v(\mathsf{T}_r).$$

Since the members of T can thus ensure that they will obtain at least

$$v(\mathsf{T}) - \frac{n_1}{n} \sum_{r=1}^{p} v(\mathsf{T}_r),$$

we conclude that

$$\bar{v}(\mathsf{T}) \geq v(\mathsf{T}) - \frac{n_1}{n} \sum_{r=1}^{p} v(\mathsf{T}_r). \tag{12}$$

From (iii) of 15.2 we see that

$$\sum_{r=1}^{p} v(\mathsf{T}_r) \leq 0. \tag{13}$$

From (12) and (13) we conclude, since n_1 and n are positive, that

$$\bar{v}(\mathsf{T}) \geq v(\mathsf{T}). \tag{14}$$

Since (14) holds for all subsets T of N, it remains true if we replace T by $-$T. Thus

$$\bar{v}(-\mathsf{T}) \geq v(-\mathsf{T}).$$

By (ii) of 15.1 this implies that

$$-\bar{v}(\mathsf{T}) \geq -v(\mathsf{T}),$$

and hence that

$$v(\mathsf{T}) \geq \bar{v}(\mathsf{T}). \tag{15}$$

From (14) and (15) we conclude that

$$\bar{v}(\mathsf{T}) = v(\mathsf{T}),$$

which completes the proof of our theorem.

REMARK 15.4. Theorem 15.3 shows that the three conditions of Theorem 15.1 are sufficient to define characteristic functions completely. Thus when we speak, henceforth, of characteristic functions, we need not refer to any game which generates them; we need only assume that they satisfy the three conditions of Theorem 15.1.

2. Reduced Form. The chief problems with which we are concerned in the theory of n-person games are, as was mentioned earlier, the questions of what will be the strength of the tendencies of the players to form various coalitions and what side-payments must be made among the several players as inducements to join a given coalition. If two games do not differ in these regards (even though their payoff functions and characteristic functions perhaps differ), they will be essentially the same from our point of view. It is natural to give a name to the relation which holds between two such games, and we shall call it *strategic equivalence.*

This notion of strategic equivalence is, of course, only an intuitive notion without any precise mathematical content, for it rests on such notions as that of a "tendency to form a coalition"—which notions have not themselves been defined mathematically. It is clear, however, that the notion is an extremely important one for our purposes and that one of the fundamental tasks of game theory is to find for it an exact mathematical definition. It is possible to present acceptable intuitive arguments to show that certain conditions are sufficient for strategic equivalence; we shall do this in the next few paragraphs.

Suppose that we are given two rectangular n-person zero-sum games, Γ and Γ', where, in both games, player i (for $i = 1, \cdots, n$) makes his choice from the set C_i; and let the payoff functions in Γ and Γ' be, respectively,

$M_1, \cdots, M_n$ and $M'_1, \cdots, M'_n$. Suppose, moreover, that there exists a positive number k such that, for $i = 1, \cdots, n$ and for any element $\| x_1 \ \cdots \ x_n \|$ of the Cartesian product of $\mathsf{C}_1, \cdots, \mathsf{C}_n$,

$$M'_i(x_1, \cdots, x_n) = k \cdot M_i(x_1, \cdots, x_n). \tag{16}$$

Then it is intuitively evident that the games Γ and Γ' are strategically equivalent. For we can think of the constant k as merely changing the monetary unit (from dollars to cents or shillings, for example), and the behavior of a rational man in playing a game does not depend on the units in which the payoff is counted. Thus we have found a sufficient condition for strategic equivalence.

Another sufficient condition is obtained if we replace condition (16) by the following: There exist n numbers $a_1, \cdots, a_n$ such that

$$a_1 + \cdots + a_n = 0$$

and, for $i = 1, \cdots, n$ and for any $\| x_1 \ \cdots \ x_n \|$ in the Cartesian product of $\mathsf{C}_1, \cdots, \mathsf{C}_n$,

$$M'_i(x_1, \cdots, x_n) = M_i(x_1, \cdots, x_n) + a_i. \tag{17}$$

To see that Γ and Γ' are strategically equivalent in this case, we notice that the payments to player i in Γ' are the same as in Γ, except that he receives in addition the amount a_i (which is, of course, possibly negative); and this amount a_i is received by i, *regardless of the course of the play.* Thus it is clear that the strategic character of Γ' would be unchanged if the payments $a_1, \cdots, a_n$ were made at the beginning of the play instead of at the end, and hence if they were not made at all—which is the case in Γ.

We see, furthermore, that in order for Γ and Γ' to be strategically equivalent, it is not really necessary to suppose, as above, that the classes from which the choices are made by the various players are actually identical; it is sufficient to suppose merely that they can be put into one-to-one correspondence in an appropriate way.

These considerations suggest that we introduce the following notion of S-equivalence, which, it follows from our intuitive arguments, is a sufficient condition for strategic equivalence.

DEFINITION 15.5. Let Γ and Γ' be two n-person zero-sum games, where the choice-sets are $\mathsf{C}_1, \cdots, \mathsf{C}_n$ and $\mathsf{C}'_1, \cdots, \mathsf{C}'_n$, respectively, and where the payoff functions are $M_1, \cdots, M_n$ and $M'_1, \cdots, M'_n$, respectively. Then Γ and Γ' are called S-*equivalent* if there exist functions $f_1, \cdots, f_n$, real numbers $a_1, \cdots, a_n$, and a positive real number k, which satisfy the following conditions:

(i) $\sum_{i=1}^{n} a_i = 0.$

(ii) For $i = 1, \cdots, n$, f_i maps C_i in a one-to-one way onto C'_i.

(iii) For $i = 1, \cdots, n$ and for any element $\|x_1 \quad \cdots \quad x_n\|$ of the Cartesian product of $\mathsf{C}_1, \cdots, \mathsf{C}_n$,

$$M_i(x_1, \cdots, x_n) = k \cdot M'_i[f_1(x_1), \cdots, f_n(x_n)] + a_i.$$

The proofs of the following two theorems will be left as exercises.

THEOREM 15.6. The relation of S-equivalence is reflexive, symmetric, and transitive.

THEOREM 15.7. If Γ and Γ' are two n-person zero-sum games, which are S-equivalent with respect to the constants $a_1, \cdots, a_n$ and k, and if v and v' are the characteristic functions of Γ and Γ', respectively, then, for any subset T of N.

$$v(\mathsf{T}) = k \cdot v'(\mathsf{T}) + \sum_{i \in \mathsf{T}} a_i.$$

In view of Theorem 15.7 the two characteristic functions v and v' are called S-*equivalent* if they satisfy the equation of this theorem, i.e., if there is a positive constant k and constants $a_1, \cdots, a_n$ (with $\sum_{i=1}^{n} a_i = 0$) such that, for every subset T of N,

$$v(\mathsf{T}) = k \cdot v'(\mathsf{T}) + \sum_{i \in \mathsf{T}} a_i.$$

In view of Theorem 15.6 the relation of S-equivalence partitions the class of all n-person zero-sum games into mutually exclusive classes, such that any two members of the same class are S-equivalent and the members of different classes are not. For some purposes it is desirable to be able to pick out of each of these classes an especially simple member; to this end we formulate the following definition.

DEFINITION 15.8. An n-person game with a characteristic function v is said to be in *reduced form* if

$$v(\{1\}) = v(\{2\}) = \cdots = v(\{n\}) = \gamma,$$

where either $\gamma = 0$ or $\gamma = -1$. When these equations hold, we also say that the characteristic function v is in *reduced form with modulus* γ.

THEOREM 15.9. If v is a characteristic function in reduced form with modulus γ, and if T is a subset of N containing p elements, then

$$p\gamma \leq v(\mathsf{T}) \leq (p - n)\gamma.$$

PROOF. From (ii) of 15.2 we see that

$$\sum_{i \in \mathsf{T}} v(\{i\}) \leq v(\mathsf{T}),$$

and thus that

$$p\gamma \leq v(\mathsf{T}). \tag{18}$$

Since (18) holds for every T, we can replace in it T by $-\mathsf{T}$; since $-\mathsf{T}$ contains $n - p$ elements, we therefore have

$$(n - p)\gamma \leq v(-\mathsf{T}). \tag{19}$$

From (19) we conclude by (ii) of 15.1 that

$$(n - p)\gamma \leq -v(\mathsf{T}),$$

and hence that

$$v(\mathsf{T}) \leq (p - n)\gamma. \tag{20}$$

Equations (18) and (20), together, give our theorem.

COROLLARY 15.10. If v is a characteristic function in reduced form with modulus 0, then, for every subset T of N,

$$v(\mathsf{T}) = 0.$$

PROOF. By 15.9 we have

$$p \cdot 0 \leq v(\mathsf{T}) \leq (p - n) \cdot 0,$$

and hence

$$v(\mathsf{T}) = 0,$$

as was to be shown.

THEOREM 15.11. If v and v' are S-equivalent characteristic functions in reduced form, then, for every subset T of N,

$$v'(\mathsf{T}) = v(\mathsf{T}).$$

PROOF. Let the moduli of v and v' be γ and γ', respectively. Since v and v' are S-equivalent, there is a positive number k and numbers $a_1, \cdots, a_n$ whose sum is 0, such that, for every subset T of N,

$$v'(\mathsf{T}) = k \cdot v(\mathsf{T}) + \sum_{i \in \mathsf{T}} a_i. \tag{21}$$

In particular, we have

$$v'(\{i\}) = kv(\{i\}) + a_i \qquad \text{for } i = 1, \cdots, n. \tag{22}$$

Adding all the equations in (22), we obtain

$$\sum_{i=1}^{n} v'(\{i\}) = k \sum_{i=1}^{n} v(\{i\}) + \sum_{i=1}^{n} a_i = k \sum_{i=1}^{n} v(\{i\});$$

and hence, since both functions are in reduced form,

$$n\gamma' = kn\gamma.$$

Thus

$$\gamma' = k\gamma,$$

and hence γ and γ' are either both 0 or both -1. If both are 0, then by 15.10 we have, for every T,

$$v'(\mathsf{T}) = 0 = v(\mathsf{T}).$$

If both are -1, then $k = 1$, and hence from (22) we have

$$(-1) = (1)(-1) + a_i,$$

so that $a_i = 0$ (for $i = 1, \cdots, n$); hence from (21), for any T, we have

$$v'(\mathsf{T}) = k \cdot v(\mathsf{T}) + \sum_{i \in \mathsf{T}} a_i = 1 \cdot v(\mathsf{T}) + 0 = v(\mathsf{T}),$$

which completes the proof.

THEOREM 15.12. Every characteristic function is S-equivalent to one, and only one, characteristic function in reduced form.

PROOF. Let v be a characteristic function. To show that v is S-equivalent to a characteristic function in reduced form, we distinguish two cases, according as $\sum_{i=1}^{n} v(\{i\}) = 0$ or $\sum_{i=1}^{n} v(\{i\}) \neq 0$.

In the first case, we take $k = 1$ and $a_i = -v(\{i\})$ (for $i = 1, \cdots, n$). Then

$$\sum_{i=1}^{n} a_i = -\sum_{i=1}^{n} v(\{i\}) = 0.$$

Moreover, by Theorem 15.7, the characteristic function v' which is S-equivalent to v with respect to these constants satisfies, for $i = 1, \cdots, n$, the condition

$$v'(\{i\}) = k \cdot v(\{i\}) + a_i = 1 \cdot v(\{i\}) - v(\{i\}) = 0.$$

In the case in which $\sum_{i=1}^{n} v(\{i\}) \neq 0$, we take

$$k = -\frac{n}{\sum_{i=1}^{n} v(\{i\})}$$

and, for $i = 1, \cdots, n$,

$$a_i = -1 + \frac{nv(\{i\})}{\sum_{i=1}^{n} v(\{i\})}.$$

Then

$$\sum_{i=1}^{n} a_i = -n + \frac{n \sum_{i=1}^{n} v(\{i\})}{\sum_{i=1}^{n} v(\{i\})} = 0,$$

and

$$\begin{aligned} v'(\{i\}) &= kv(\{i\}) + a_i \\ &= -\frac{n}{\sum_{i=1}^{n} v(\{i\})} \cdot v(\{i\}) - 1 + \frac{nv(\{i\})}{\sum_{i=1}^{n} v(\{i\})} = -1. \end{aligned}$$

That a characteristic function is not S-equivalent to two distinct characteristic functions in reduced form follows from Theorem 15.11, together with the fact that the relation of S-equivalence is transitive.

REMARK 15.13. From Theorems 15.12 and 15.7 we see that in order to give an adequate theory of all n-person zero-sum games, it suffices to consider only games in reduced form.

Corollary 15.10 shows that games in reduced form with modulus 0 differ very markedly from those with modulus -1. In fact, from 15.10 we see that when the modulus is 0, every set of players gets paid 0; thus there is no point in forming coalitions in such a game, and there can be no question of side-payments to induce a player to enter a coalition; hence no theory is required for such games. Thus we can henceforth restrict our attention to games in reduced form with modulus -1. We shall call such games *essential*, in contradistinction to games with modulus 0, which will be called *inessential*. If games are not in reduced form, we call them essential or inessential according as they are S-equivalent to essential or inessential games in reduced form.

The following two theorems give conditions that games (not necessarily in reduced form) be inessential; their proofs will be left as exercises.

THEOREM 15.14. A game is inessential if, and only if, its characteristic function v satisfies the condition

$$\sum_{i=1}^{n} v(\{i\}) = 0.$$

THEOREM 15.15. A game is inessential if, and only if, its characteristic function v is such that, whenever $\mathsf{R} \cap \mathsf{T} = \Lambda$, we have

$$v(\mathsf{R} \cup \mathsf{T}) = v(\mathsf{R}) + v(\mathsf{T}).$$

Thus in an essential game some coalition has a positive tendency to form.

From Corollary 15.10 we see that for each n there is, to within S-equivalence, but one inessential n-person game. It is interesting to notice that for $n = 3$ there is also but one essential n-person game in reduced form. For if v is the characteristic function of such a game, then we have

$$v(\{1\}) = v(\{2\}) = v(\{3\}) = -1,$$

and hence by (ii) of 15.1,

$$v(\{2, 3\}) = v(\{1, 3\}) = v(\{1, 2\}) = +1;$$

since we also have

$$v(\Lambda) = v(\{1, 2, 3\}) = 0,$$

we therefore see that the value of $v(\mathsf{T})$ is determined for every T. Thus it is possible to speak of *the* essential three-person game in reduced form.

On the other hand, for $n > 3$ there are infinitely many essential n-person games in reduced form. We shall examine this question a little more closely for $n = 4$.

If v is the characteristic function of an essential four-person game in reduced form, then we see immediately that

$$v(\mathsf{T}) = \begin{cases} 0 \\ -1 \\ 1 \\ 0 \end{cases} \quad \text{when } \mathsf{T} \text{ has } \begin{cases} 0 \\ 1 \\ 3 \\ 4 \end{cases} \text{ elements.} \tag{23}$$

Thus the value of $v(\mathsf{T})$ is determined, except for the case in which T contains just two elements. Since

$$v(\{2, 3\}) = -v(\{1, 4\}),$$
$$v(\{1, 3\}) = -v(\{2, 4\}),$$
$$v(\{1, 2\}) = -v(\{3, 4\}),$$

however, we see that the values of v will be completely determined if we assign values to $v(\{1, 4\})$, $v(\{2, 4\})$, and $v(\{3, 4\})$. Moreover, by Theorem 15.9 we see that if T is any set with two elements, then

$$-2 \leq v(\mathsf{T}) \leq 2.$$

Thus if we set

$$\left.\begin{aligned} v(\{1, 4\}) &= 2x_1, \\ v(\{2, 4\}) &= 2x_2, \\ v(\{3, 4\}) &= 2x_3, \end{aligned}\right\} \tag{24}$$

then we shall have

$$\left.\begin{aligned} v(\{2, 3\}) &= -2x_1, \\ v(\{1, 3\}) &= -2x_2, \\ v(\{1, 2\}) &= -2x_3, \end{aligned}\right\} \tag{25}$$

and

$$-1 \leq x_i \leq 1 \qquad \text{for } i = 1, 2, 3. \tag{26}$$

Moreover, it is easily verified that if x_1, x_2, and x_3 are any numbers which satisfy (26), and if we define $v(\mathsf{T})$ by (23), (24), and (25), then v is the characteristic function of an essential four-person game in reduced form. To do this, it is only necessary to show that the function v defined in this way satisfies the conditions of Theorem 15.1.

Thus if we regard the numbers x_1, x_2, and x_3 as the Cartesian coordinates of a point in 3-dimensional Euclidean space, then the totality of essential four-person games in reduced form is seen to correspond in a one-to-one way to the points $\| x_1 \quad x_2 \quad x_3 \|$ whose coordinates satisfy the condition in (26); these points are simply the points of a certain cube.

Using the terminology of the geometers, we can say that while there is but one essential three-person game in reduced form, there is a triple infinity of essential four-person games in reduced form.

EXAMPLE 15.16. The game called "odd man out" is played as follows: Each of the players (there are three of them) writes down either "heads" or "tails" on a piece of paper. After each player has chosen his word, the umpire compares the three: if all the players have written down the same word, then no payments are made; otherwise, the odd man pays each of the others a dollar.

Thus if M_1, M_2, and M_3 are the payoff functions for the three players, and if we represent "heads" by "1" and "tails" by "2," then

$$M_1(1, 1, 1) = M_1(2, 2, 2) = 0,$$
$$M_1(1, 1, 2) = M_1(2, 2, 1) = M_1(1, 2, 1) = M_1(2, 1, 2) = 1,$$
$$M_1(1, 2, 2) = M_1(2, 1, 1) = -2,$$

and similarly for M_2 and M_3.

Now suppose that players 2 and 3 make a coalition against player 1. Player 1 has, of course, only two strategies: he can choose 1, or he can choose 2. The coalition $\{2, 3\}$, on the other hand, has four strategies: both players can choose 1; both can choose 2; player 2 can choose 1, while player 3 chooses 2; or player 2 can choose 2, while player 3 chooses 1. From the definition of M_1, we now can easily compute the payoff matrix to player 1 for various choices of strategies. Thus if player 1 chooses 1, for instance, while the coalition $\{2, 3\}$ chooses $\| 1 \quad 2 \|$, then, since $M_1(1, 1, 2) = 1$, the payoff to player 1 will be 1. The complete matrix is Matrix 1.

MATRIX 1

	$\|1 \quad 1\|$	$\|1 \quad 2\|$	$\|2 \quad 1\|$	$\|2 \quad 2\|$
1	0	1	1	-2
2	-2	1	1	0

Since the second and third columns of Matrix 1 dominate the first, we see that when the coalition $\{2, 3\}$ plays an optimal strategy, the players will never use either of the strategies $\| 1 \quad 2 \|$ and $\| 2 \quad 1 \|$. Solving the matrix, the value of the game (to player 1) is -1. Thus

$$v(\{1\}) = -1,$$

and hence

$$v(\{2, 3\}) = 1.$$

From symmetry we conclude that

$$v(\{2\}) = v(\{3\}) = -1,$$

and

$$v(\{1, 3\}) = v(\{1, 2\}) = 1.$$

Moreover, as for any three-person zero-sum game,

$$v(\Lambda) = v(\{1, 2, 3\}) = 0.$$

Thus we have completely determined the characteristic function of this game. The payoff functions happen, in this case, to be such that the characteristic function is already in reduced form.

BIBLIOGRAPHICAL REMARKS

The material in this chapter is mostly to be found in von Neumann and Morgenstern [1]. A complete discussion of **S**-equivalence is to be found in McKinsey [1].

EXERCISES

1. In a rectangular three-person zero-sum game each player makes a choice from the set $\{1, 2\}$. The payoff functions are defined as follows:

$$\begin{aligned}
&M_1(1, 1, 1) = +1, \quad &&M_2(1, 1, 1) = +1, \quad &&M_3(1, 1, 1) = -2,\\
&M_1(1, 1, 2) = -1, &&M_2(1, 1, 2) = -1, &&M_3(1, 1, 2) = +2,\\
&M_1(1, 2, 1) = +1, &&M_2(1, 2, 1) = +1, &&M_3(1, 2, 1) = -2,\\
&M_1(1, 2, 2) = -1, &&M_2(1, 2, 2) = -1, &&M_3(1, 2, 2) = +2,\\
&M_1(2, 1, 1) = -1, &&M_2(2, 1, 1) = -1, &&M_3(2, 1, 1) = +2,\\
&M_1(2, 1, 2) = +1, &&M_2(2, 1, 2) = +1, &&M_3(2, 1, 2) = -2,\\
&M_1(2, 2, 1) = -1, &&M_2(2, 2, 1) = -1, &&M_3(2, 2, 1) = +2,\\
&M_1(2, 2, 2) = +1, &&M_2(2, 2, 2) = +1, &&M_3(2, 2, 2) = -2.
\end{aligned}$$

Find the characteristic function of the game.

2. Find the characteristic function in reduced form which is **S**-equivalent to the following characteristic function v for a four-person game:

$$v(\Lambda) = 0, \qquad v(\{2, 3\}) = 0,$$
$$v(\{1\}) = -1, \qquad v(\{2, 4\}) = 0,$$
$$v(\{2\}) = -2, \qquad v(\{3, 4\}) = 0,$$
$$v(\{3\}) = -2, \qquad v(\{1, 2, 3\}) = 0,$$
$$v(\{4\}) = 0, \qquad v(\{1, 2, 4\}) = 2,$$
$$v(\{1, 2\}) = 0, \qquad v(\{1, 3, 4\}) = 2,$$
$$v(\{1, 3\}) = 0, \qquad v(\{2, 3, 4\}) = 1,$$
$$v(\{1, 4\}) = 0, \qquad v(\{1, 2, 3, 4\}) = 0.$$

3. Define an "odd-man-out" game for four players (in analogy to the three-player version of Example 15.16), and find its characteristic function.

4. Prove Theorem 15.6.

5. Prove Theorem 15.7.

6. Prove Theorem 15.14.

7. Prove Theorem 15.15.

8. An n-person game is called *symmetric* if, whenever T_1 and T_2 are two subsets of N with the same number of elements, $v(\mathsf{T}_1) = v(\mathsf{T}_2)$. Show that there is only one symmetric essential four-person game in reduced form, and find its characteristic function.

9. Show that there are infinitely many symmetric essential five-person games in reduced form.

10. Show that, for the game defined in the proof of Theorem 15.3, the distinguished subsets of N are mutually exclusive, and their union is N.

CHAPTER 16

SOLUTIONS OF *n*-PERSON GAMES

1. Imputations. As has been mentioned, we are interested, in the case of n-person games, in the questions of what coalitions will tend to form and what each player will be paid (after all side-payments are made) in the event a given coalition forms. The payments to the several players, for given coalitions and side-payments, can be represented as a vector (i.e., as an ordered n-tuple) of real numbers $\| x_1 \quad x_2 \quad \cdots \quad x_n \|$, where x_i (for $i = 1, \cdots, n$) is the amount the ith player receives.

Since the game is zero-sum, however, so that no money is created or destroyed, it is clear that the payments to the various players must add up to zero, i.e.,

$$\sum_{i=1}^{n} x_i = 0.$$

Moreover, we notice that no such vector $\| x_1 \quad \cdots \quad x_n \|$ could ever be realized unless, for each i, we have

$$x_i \geq v(\{i\});$$

for player i can see to it that he obtains $v(\{i\})$ all by himself (i.e., even though he cannot persuade any other player to cooperate with him), and hence he would certainly reject any system of distribution which would give him less than $v(\{i\})$.

Since we shall frequently be talking about vectors which satisfy the above two conditions, it is convenient to give them a name; accordingly, we give the following definition.

DEFINITION 16.1. By an *imputation* for an n-person game with characteristic function v, we shall mean a vector

$$\| x_1 \quad \cdots \quad x_n \|$$

such that

$$\text{(i)} \qquad \sum_{i=1}^{n} x_i = 0,$$

and

$$\text{(ii)} \qquad x_i \geq v(\{i\}) \qquad \text{for } i = 1, \cdots, n.$$

REMARK 16.2. It might be thought that the second condition of our definition should be strengthened so as to say that if T is any subset of N, then

$$\sum_{i \in \mathsf{T}} x_i \geq v(\mathsf{T}).$$

But the intuitive grounds for this stronger condition are not so clear as for the weaker condition. For, although the players of set T can, by making a coalition, be sure that collectively they will obtain $v(\mathsf{T})$, it is by no means clear that they will be willing to join each other in such a way. Moreover, from the formal point of view, it can readily be shown that such a condition would, in general, make the class of imputations empty. Thus, for instance, if this condition were satisfied by an imputation $\| x_1 \quad x_2 \quad x_3 \|$ for the essential three-person game in reduced form, then we should have

$$x_1 + x_2 \geq v(\{1, 2\}) = 1$$

and hence, since $x_1 + x_2 + x_3 = 0$,

$$-x_3 \geq 1$$

or

$$x_3 \leq -1;$$

since

$$x_3 \geq v(\{3\}) = -1,$$

we should therefore have

$$x_3 = -1.$$

In a similar fashion we could obtain

$$x_2 = -1$$

and

$$x_1 = -1,$$

contrary to the assumption that $x_1 + x_2 + x_3 = 0$; thus there would exist no imputation for this game.

REMARK 16.3. It is easily verified that the set of all imputations for an

n-person game is a convex subset of n-dimensional Euclidean space. Thus if there are two distinct imputations for a given game, then there are infinitely many. We notice that an inessential game has only one imputation, namely, $\| v(\{1\}) \quad \cdots \quad v(\{n\}) \|$. On the other hand, an essential game has infinitely many imputations.

An imputation $\| y_1 \quad \cdots \quad y_n \|$ will clearly be preferred by player i to the imputation $\| x_1 \quad \cdots \quad x_n \|$ if it gives him more—that is to say, if

$$y_i > x_i .$$

Similarly, $\| y_1 \quad \cdots \quad y_n \|$ will be preferred by a subset T of the set N of all players if

$$y_i > x_i , \qquad \text{for all } i \text{ in } \mathsf{T}.$$

Unless

$$v(\mathsf{T}) \geq \sum_{i \in \mathsf{T}} y_i ,$$

however, the preference of T for $\| y_1 \quad \cdots \quad y_n \|$ will be idle, since the players would find themselves unable (without outside help) to ensure that they would get the amount that they would be allowed by this imputation.

(In speaking of a set T of players' preferring one imputation to another, it is, of course, natural to restrict ourselves to the case in which T is not empty, for it would be silly to speak of an empty set of people having a preference for one thing as opposed to another.)

These considerations lead us to the notion of dominance, which is defined as a non-idle preference by a nonempty set of players.

DEFINITION 16.4. Let $\| y_1 \quad \cdots \quad y_n \|$ and $\| x_1 \quad \cdots \quad x_n \|$ be imputations for a game whose characteristic function is v, and let T be a subset of the players. Then we say that $\| y_1 \quad \cdots \quad y_n \|$ *dominates* $\| x_1 \quad \cdots \quad x_n \|$ *with respect to* T if the following conditions are satisfied:

$$\text{(i)} \qquad \mathsf{T} \neq \Lambda .$$

$$\text{(ii)} \qquad v(\mathsf{T}) \geq \sum_{i \in \mathsf{T}} y_i .$$

$$\text{(iii)} \qquad y_i > x_i \qquad \text{for all } i \text{ in } \mathsf{T}.$$

When $\| y_1 \quad \cdots \quad y_n \|$ dominates $\| x_1 \quad \cdots \quad x_n \|$ with respect to T, we write

$$\| y_1 \quad \cdots \quad y_n \| \underset{\mathsf{T}}{\succ} \| x_1 \quad \cdots \quad x_n \| .$$

If there exists any set T with respect to which $\| y_1 \quad \cdots \quad y_n \|$ dominates

$\| x_1 \quad \cdots \quad x_n \|$, then we say that $\| y_1 \quad \cdots \quad y_n \|$ *dominates* $\| x_1 \quad \cdots \quad x_n \|$, and we write

$$\| y_1 \quad \cdots \quad y_n \| \succ \| x_1 \quad \cdots \quad x_n \|.$$

REMARK 16.5. It is of interest to examine some of the general properties of the relation of dominance.

It is clear from (i) of 16.4 that no imputation can dominate another with respect to the empty set. Moreover, we see from (ii) of 16.1 and from 16.4 that an imputation cannot dominate another with respect to a one-element set, for if we had

$$\| y_1 \quad \cdots \quad y_n \| \underset{\{i\}}{\succ} \| x_1 \quad \cdots \quad x_n \|,$$

then we could conclude from (ii) of 16.1 and (ii) of 16.4 that

$$y_i = v(\{i\}),$$

and hence from (iii) of 16.4 that

$$v(\{i\}) > x_i,$$

which would contradict (ii) of 16.1. In a similar fashion we see, using (i) of 16.1, that an imputation cannot dominate another with respect to the set N of all players. Thus if one imputation dominates another with respect to a set T, then T must contain at least 2 and, at most, $n - 1$ members.

It is seen immediately from (iii) of 16.4 that no imputation can dominate itself (with respect to any set T). Moreover, for any fixed set T the relation of dominance with respect to T is transitive; i.e., if

$$\| z_1 \quad \cdots \quad z_n \| \underset{\mathsf{T}}{\succ} \| y_1 \quad \cdots \quad y_n \|$$

and

$$\| y_1 \quad \cdots \quad y_n \| \underset{\mathsf{T}}{\succ} \| x_1 \quad \cdots \quad x_n \|,$$

then also

$$\| z_1 \quad \cdots \quad z_n \| \underset{\mathsf{T}}{\succ} \| x_1 \quad \cdots \quad x_n \|.$$

Thus the relation of dominance with respect to a fixed set T is what is called a "partial ordering." It is not a "complete ordering," since there will ordinarily exist imputations neither of which will dominate the other; thus, for example, in the essential three-person game in reduced form, neither of the imputations

$$\| .1 \quad -.1 \quad 0 \|,$$
$$\| -.1 \quad .1 \quad 0 \|$$

dominates the other with respect to the set $\{1, 2\}$.

When we consider the relation of dominance in general, however (i.e., when dominance is no longer merely considered with respect to a fixed set T), then the situation is more complicated. It is still true that nothing dominates itself, but the relation of dominance is no longer transitive. Thus, for example, consider the following three imputations

$$\| .1 \quad .1 \quad -.2 \|,$$
$$\| 0 \quad 0 \quad 0 \|,$$
$$\| -.1 \quad .2 \quad -.1 \|$$

for the essential three-person game in reduced form. We see that

$$\| .1 \quad .1 \quad -.2 \| \underset{\{1, 2\}}{\succ} \| 0 \quad 0 \quad 0 \|$$

and

$$\| 0 \quad 0 \quad 0 \| \underset{\{1, 3\}}{\succ} \| -.1 \quad .2 \quad -.1 \|,$$

so that

$$\| .1 \quad .1 \quad -.2 \| \succ \| 0 \quad 0 \quad 0 \|$$

and

$$\| 0 \quad 0 \quad 0 \| \succ \| -.1 \quad .2 \quad -.1 \|;$$

but, on the other hand, we see that

$$\| .1 \quad .1 \quad -.2 \| \succ \| -.1 \quad .2 \quad -.1 \|$$

is false. Indeed, one can even give an example of a five-person game having two imputations, each of which dominates the other!

2. Definition of a Solution. Now we want to consider the question of what imputations are likely to arise from actual plays of a game. The first answer which we are tempted to give is the following: An imputation $\alpha = \| a_1 \quad \cdots \quad a_n \|$ can be realized in an actual play if there exists no imputation β which dominates α. For no set of players would have any motive for changing from α to any other imputation; thus if, in the course of the bargaining, the imputation α were to be suggested, then everyone would realize that he could not possibly get more than the α promised him; and thus an end of bargaining would be reached.

Unfortunately, however, there does not, in general, exist such an imputation. Indeed we can show that (except in the case of an inessential game, where there is but one imputation) every imputation is dominated by some other imputation. Let $\| x_1 \ \cdots \ x_n \|$ be any imputation for an essential n-person game. Then there exists an integer k such that

$$x_k > v(\{k\});$$

for otherwise we should have, by Theorem 15.14,

$$\sum_{i=1}^{n} x_i = \sum_{i=1}^{n} v(\{i\}) < 0,$$

which contradicts (i) of 16.1. If we now set

$$y_k = v(\{k\})$$

and

$$y_i = x_i + \frac{x_k - v(\{k\})}{n-1} \qquad \text{for } i \neq k,$$

then it is easily verified that $\| y_1 \ \cdots \ y_n \|$ is an imputation which dominates $\| x_1 \ \cdots \ x_n \|$ with respect to the set $\{1, \cdots, k-1, k+1, \cdots, n\}$.

Since we thus seem to have strayed into a blind alley, let us turn back and consider the essential three-person game in reduced form. There appear to be just three possibilities here: the three ways in which a coalition of two can be formed. If players 1 and 2 form a coalition, then it appears reasonable to suppose that they will, together, take as much as they possibly can (namely, $+1$) and that player 3 will accordingly be given -1. Moreover, from the fact that the game is completely symmetrical (so that neither player 1 nor player 2 is in a more advantageous position), we would intuitively expect them to divide their winnings equally; therefore, in case players 1 and 2 cooperate, it appears that we should arrive at the imputation $\| \tfrac{1}{2} \ \ \tfrac{1}{2} \ \ -1 \|$. Similarly, in case players 1 and 3 cooperate, we should expect the imputation $\| \tfrac{1}{2} \ \ -1 \ \ \tfrac{1}{2} \|$, and in case players 2 and 3 cooperate, we should expect the imputation $\| -1 \ \ \tfrac{1}{2} \ \ \tfrac{1}{2} \|$. Thus we obtain the following set of three imputations:

$$\mathsf{A} = \left\{ \left\| \frac{1}{2} \quad \frac{1}{2} \quad -1 \right\|, \left\| \frac{1}{2} \quad -1 \quad \frac{1}{2} \right\|, \left\| -1 \quad \frac{1}{2} \quad \frac{1}{2} \right\| \right\}.$$

This set of three imputations has the property that no one of them dominates another. Moreover, every other imputation is dominated by at least one member of A. For suppose, if possible, that $\| x_1 \ \ x_2 \ \ x_3 \|$ is an imputation not in **A** and is not dominated by any member of **A**. Since $\| x_1 \quad x_2 \quad x_3 \|$ is not dominated by $\| \frac{1}{2} \quad \frac{1}{2} \quad -1 \|$, and since $\frac{1}{2} + \frac{1}{2} = 1 = v(\{1, 2\})$, we conclude that we must have either

$$x_1 \geq \frac{1}{2}$$

or

$$x_2 \geq \frac{1}{2}.$$

Let us suppose that the first of these inequalities holds (in case the second holds, the argument is similar). Thus

$$x_1 \geq \frac{1}{2}.$$

Moreover, since $\| x_1 \quad x_2 \quad x_3 \|$ is not dominated by $\| -1 \quad \frac{1}{2} \quad \frac{1}{2} \|$, we see, similarly, that either

$$x_2 \geq \frac{1}{2}$$

or

$$x_3 \geq \frac{1}{2}.$$

Suppose, again, that the first of these inequalities holds, since a similar argument can take care of the other case. Then we have

$$x_1 \geq \frac{1}{2},$$

and

$$x_2 \geq \frac{1}{2}.$$

Since

$$x_1 + x_2 = -x_3 \leq 1,$$

we conclude that

$$x_1 = x_2 = \frac{1}{2}$$

and that

$$x_3 = -1,$$

so that $\| x_1 \quad x_2 \quad x_3 \| \in$ **A**, contrary to hypothesis.

The two properties just established for the set A are sufficiently important that it is convenient to give a name to sets of imputations which enjoy them.

DEFINITION 16.6. A set A of imputations for a given n-person game is called a *solution* of the game if

(i) no member of A dominates another member of A;
(ii) every imputation not in A is dominated by some member of A.

REMARK 16.7. In von Neumann's development of the theory of n-person games, the notion of "solutions," as the term has just been defined, occupies the central position one would expect from the name. The theory consists essentially in a search for solutions and a discussion of their properties.

The intuitive justification of the use of the word "solution" is, of course, now very different from what it was in the case of two-person games. In the former case, a solution meant a set of probabilities with which the player should play his various pure strategies in order to maximize his expectation of gain. But in the case of n-person games (for $n > 2$), a solution purports merely to give a set of possible ways in which winnings can be divided at the end of a play.

Some people have felt dissatisfied with the intuitive basis of this notion, however; and the question has been raised as to whether knowing a solution of a given n-person game would enable a person to play it with greater expectation of profit than if he were quite ignorant of this theory. If it were possible to give an example of a game which possessed no solution in the sense of von Neumann, then it is clear that those who maintain that this notion of a solution does not constitute an adequate foundation for the theory of n-person games would be quite justified in their criticism. Thus it becomes important to show that every n-person game has a solution (in the given sense), but unfortunately this has not been accomplished; we know only that certain special games have solutions (for example, it is known that all three-person games and all four-person games have solutions, but it has never been shown that all five-person games have solutions). We shall see that this general problem is reducible to the corresponding problem for games in reduced form.

3. Isomorphic Games.

DEFINITION 16.8. Two n-person games v and v' are called *isomorphic* if there exists a one-to-one correspondence, $\leftrightarrow$, between the imputations of v and the imputations of v' such that, for every subset T of the players, if α and β are imputations for v and α' and β' are imputations for v' and $\alpha \leftrightarrow \alpha'$ and $\beta \leftrightarrow \beta'$, then

$$\alpha \underset{\mathsf{T}}{\succ} \beta \qquad \text{in game } v$$

if, and only if,

$$\alpha' \underset{\mathsf{T}}{\succ} \beta' \qquad \text{in game } v'.$$

It is easily shown that the relation of isomorphism of games is reflexive, symmetric, and transitive. Moreover, we have the following obvious theorem.

THEOREM 16.9. Suppose that the games v and v' are isomorphic under the relation $\leftrightarrow$, and that A is a solution of v. Let A' be the set of all imputations α' such that, for some α in A, $\alpha \leftrightarrow \alpha'$. Then A' is a solution of v.

The following theorem amounts to saying that S-equivalence is a sufficient condition for isomorphism. It can also be shown that it is a necessary condition, but we shall omit the proof of this.

THEOREM 16.10. Let v and v' be two games which are S-equivalent with the constants k and $a_1, \cdots, a_n$. If $\| x_1 \quad \cdots \quad x_n \|$ is any imputation for v, we let

$$\| x_1 \quad \cdots \quad x_n \| \leftrightarrow \| kx_1 + a_1 \quad \cdots \quad kx_n + a_n \|.$$

Then the relation $\leftrightarrow$ establishes an isomorphism between v and v'.

PROOF. First we want to show that if $\| x_1 \quad \cdots \quad x_n \|$ is any imputation for v, then $\| kx_1 + a_1 \quad \cdots \quad kx_n + a_n \|$ is an imputation for v'. But, since $\sum_{i=1}^{n} a_i = 0$, we have

$$\sum_{i=1}^{n} (kx_i + a_i) = k \sum_{i=1}^{n} x_i + \sum_{i=1}^{n} a_i = k \cdot 0 + 0 = 0.$$

Moreover, for each i, we have (from the S-equivalence)

$$v'(\{i\}) = kv(\{i\}) + a_i;$$

and since (from the fact that $\| x_1 \quad \cdots \quad x_n \|$ is an imputation for v)

$$x_i \geq v(\{i\}),$$

we conclude that

$$kx_i + a_i \geq v'(\{i\}),$$

as was to be shown.

To complete our proof, it is sufficient to show that if T is any subset of the players, and if $\| x_1 \quad \cdots \quad x_n \|$ and $\| y_1 \quad \cdots \quad y_n \|$ are imputations (for v) such that

$$\| x_1 \quad \cdots \quad x_n \| \underset{\mathsf{T}}{\succ} \| y_1 \quad \cdots \quad y_n \| \qquad \text{in game } v,$$

then

$$\| kx_1 + a_1 \quad \cdots \quad kx_n + a_n \| \underset{\mathsf{T}}{\succ} \| ky_1 + a_1 \quad \cdots \quad ky_n + a_n \| \qquad \text{in game } v'.$$

But, for any member i of T, we have

$$x_i > y_i,$$

and hence, since k is positive,

$$kx_i > ky_i,$$

and hence

$$kx_i + a_i > ky_i + a_i.$$

Moreover, since

$$\sum_{i \in \mathsf{T}} x_i \leq v(\mathsf{T}),$$

we see by means of Theorem 15.7 that

$$\begin{aligned} \sum_{i \in \mathsf{T}} (kx_i + a_i) &= k \sum_{i \in \mathsf{T}} x_i + \sum_{i \in \mathsf{T}} a_i \\ &\leq kv(\mathsf{T}) + \sum_{i \in \mathsf{T}} a_i = v'(\mathsf{T}), \end{aligned}$$

which completes the proof.

COROLLARY 16.11. Let v and v' be two games which are S-equivalent with the constants k and $a_1, \cdots, a_n$, let A be a solution of v, and let A' be the set of all imputations having the form

$$\| kx_1 + a_1 \quad \cdots \quad kx_n + a_n \|,$$

where $\| x_1 \quad \cdots \quad x_n \| \in \mathsf{A}$. Then A' is a solution of v'.

PROOF. By Theorems 16.10 and 16.9.

COROLLARY 16.12. If every game in reduced form has a solution, then every game has a solution.

PROOF. By Theorem 15.12 and Corollary 16.11.

REMARK 16.13. Thus in order to show that every game has a solution, it would suffice to show that every game in reduced form has a solution (and Corollary 16.11 would even enable us to find solutions of arbitrary games if we could find solutions of all games in reduced form). We have seen, moreover, that there is always only one imputation for an inessential game in reduced form, namely, the imputation $\| 0 \quad 0 \quad \cdots \quad 0 \|$, and that the set consisting of this single imputation is (trivially) a solution. Thus there remains only the problem of showing that every essential game in reduced form has a solution; this problem appears to be difficult, however. It should be noticed that since we have seen that the essential three-person game in reduced form has a solution, it follows that every three-person game has a solution.

REMARK 16.14. It might be thought that in order for the definition of a solution to be regarded as satisfactory, it would also be necessary that a given game have a unique solution. However, von Neumann dissents from this view and holds that a given solution represents merely a socially accepted standard of behavior and consequently that there can, and should, be many solutions—corresponding to the many possible stable organizations of society. We shall be able to make his position in this regard somewhat clearer after we have solved the problem of finding all solutions of three-person games.

4. Three-person Games. In order to find all the solutions of the essential three-person game in reduced form, it is convenient to introduce a new coordinate system for the Euclidean plane.

We take as axes three concurrent straight lines which make angles of 60° with each other; and by the coordinates of an arbitrary point we mean the perpendicular distances from it to these three lines, the distances being called positive or negative as indicated in Fig. 1 (thus, for example, x_1 is positive for points above the horizontal line and is negative for points below this line).

Since, as is well known, the points of the Euclidean plane can be represented by using only two coordinates (as in the usual Cartesian system, for example), we should expect these three coordinates not to be mutually independent. And indeed, it can be easily shown (using a little trigonometry) that for every point $\| x_1 \quad x_2 \quad x_3 \|$ we have

$$x_1 + x_2 + x_3 = 0.$$

Moreover, if x_1, x_2, and x_3 are any three numbers whose sum is zero, then there exists a point in the plane whose coordinates are $\| x_1 \quad x_2 \quad x_3 \|$. Thus we have a coordinate system which automatically satisfies one of the conditions for an imputation (for a three-person game). The other condition

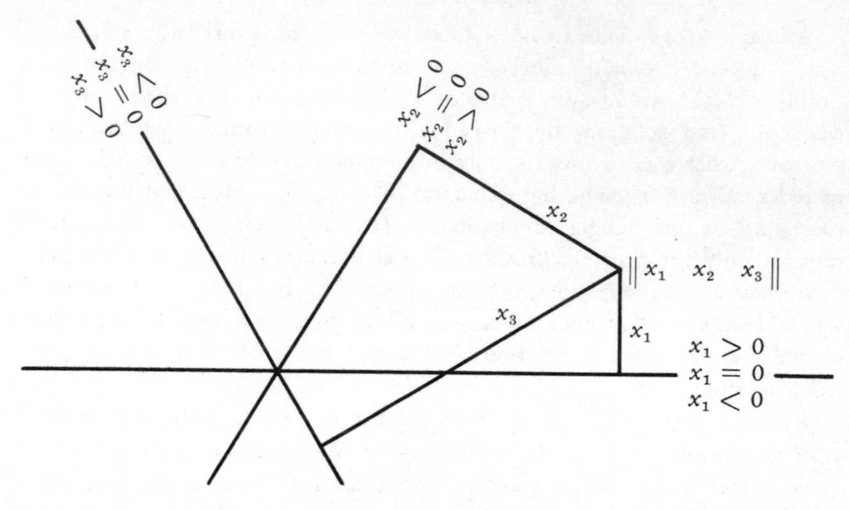

Fig. 1

amounts to saying that the point corresponding to the imputation shall lie within the shaded triangle as shown in Fig. 2. Thus this shaded area, which we shall call the *fundamental triangle*, represents all imputations for this game.

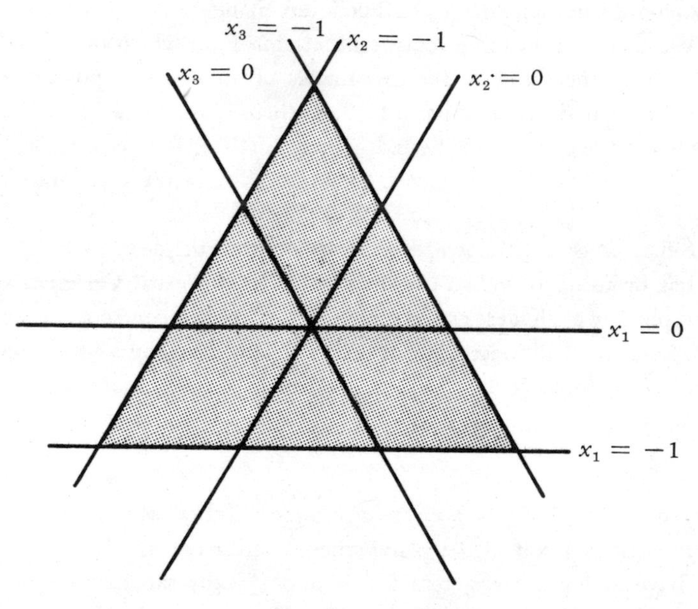

Fig. 2

We turn now to the geometrical representation of the relation of domination. Since we have seen that one imputation cannot dominate another with respect to the empty set, or a one-element set, or the set of all players, we need consider, in the case of three-person games, only domination with respect to two-element sets.

Moreover, we notice that if $\| x_1 \quad x_2 \quad x_3 \|$ is any imputation, then

$$x_1 + x_2 \leq v(\{1, 2\}).$$

For if we had

$$x_1 + x_2 > v(\{1, 2\}),$$

then we could conclude that

$$-x_3 > -v(\{3\}),$$

and hence that

$$x_3 < v(\{3\}),$$

contrary to the definition of an imputation. Similarly, we conclude that

$$x_1 + x_3 \leq v(\{1, 3\})$$

and that

$$x_2 + x_3 \leq v(\{2, 3\}).$$

Thus (ii) of 16.4 is always satisfied by a two-element T in our three-person game; and we conclude that an imputation $\| x_1 \quad x_2 \quad x_3 \|$ dominates an imputation $\| y_1 \quad y_2 \quad y_3 \|$ if, and only if, either

$$x_1 > y_1 \quad \text{and} \quad x_2 > y_2,$$

or

$$x_1 > y_1 \quad \text{and} \quad x_3 > y_3,$$

or

$$x_2 > y_2 \quad \text{and} \quad x_3 > y_3.$$

From this we conclude that an imputation $\| x_1 \quad x_2 \quad x_3 \|$ dominates just those imputations which lie in the shaded areas of Fig. 3 (the shaded areas exclusive of the three boundary lines passing through $\| x_1 \quad x_2 \quad x_3 \|$). We see, moreover, that every point in the unshaded areas represents an imputation which dominates $\|x_1 \quad x_2 \quad x_3\|$. Thus, if $\|x_1 \quad x_2 \quad x_3\|$ and $\|y_1 \quad y_2 \quad y_3\|$ are two imputations, neither one of which dominates the other, then the corresponding points lie on a line parallel to one of the coordinate axes.

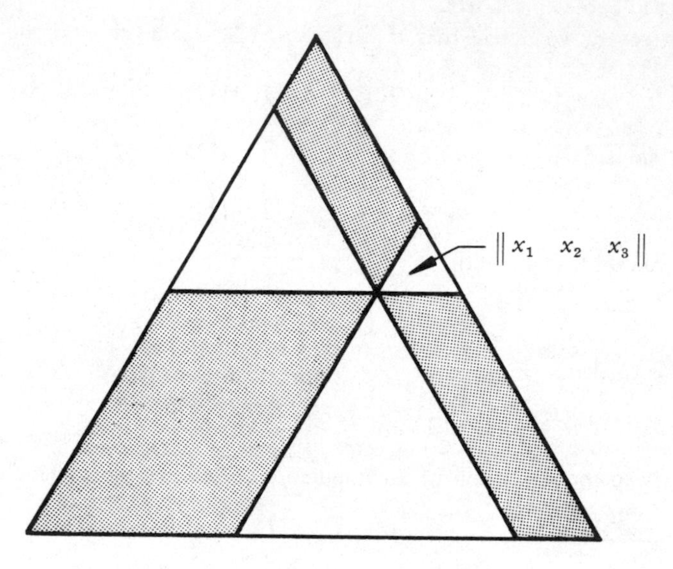

Fig. 3

We turn now to the problem of determining all solutions of this game. Since the game is essential, we see from the construction in 16.5 that every solution **A** must contain at least two imputations. Moreover, we have seen that every two members of **A** must lie on a line parallel to one of the coordinate axes (for otherwise, one of them would dominate the other, contrary to (i) of 16.6).

We now distinguish two cases, according as all points of **A** lie on the same line or not. In the second case, we are led to the solution

$$\mathbf{A} = \left\{ \left\| \frac{1}{2} \quad \frac{1}{2} \quad -1 \right\|, \left\| \frac{1}{2} \quad -1 \quad \frac{1}{2} \right\|, \left\| -1 \quad \frac{1}{2} \quad \frac{1}{2} \right\| \right\},$$

which was considered previously. In the first case, we conclude that **A** must contain all points within the fundamental triangle which lie on the given line and, moreover, that one of the following three conditions must be satisfied. Either

(i) **A** consists of all imputations $\|c \quad x_2 \quad x_3\|$, where c is fixed and satisfies the inequalities $-1 \leq c < ½$; or

(ii) A consists of all imputations $\| x_1 \quad c \quad x_3 \|$, where c is fixed and satisfies the inequalities $-1 \leq c < \frac{1}{2}$; or

(iii) A consists of all imputations $\| x_1 \quad x_2 \quad c \|$, where c is fixed and satisfies the inequalities $-1 \leq c < \frac{1}{2}$.

REMARK 16.15. Therefore in the case of the essential three-person game we have an embarrassing richness of solutions. In addition to the finite solution (consisting of three imputations) with which we started our discussion, we have also found an infinite collection of infinite solutions. We notice, moreover, that every point of the fundamental triangle lies in at least one of the sets described, so that every imputation belongs to at least one solution; thus no possible imputation is ruled out.

Von Neumann accounts for this situation, as was mentioned earlier, by regarding the various solutions as representing various standards of social behavior. In the case of the three-person game in question, he calls the unique finite solution the "nondiscriminatory" solution and the others "discriminatory." Thus, for instance, a solution of the form (iii), above, represents a social arrangement where players 1 and 2 have decided to exclude player 3 from their negotiations but to allow him a fixed amount c (the smaller the c, of course, the worse for player 3). How players 1 and 2 will divide the "spoils," $-c$, between them is not decided by the theory; this would presumably be determined by such extraneous factors as the relative persuasive powers of the players and their relative stubbornness.

Putting all of this in the most favorable light possible, it seems that our set of solutions tells us that when three people play the essential three-person game in reduced form, then either (1) two of them should decide to exclude the third from all negotiations and arbitrarily assign to him a fixed amount c, deciding between themselves (by some method not specified) how to divide the amount $-c$; or (2) no person should be excluded from the negotiations, but some two should end up by giving the third -1 and dividing the amount $+1$ equally between themselves.

It seems hard to believe that knowing this would enable a man to play the game more smoothly or profitably, especially if the other two players were unacquainted with this "correct" way of doing things! Thus it appears that the theory of n-person games is not yet in a completely satisfactory form.

BIBLIOGRAPHICAL REMARKS

An extremely detailed and painstaking treatment of solutions of n-person games is given in von Neumann and Morgenstern [1].

Another approach to the subject of n-person games is developed in Nash [2].

For other developments, which are too recent to be discussed in the text of this book, see Bott [1] and Shapley [1] and [2].

EXERCISES

1. The characteristic function v of a certain five-person game is defined as follows:

$$
\begin{aligned}
v(\mathsf{T}) &= 0 && \text{if } \mathsf{T} \text{ contains 0 elements,}\\
v(\mathsf{T}) &= -1 && \text{if } \mathsf{T} \text{ contains 1 element,}\\
v(\mathsf{T}) &= 0 && \text{if } \mathsf{T} \text{ contains 2 elements,}\\
v(\mathsf{T}) &= 0 && \text{if } \mathsf{T} \text{ contains 3 elements,}\\
v(\mathsf{T}) &= +1 && \text{if } \mathsf{T} \text{ contains 4 elements,}\\
v(\mathsf{T}) &= 0 && \text{if } \mathsf{T} \text{ contains 5 elements.}
\end{aligned}
$$

Show that each of the two imputations

$$
\begin{aligned}
\alpha &= \| -.1 \quad -.1 \quad -.2 \quad -.2 \quad +.6 \|,\\
\beta &= \| -.2 \quad -.2 \quad -.1 \quad -.1 \quad +.6 \|
\end{aligned}
$$

dominates the other.

2. Show that if there are two imputations for an n-person game, each of which dominates the other, then $n \geq 5$.

3. Show that, in the case of the essential three-person game in reduced form, there exist exactly three imputations which do not dominate any imputations at all. Generalize this to the case of an essential n-person game.

4. Find all solutions of the three-person game whose characteristic function v is defined as follows:

$$
\begin{aligned}
v(\Lambda) &= 0, & v(\{1, 2\}) &= 8,\\
v(\{1\}) &= -4, & v(\{1, 3\}) &= 3,\\
v(\{2\}) &= -3, & v(\{2, 3\}) &= 4,\\
v(\{3\}) &= -8, & v(\{1, 2, 3\}) &= 0.
\end{aligned}
$$

5. Prove Theorem 16.9.

6. Show that the relation of isomorphism of games is reflexive, symmetric, and transitive.

7. The characteristic function v of a certain four-person game is defined as follows:

$$
v(\mathsf{T}) = 0 \qquad \text{if } \mathsf{T} \text{ contains 0 elements,}
$$

$$
\begin{aligned}
v(\mathsf{T}) &= -1 && \text{if } \mathsf{T} \text{ contains 1 element,}\\
v(\mathsf{T}) &= 0 && \text{if } \mathsf{T} \text{ contains 2 elements,}\\
v(\mathsf{T}) &= +1 && \text{if } \mathsf{T} \text{ contains 3 elements,}\\
v(\mathsf{T}) &= 0 && \text{if } \mathsf{T} \text{ contains 4 elements.}
\end{aligned}
$$

Show that the following set of thirteen imputations is a solution of this game:

$$
\left\| 0 \quad 0 \quad 0 \quad 0 \right\|, \quad \left\| \tfrac{1}{2} \quad \tfrac{1}{2} \quad 0 \quad -1 \right\|, \quad \left\| \tfrac{1}{2} \quad \tfrac{1}{2} \quad -1 \quad 0 \right\|,
$$

$$
\left\| \tfrac{1}{2} \quad 0 \quad \tfrac{1}{2} \quad -1 \right\|, \quad \left\| \tfrac{1}{2} \quad -1 \quad \tfrac{1}{2} \quad 0 \right\|, \quad \left\| \tfrac{1}{2} \quad 0 \quad -1 \quad \tfrac{1}{2} \right\|,
$$

$$
\left\| \tfrac{1}{2} \quad -1 \quad 0 \quad \tfrac{1}{2} \right\|, \quad \left\| 0 \quad \tfrac{1}{2} \quad \tfrac{1}{2} \quad -1 \right\|, \quad \left\| -1 \quad \tfrac{1}{2} \quad \tfrac{1}{2} \quad 0 \right\|,
$$

$$
\left\| 0 \quad \tfrac{1}{2} \quad -1 \quad \tfrac{1}{2} \right\|, \quad \left\| -1 \quad \tfrac{1}{2} \quad 0 \quad \tfrac{1}{2} \right\|, \quad \left\| 0 \quad -1 \quad \tfrac{1}{2} \quad \tfrac{1}{2} \right\|,
$$

$$
\left\| -1 \quad 0 \quad \tfrac{1}{2} \quad \tfrac{1}{2} \right\|.
$$

8. Find another solution of the game defined in Exercise 7.

CHAPTER 17

GAMES WITHOUT ZERO-SUM RESTRICTION: THE VON NEUMANN–MORGENSTERN THEORY

1. Characteristic Functions. So far we have considered only zero-sum games, i.e., games in which the sum of the expectations of the various players is zero. Although parlor games are ordinarily of this sort, games which do not satisfy the zero-sum condition are (as was mentioned in Chap. 1) very important from the point of view of applications to economic theory. Thus if we consider, for example, the mutual interactions of a labor union with an industrial company as a two-person game, it is clear that this game is not zero-sum, for it may be that certain actions (agreement on a contract, for instance) are advantageous to both parties, while other actions (for instance, a strike which causes the plant to shut down) hurt both sides—though not usually to the same extent, of course. Thus the theory of games which are not zero-sum occupies an exceedingly important position so far as regards the development of the social sciences.

This chapter will be devoted to a study of games where the zero-sum restriction is not necessarily satisfied. Since we do not want our theory to exclude zero-sum games from consideration, however, we shall use the expression "general games" to include both zero-sum games and games which are not zero-sum; when there is no danger of misunderstanding, we shall say simply "game" instead of "general game."

Unfortunately it happens that, despite the great importance of general games for the social sciences, there is not available so far any treatment of such games which can be regarded as even reasonably satisfactory. We are not going to attempt to give a complete account of the various theories which have been developed in this domain. We shall confine ourselves to a brief sketch of the von Neumann and Morgenstern theory.

In considering general games, it is clear, in the first place, that we need consider only games in rectangular form. For, by the introduction of the notion of a strategy, we can always reduce every general game to a general game in rectangular form. Thus a general n-person game with the players $\{1, \cdots, n\}$ is completely specified when we describe n choice-sets $\{\mathsf{C}_1, \cdots, \mathsf{C}_n\}$ and n payoff functions $\{M_1, \cdots, M_n\}$. A play of the game consists in the following: Player i (for $i = 1, \cdots, n$) chooses an element x_i

from the set C_i and communicates his choice to the umpire (but not to the other players); when the choices have all been made, the umpire pays player i (for $i = 1, \cdots, n$) the amount

$$M_i(x_1, \cdots, x_n).$$

If each of the sets C_i is finite, we call the game itself *finite*. The game is zero-sum if, whenever $\| x_1 \quad \cdots \quad x_n \|$ belongs to the Cartesian product $C_1 \times \cdots \times C_n$ of the sets $C_1, \cdots, C_n$, we have

$$\sum_{i=1}^{n} M_i(x_1, \cdots, x_n) = 0.$$

Henceforth, of course, we shall not ordinarily assume that the above equation is satisfied.

From a certain point of view it is possible to regard a general n-person game as a special kind of $(n + 1)$-person game. For suppose that we have a general n-person game with the players $\{1, \cdots, n\}$, the choice-sets $\{C_1, \cdots, C_n\}$, and the payoff functions $\{M_1, \cdots, M_n\}$; and suppose that we take C_{n+1} to be an arbitrary set and that we define the function M_{n+1} by setting, for any member $\| x_1 \quad \cdots \quad x_n \|$ of $C_1 \times \cdots \times C_n$,

$$M_{n+1}(x_1, \cdots, x_n) = - \sum_{i=1}^{n} M_i(x_1, \cdots, x_n).$$

Then we can regard $\{C_1, \cdots, C_n, C_{n+1}\}$ and $\{M_1, \cdots, M_n, M_{n+1}\}$ as choice-sets and payoff functions of an $(n + 1)$-person game with the players $\{1, \cdots, n, n + 1\}$; moreover, from the way in which M_{n+1} was defined, we see immediately that this new game is zero-sum.

It should not be concluded, of course, that this construction enables us in one step to reduce the theory of general games to the theory of zero-sum games, for the $(n + 1)$-person game constructed above has some special properties. In the first place, the values assumed by the payoff functions are independent of the choice made by the player $n + 1$; and, what is more important, the player $n + 1$, since he is only a mathematical fiction so far as regards the original game, must not be conceived as entering into coalitions or making side-payments. Nevertheless, this extension of a general n-person game to a zero-sum $(n + 1)$-person game has a certain formal usefulness.

Let Γ be a general n-person game and let Γ' be its zero-sum $(n + 1)$-person extension as introduced above. From the results of Chap. 15, we see that Γ' possesses a characteristic function; i.e., there is a real-valued function v, defined over all subsets of the set $N_{n+1} = \{1, \cdots, n, n + 1\}$ of players of Γ', which is such that $v(T)$, for every subset T of N_{n+1}, represents the amount that the players in T can expect to win if they combine into a coalition. This function v is, a fortiori, defined over the set $\mathsf{N}_n = \{1, \cdots, n\}$ of players of the original game, and, for any subset T of N_n, $v(\mathsf{T})$ still represents the amount the players of T can expect to win if they make a coalition. We also call v (when restricted in its arguments to subsets of N_n) the *characteristic function* of the original game Γ.

It can easily be shown that, in case Γ is already a zero-sum game, the characteristic function of Γ, as defined above, is identical with the characteristic function as defined in Chap. 15. Thus our new definition is consistent with the old one, and we can speak quite generally of the characteristic functions of games, without regard to the question as to whether they are zero-sum or not.

We now have the following theorem:

> THEOREM 17.1. If v is the characteristic function of a (general) game whose players are $\mathsf{N}_n = \{1, \cdots, n\}$, then
>
> (i) $v(\Lambda) = 0$;
>
> (ii) if R and T are mutually exclusive subsets of N_n, then $v(\mathsf{R} \cup \mathsf{T}) \geq v(\mathsf{R}) + v(\mathsf{T})$.
>
> Moreover, if v is any real-valued function, defined over the class of all subsets of N_n, which satisfies conditions (i) and (ii), then there exists a (general) game Γ which has v for its characteristic function.

PROOF. If v is the characteristic function of a general game Γ whose players are N_n, then by definition there exists a zero-sum game Γ', with players N_{n+1}, such that the characteristic function v' of Γ' satisfies

$$v(\mathsf{T}) = v'(\mathsf{T}) \qquad \text{for } \mathsf{T} \subseteq \mathsf{N}_n. \tag{1}$$

By (i) of Lemma 15.2 we see that

$$v'(\Lambda) = 0; \tag{2}$$

and from (iii) of Theorem 15.1, if R and T are mutually exclusive subsets of N_{n+1} (and hence, a fortiori, if they are mutually exclusive subsets of N_n), we have

$$v'(\mathsf{R} \cup \mathsf{T}) \geq v'(\mathsf{R}) + v'(\mathsf{T}). \tag{3}$$

The first part of our theorem follows from (2) and (3), by means of (1).

To prove the second part of the theorem, let v be any real-valued function which is defined over the class of all subsets of N_n and which satisfies conditions (i) and (ii). We define the function v' over the class of all subsets of N_{n+1} as follows:

$$v'(\mathsf{T}) = v(\mathsf{T}) \qquad \text{if } n+1 \notin \mathsf{T}, \tag{4}$$

$$v'(\mathsf{T}) = -v(\mathsf{N}_{n+1} - \mathsf{T}) \qquad \text{if } n+1 \in \mathsf{T}. \tag{5}$$

By definition of the characteristic function of a general game, we see from (4) that in order to complete the proof it suffices to show that v' is the characteristic function of a zero-sum $(n+1)$-person game. From Theorem 15.3, we conclude, indeed, that it suffices to show that v' satisfies the three conditions of Theorem 15.1. But by (i) and (5) it is immediately seen that (i) of 15.1 is satisfied by v'; and the fact that (ii) of 15.1 is satisfied by v' is an immediate consequence of (5). Moreover, if neither R nor T contains $n+1$, then (iii) of 15.1 follows directly from (ii) and (4); and if both R and T contain $n+1$, then (iii) of 15.1 is vacuously true. Thus it suffices to show that (iii) of 15.1 is true in case $n+1$ belongs to one of the sets R or T, but not to the other. Without loss of generality, we can suppose that

$$n+1 \in \mathsf{R}, \tag{6}$$

$$n+1 \notin \mathsf{T}. \tag{7}$$

The two sets T and $\mathsf{N}_{n+1} - (\mathsf{R} \cup \mathsf{T})$ are mutually exclusive subsets of N_n; hence by (ii)

$$v(\mathsf{T} \cup [\mathsf{N}_{n+1} - (\mathsf{R} \cup \mathsf{T})]) \geq v(\mathsf{T}) + v[\mathsf{N}_{n+1} - (\mathsf{R} \cup \mathsf{T})]. \tag{8}$$

Since R and T are mutually exclusive, we have $\mathsf{T} \subseteq \mathsf{N}_{n+1} - \mathsf{R}$, and hence

$$\mathsf{T} \cup [\mathsf{N}_{n+1} - (\mathsf{R} \cup \mathsf{T})] = \mathsf{T} \cup [\mathsf{N}_{n+1} - \mathsf{R}] = \mathsf{N}_{n+1} - \mathsf{R}. \tag{9}$$

From (8) and (9)

$$v(\mathsf{N}_{n+1} - \mathsf{R}) \geq v(\mathsf{T}) + v[\mathsf{N}_{n+1} - (\mathsf{R} \cup \mathsf{T})],$$

and hence by (5) and (4)

$$-v'(\mathsf{R}) \geq v'(\mathsf{T}) - v'(\mathsf{R} \cup \mathsf{T}). \tag{10}$$

From (10) we conclude by transposition that

$$v'(\mathsf{R} \cup \mathsf{T}) \geq v'(\mathsf{R}) + v'(\mathsf{T}),$$

as was to be shown.

The notion of S-equivalence, and the intuitive arguments to justify it, can now be extended from zero-sum games to general games.

DEFINITION 17.2. Two n-person characteristic functions (for general n-person games) v and v' are called S-*equivalent* if there is a positive constant k and n numbers $a_1, \cdots, a_n$ such that, for every subset T of N_n,

$$v(\mathsf{T}) = k \cdot v'(\mathsf{T}) + \sum_{i \in \mathsf{T}} a_i .$$

REMARK 17.3. Since we are no longer restricting ourselves to zero-sum games, it is no longer necessary to make the restriction, in defining S-equivalence, that

$$\sum_{i=1}^{n} a_i = 0 .$$

Thus it should be remarked that Definition 17.2 makes a somewhat wider class of games S-equivalent to a given zero-sum game than does Definition 15.5 (and the remark following Theorem 15.7). Indeed, zero-sum games, under the new definition, can now be S-equivalent to *constant-sum* games, i.e., to games whose characteristic functions satisfy the condition (for every subset T of N_n)

$$v(\mathsf{T}) + v(\mathsf{N}_n - \mathsf{T}) = v(\mathsf{N}_n) .$$

To avoid any misunderstandings in this connection, we shall henceforth use the term S-equivalence always in the sense of 17.2.

The following notion of reduced forms clearly coincides with the notion defined in 15.8 for the case of zero-sum games.

DEFINITION 17.4. A general n-person game with characteristic function v is said to be in *reduced form* if

$$v(\{1\}) = v(\{2\}) = \cdots = v(\{n\}) = \gamma ,$$

where either $\gamma = 0$ or $\gamma = -1$, and, in addition,

$$v(\mathsf{N}_n) = 0 .$$

We call γ the *modulus* of v.

The proof of the following theorem is similar to that of Theorem 15.12.

> THEOREM 17.5. If v is the characteristic function of any general game, then v is S-equivalent to one, and only one, characteristic function in reduced form.

As in the case of zero-sum games, we call a general game *essential* if it is S-equivalent to a game in reduced form of modulus -1; otherwise, it is called *inessential*.

2. Imputations and Solutions. As in the case of zero-sum games, we now wish to introduce the notion of an ordered sequence $\| x_1 \quad \cdots \quad x_n \|$ of real numbers to represent a possible division of money among the players at the end of a play. As before, we need not consider divisions which give some one of the players less than he could gain by his own unaided efforts; so we wish to impose the condition

$$x_i \geq v(\{i\}) \qquad \text{for } i = 1, \cdots, n.$$

Moreover, since we allow agreements and side-payments among the players, it seems natural to consider only divisions $\| x_1 \quad \cdots \quad x_n \|$ such that

$$\sum_{i=1}^{n} x_i = v(\{1, \cdots, n\}).$$

For clearly the players can always see to it that, together, they get $v(\{1, \cdots, n\})$—by making a coalition, so to speak, "against nature." And if a division were proposed such that

$$\sum_{i=1}^{n} x_i < v(\{1, \cdots, n\}),$$

then the players could be shown a method of playing and a way of dividing the payments at the end of the play which would give player i (for $i = 1, \cdots, n$) more than x_i; namely, there would be a way of playing such that each player would get

$$x_i + \frac{1}{n}\left[v(\{1, \cdots, n\}) - \sum_{i=1}^{n} x_i \right] > x_i.$$

Hence we define an imputation for a general game as follows:

DEFINITION 17.6. By an *imputation* for a general game whose characteristic function is v, we mean a vector

$$\| x_1 \quad \cdots \quad x_n \|$$

such that

$$\text{(i)} \qquad \sum_{i=1}^{n} x_i = v(\{1, \cdots, n\})$$

and

$$\text{(ii)} \qquad x_i \geq v(\{i\}) \qquad \text{for } i = 1, \cdots, n.$$

REMARK 17.7. In case the general game happens to be a zero-sum game, then we obtain the same notion of an imputation by Definition 17.6 as by Definition 16.1. For (ii) of 17.6 is the same as (ii) of 17.1; and, in case the game is zero-sum, we have

$$v(\{1, \cdots, n\}) = 0,$$

so that (i) of 17.6 is the same as (i) of 16.1.

We can now introduce the notions of domination and solution, exactly as in the case of zero-sum games (see Definitions 16.4 and 16.6). The remarks made about these notions in Chap. 16 apply *mutatis mutandis* to general games. It remains true, in particular, that games which are S-equivalent are isomorphic with respect to dominance; so we again need consider only the reduced forms of games.

To make clearer the notion of a solution, we shall find all solutions of two-person games in reduced form. If v is the characteristic function of such a game, then we have:

$$v(\mathsf{T}) = 0 \quad \text{when } \mathsf{T} \text{ contains 0 elements},$$
$$v(\mathsf{T}) = \gamma \quad \text{when } \mathsf{T} \text{ contains 1 element},$$
$$v(\mathsf{T}) = 0 \quad \text{when } \mathsf{T} \text{ contains 2 elements}.$$

Thus the characteristic function is completely determined when γ is given; and we need distinguish only two cases, according as $\gamma = 0$ or $\gamma = -1$.

If $\gamma = 0$, then v is identically zero, and the game is inessential. In this case there is only one imputation, namely, $\|0 \quad 0\|$, and the set consisting of this imputation is a solution, and, of course, the only possible one.

If $\gamma = -1$, then there are infinitely many imputations, namely, the set of all couples $\|x_1 \quad x_2\|$ such that

$$-1 \leq x_1, \qquad -1 \leq x_2$$

and

$$x_1 + x_2 = 0.$$

Thus an imputation is any ordered couple of the form $\|x \quad -x\|$, where $-1 \leq x \leq 1$. Now we notice that no imputation can dominate another; this can be seen by considering the various possible sets of players. Thus, for instance, if we had

$$\|x \quad -x\| \underset{\{1\}}{\succ} \|y \quad -y\|,$$

then we should have

$$y < x \leq v(\{1\}),$$

and hence

$$y < v(\{1\}),$$

which contradicts the hypothesis that $\| y \quad -y \|$ is an imputation. Thus the set of all imputations is again a solution; and it is clearly the only solution.

It follows that the solution of the essential two-person (general) game which is not in reduced form is the set of all couples $\| x_1 \quad x_2 \|$, where x_1 and x_2 satisfy the conditions

$$v(\{1\}) \leq x_1,$$
$$v(\{2\}) \leq x_2,$$
$$x_1 + x_2 = v(\{1, 2\}).$$

The interpretation of this result, in terms of the way two people should (or, perhaps, merely "do") play a non-zero-sum two-person game, is as follows: They should find a way to play the game which will maximize the sum of their gains and then divide this sum between the two players—in such a way that each gets at least what he could get if he were playing "on his own," but with the other player trying to do him as much harm as possible. Aside from this last condition, the theory offers no way of deciding how the profits will be divided.

EXAMPLE 17.8. Consider the two-person game in which there are two strategies available for each player and in which the payoff matrices for the two players are as follows:

$$\begin{Vmatrix} 1 & -2 \\ -1 & 1 \end{Vmatrix}, \quad \begin{Vmatrix} 1 & 3 \\ 4 & -1 \end{Vmatrix}.$$

(Thus, for example, if player 1 chooses his first strategy and player 2 chooses his second strategy, then the first player gets -2 and the second player gets 3.)

Now $v(\{1\})$ is the value of the zero-sum game whose matrix is

$$\begin{Vmatrix} 1 & -2 \\ -1 & 1 \end{Vmatrix}.$$

Thus

$$v(\{1\}) = -\frac{1}{5};$$

player 1 can be sure of getting at least $-1/5$ by using the mixed strategy $\| 2/5 \quad 3/5 \|$. Similarly, $v(\{2\})$ is the value of the zero-sum game whose matrix is

$$\begin{Vmatrix} 1 & 4 \\ 3 & -1 \end{Vmatrix}.$$

Thus

$$v(\{2\}) = \frac{13}{7};$$

player 2 can be assured of getting at least $13/7$ by using the mixed strategy $\| 4/7 \quad 3/7 \|$. To find $v(\{1, 2\})$, we merely take the maximum of the sums of corresponding elements of the two originally given matrices; thus

$$v(\{1, 2\}) = 3.$$

Finally we have

$$v(\Lambda) = 0.$$

The solution of this game is therefore the set of all couples $\| x_1 \quad x_2 \|$ such that

$$x_1 + x_2 = 3,$$

$$x_1 \geq -\frac{1}{5},$$

$$x_2 \geq \frac{13}{7}.$$

Thus the two players can expect to obtain $x_1 = -1/5 + 47/35\,\theta$ and $x_2 = 13/7 + 47/35(1 - \theta)$, where θ is a number between 0 and 1 to be determined by negotiation between the two players.

It is possible to find also all solutions of non-zero-sum three-person games. This investigation is not especially difficult, but we shall omit it in the interest of brevity. We turn, instead, to a criticism of one of the fundamental assumptions of von Neumann's theory.

It should be noticed that von Neumann's whole theory of general games is based on the notion of the characteristic function. This implies that if two games have identical characteristic functions, then they will have identical solutions. It is, to say the least, debatable, however, whether this is satisfactory from the point of view of intuition. The intuitive difficulties involved here can be brought into focus by the following example.

EXAMPLE 17.9. Consider a two-person game in rectangular form, in which the first player has available only one (pure) strategy, while the second player has two strategies. The payoff matrices are as follows:

$$\| 0 \quad 10 \|, \qquad \| -1000 \quad 0 \|.$$

Thus if the second player plays strategy 1, then the first player gets 0, and the second player gets -1000; if he plays strategy 2, then the first player gets 10 and the second player gets 0 (the game is, of course, not zero-sum).

Now it seems intuitively reasonable to suppose that player 1 is here in a better position than player 2. For, if player 2 behaves in such fashion as to maximize his own gain, then he will choose strategy 2 so as to get 0 instead of -1000, and in this case player 1 will get 10. It might, of course, be thought that player 2 would be able to get some of the 10 from player 1 by threatening to play strategy 1, which would reduce the gain of player 1 from 10 down to 0; but, in view of the fact that player 2 would himself lose so heavily if he were to carry out such a threat (in fact, he would lose much more than would player 1), it appears unlikely that player 1 would take him seriously. In making such a threat, player 2 would appear to be in a position analogous to that of a workman who would say to his employer, "Since you make a profit from my labor, I demand that you share this profit with me; if you refuse to do so, I shall maim myself in such a way that I shall henceforth be unable to work, and you will get no profit at all from me"; the workman could hardly expect anything but rude and stubborn resistance to such a demand.

For our present purposes, however, it is not even necessary to say that player 1 would be able to keep all the 10; our point will be made if it is admitted even that player 1 and player 2 are not in equally advantageous positions, so that player 1 would be able to keep more than 5. Perhaps the correct way to put this intuitive problem is to ask ourselves whether, in case we had to play this game, it would be a matter of indifference whether we were to take the role of player 1 or player 2. Most people seem to feel that they would rather be player 1; in fact, most people would pay money to be player 1 instead of player 2.

If it is admitted that player 1 is in a better position in this game than is player 2, then it appears that any "solution" of the game should be defined in such a way as to reflect this asymmetry. This is not true, however, of the solution in the sense of von Neumann; for we have seen that the solution is defined entirely in terms of the characteristic function, and it happens that the characteristic function of this game is symmetric in 1 and 2; indeed, it is easily verified that the characteristic function v is as follows:

$$v(\Lambda) = 0,$$

$$v(\{1\}) = v(\{2\}) = 0,$$

$$v(\{1, 2\}) = 10.$$

Thus the solution of this game, in the sense of von Neumann, is the set of all ordered couples $\|x \quad y\|$, where x and y are non-negative numbers whose sum is 10.

BIBLIOGRAPHICAL REMARK

The material in this chapter is taken largely from von Neumann and Morgenstern [1], Chaps. 5 and 6.

EXERCISES

1. The game of "Russian roulette" is played as follows: A bullet is put into one chamber of a pistol (the other chambers being left empty); each of several men then spins the cylinder, points the pistol at his head, and pulls the trigger. The players play in turn until they have all had one play, or until one has shot himself, whichever happens first. The payoff to each player is either death, or the feeling of relief that comes from knowing that he has escaped death. Is Russian roulette a zero-sum game?

2. Consider a two-person game in which there are two strategies available for each player and in which the payoff matrices for the two players are as follows:

$$\left\|\begin{matrix} 1 & 3 \\ 0 & 10 \end{matrix}\right\|, \qquad \left\|\begin{matrix} -1 & 2 \\ 5 & 1 \end{matrix}\right\|.$$

Find the characteristic function of this game.

3. Find the reduced form of the following characteristic function for a three-person game:

$$\begin{aligned} v(\Lambda) &= 0, & v(\{1, 2\}) &= 3, \\ v(\{1\}) &= 0, & v(\{1, 3\}) &= 6, \\ v(\{2\}) &= 1, & v(\{2, 3\}) &= 7, \\ v(\{3\}) &= 4, & v(\{1, 2, 3\}) &= 8. \end{aligned}$$

4. Find a solution (in the sense of von Neumann) for the three-person game whose characteristic function is as follows:

$$v(\Lambda) = 0,$$

$$v(\{1\}) = v(\{2\}) = v(\{3\}) = -1,$$

$$v(\{1, 2\}) = v(\{1, 3\}) = v(\{2, 3\}) = v(\{1, 2, 3\}) = 0.$$

5. Prove Theorem 17.5.

6. Prove Theorem 15.9 for the case of general games.

7. Prove Theorem 15.15 for the case of general games.

CHAPTER 18
SOME OPEN PROBLEMS

1. Two Types of Problems. As the student will doubtless realize by this time, the theory of games is still in a far from satisfactory state of development. Of course, every branch of mathematics, no matter how old, continues to present difficult problems; but in the more anciently established disciplines (such as the theory of numbers, for example), the unsolved problems already have a sharp and definite character; we seem to know exactly what is meant, let us say, by Fermat's Last Problem; it is only a matter of discovering a proof, or a disproof, of this clearly formulated statement. In the theory of games, however, we are also confronted with difficult problems of what might be called a "conceptual" sort. By this is meant such problems as: what formal extensions of the mathematical theory are likely to be useful in practical applications and what modifications of the existing theory are necessary in order to permit application to given practical situations. In the most general sense, these problems amount to asking what sort of formal mathematical apparatus is best fitted as a tool for dealing with situations involving conflict among rational agents; thus conceptual problems, as opposed to what we might call "technical" problems, are concerned with the question of what sort of science the theory of games ought to be, rather than with questions as to what theorems can be established in portions of the theory already agreed upon.

This distinction between technical and conceptual problems is, of course, not very sharp: many problems are partly technical and partly conceptual. In this chapter we shall discuss three of the most important unsolved problems of the theory of games; the first of these problems is partly technical, and the other two are largely conceptual.

2. Games Played over Function Space. When we discussed infinite games in Chap. 7, we very quickly turned to the special case of a continuous game, where each player chooses a number from the closed unit interval; and our whole discussion in the next few chapters was confined to this case and to trivial modifications of it. Now, since continuous games (for the infinite case) are analogous to rectangular games (for the finite case), it might be thought that by introducing strategies we could reduce every infinite game to a continuous game, just as every finite game can, by the introduction of strategies, be reduced to a rectangular game. Unfortunately, however, this is not the case; for it can easily happen, for instance, that the number of strategies is so large that the strategies cannot even be put into one-to-one correspondence with the real numbers in the closed unit interval.

Thus suppose, for instance, that a game has four moves: in the first move, P_1 chooses a real number x_1; in the second move, P_2, knowing x_1, chooses a real number y_1; in the third move, P_1, knowing y_1, but having forgotten x_1, chooses a real number x_2; and in the last move, P_2, knowing y_1 and x_2, but not knowing x_1, chooses a real number y_2. (The payoff is then some function of the four variables x_1, x_2, y_1, and y_2.) A pure strategy for P_1 is now an ordered couple $\| a \quad f \|$, where a is a real number and f is a function of one real variable (it depends on y_1); and a pure strategy for P_2 is an ordered couple $\| g \quad h \|$, where g is a function of one real variable (it depends on x_1) and h is a function of two real variables (it depends on y_1 and x_2). Since there are more functions of a real variable than there are real numbers (indeed, if c is the number of real numbers, then there are 2^c functions of a real variable), it is clear that the strategies of neither P_1 nor P_2 can be put into one-to-one correspondence with the points of the closed unit interval, and hence that this game cannot be reduced to a continuous game.

It is easy, as a matter of fact, to describe games where each player has but one move and where neither is informed of the other's choice, but which are not equivalent to continuous games. To facilitate the description of a simple game of this sort, let us denote by F the set of all functions f which are defined over the closed unit interval and are such that $0 \leq f(x) \leq 1$ (for $0 \leq x \leq 1$) and the integral $\int_0^1 f(x)\,dx$ exists. The game is now as follows: P_1 chooses a member f of F, and P_2, without being informed what choice P_1 has made, chooses a member g of F; P_2 then pays P_1 the amount

$$M(f, g) = \int_0^1 [f(x) - g(x)]^2\,dx.$$

Since it follows from well-known results that F has more members than there are real numbers, we again see that this game cannot be reduced to a continuous game by relabeling the elements of F.

It is clear that the payoff function for a game of the type just described need not necessarily have a saddle-point, and hence it is natural to suppose that the players will make use of mixed strategies. Here a mixed strategy will be a distribution function over the set F, or, as we sometimes say, a distribution function *over function space.* But now the perplexing question arises: to what class A of subsets of F are the distribution functions to be regarded as assigning probability. It can easily be shown that we cannot obtain intuitively acceptable results by supposing that A is the class of all subsets of F; and, on the other hand, once we start leaving subsets out of F, it is not easy to know where to stop. This question is, of course, connected with the question of how we are going to define the expectation of P_1 in case P_1 uses the distribution function F over F and P_2 uses the distribution function G over F.

For purposes of comparison here, it may be remarked that our definition (in Chap. 8) of distribution functions over the unit interval amounts to assuming that a probability is assigned to every Lebesgue measurable set. But it is not clear that there is an equally natural class of subsets of function space.

This problem, as described up to this point, is almost entirely conceptual. The problem of finding an intuitively satisfactory way of introducing distribution functions over function space, however, appears to be extremely difficult, and it may very well happen that a solution will not be found for it. If this turns out to be the case, we are still left with the more technical problem of picking out some large subclass of games over function space which can be solved without making use of the notion of a distribution function; a reasonable candidate for such a treatment would appear to be the class of games over function space which are convex for the minimizing player (the definition of convex functions given in Chap. 12 can readily be generalized to functions having functions as arguments).

3. Pseudo-games. Another important question is the problem of finding some rational way of dealing with conflict situations which are not technically zero-sum games, though they closely resemble such games. Just to have a term, let us agree to call such situations *pseudo-games.*

One way in which a conflict situation can fail to be a game is for the players to be forbidden to use mixed strategies. Situations of this sort are especially apt to arise in connection with military engagements. Thus suppose that two members of a combat team are forced to separate and that they cannot communicate with each other, or with their home base, because of the danger of revealing their positions to the enemy. Then even though it may be desirable for them jointly to play a mixed strategy, they may be unable to do so. (It might be thought that they could avoid the difficulty merely by each taking along an appropriate random list of strategies to be played successively; but even this device may be impracticable because of the very great loss that would result if the enemy were to capture such a list.) In such a case it can happen that the "game" has no "value"; and the ordinary theory of zero-sum two-person games is no longer applicable. Apparently the players are reduced to using behavior strategies, and the game may fail to possess optimal behavior strategies.

Somewhat similar circumstances can lead to pseudo-games which violate assumption 5 (d) of Chap. 6, i.e., which have plays intersecting the same information set more than once. The problem of dealing with such pseudo-games appears to be related to the problem of defining optimal ways of playing a game in normalized form, where the values of the elements of the payoff matrix are not known exactly but are merely required to satisfy certain inequalities.

In this connection we mention finally situations which technically fail to be games because of peculiarities in the nature of the payoff. We have always assumed that the values of the payoff functions are things which can be transferred from one player to another and which have the same utility to all. But in practice, of course, it can happen that the values of the payoff function are, for instance, sums of money and that the various players differ so greatly in their financial status that a dollar is much more important to some of them than to others; it can even happen that the values of the payoff are things such as the glow of satisfaction that comes from winning a chess game or the death that comes from losing a game of Russian roulette, which are not transferable at all. Thus we are left with the difficult problem of how to define a solution of a game where restrictions are placed on the transferability of goods.

4. Non-zero-sum Games and *n*-Person Games. The most crying need in game theory at the present time, however, is for a more satisfactory theory of non-zero-sum games and n-person games (for $n > 2$).

The theory of von Neumann, at best, deals with only a rather special type of such games: the type where agreements, negotiations, and side-payments among the players are allowed. Actually, of course, many situations to which we should like to apply the theory of games are not of this sort. Thus, if we wish to consider the behavior of three corporations as a game, we are faced with the fact that the antitrust laws prohibit their making a coalition. (It is sometimes said that such a situation should properly be regarded as a four-person game, with the government constituting the fourth player; but to treat this larger situation as a game would necessitate, among other things, the strange assumption that the government is prepared to enter into a coalition with a private corporation.) Coalitions and side-payments are forbidden, indeed, even in most parlor games. (I once knew of two members of a bridge club who had made the private agreement that whenever they played at the same table without being partners, they would bid very high and double and re-double. By this means they were assured that at least one of them would have an extremely high score for the afternoon's play; but the other ladies took a dim view of this little arrangement.)

In this connection it should be remarked that Nash has distinguished between *noncooperative* and *cooperative* games. In the former, no communication is allowed between the players and, in particular, they are not allowed to make agreements about side-payments; in a cooperative game, on the other hand, communication is allowed.

Nash regards the noncooperative games as the more fundamental and attempts to reduce the cooperative games to noncooperative games in the following way: The negotiations of the cooperative game are included as formal moves in a noncooperative game (these moves consist in such procedures as, for example, one player's offer of a side-payment to another). He deals with the noncooperative games, in turn, by the introduction of the notion of an equilibrium point (this notion was explained in Chap. 6 of this book).

It must be remarked that Nash's theory—although it represents a considerable advance—has some serious inadequacies and certainly cannot be regarded as a definitive solution of the conceptual problems of this domain. In the first place, so far as regards the noncooperative game, it does not appear that knowing the position of the equilibrium points will necessarily be of much help in playing the game. Thus consider the two-person (non-zero-sum) game whose matrices are as follows:

$$\begin{Vmatrix} 4 & -3 \\ 1 & -2 \end{Vmatrix}, \qquad \begin{Vmatrix} -20 & -30 \\ 10 & 40 \end{Vmatrix}.$$

Since $4 > 1$ and $-20 > -30$, we see that there is an equilibrium point in the upper left-hand corner (thus if P_1 plays the first row, then P_2 cannot do better than to play the first column; and, conversely, if P_2 plays the first column, then P_1 cannot do better than to play the first row). Similarly, since $-2 > -3$ and $40 > 10$, there is also an equilibrium point in the lower right-hand corner. Here P_1 would, of course, prefer the equilibrium point in the upper left-hand corner, and P_2 would prefer the equilibrium point in the lower right-hand corner. The theory of Nash seems to throw little light on the question of how to play a game having such a pair of payoff matrices.

In the second place, even if the theory of noncooperative games were in a completely satisfactory state, there appear to be difficulties in connection with the reduction of cooperative games to noncooperative games. It is extremely difficult in practice to introduce into the cooperative games the moves corresponding to negotiations in a way which will reflect all the infinite variety permissible in the cooperative game, and to do this without giving one player an artificial advantage (because of his having the first chance to make an offer, let us say).

Thus it seems that, despite the great ingenuity that has been shown in the various attacks on the problem of n-person and non-zero-sum games, we have not yet arrived at a satisfactory notion of a solution of such a game. This whole aspect of the theory presents a challenging problem to the mathematician—and an extremely important one—since the application of game theory to a very wide class of practical situations must wait for such a definition.

BIBLIOGRAPHICAL AND HISTORICAL REMARKS

The first and second problems of this chapter were posed in Helmer [1]. For additional unsolved problems in game theory, see Kuhn and Tucker [1].

For more recent work in n-person and non-zero-sum games, see Bott [1] and Shapley [2]; the latter paper gives a solution to problem (10) of Kuhn and Tucker [1]. See also Nash [1] and Raiffa [1].

Bibliography

ANDERSON, O. [1], "Theorie der Glücksspiele und ökonomisches Verhalten," *Schweizerische Zeitschrift für Volkswirtschaft und Statistik,* Vol. 85, 1949, pp. 46–53.

ARROW, K. J., E. W. BARANKIN, AND D. BLACKWELL [1], "Admissible Points of Convex Sets," *Contributions to the Theory of Games—II,* Annals of Mathematics Study No. 28, Princeton University Press, Princeton, N.J. (to be published).

ARROW, K. J., D. BLACKWELL, AND M. A. GIRSHICK [1], "Bayes and Minimax Solutions of Sequential Decision Problems," *Econometrica,* Vol. 17, 1949, pp. 213–244.

BELLMAN, R. [1], "Games of Bluffing," *Rendiconti del Palermo* (to be published).

BELLMAN, R., AND D. BLACKWELL [1], "Some Two-person Games Involving Bluffing," *Proceedings, National Academy of Sciences, U.S.A.,* Vol. 35, 1949, pp. 600–605.

BITTER, R. [1], "The Mathematical Formulation of Strategic Problems," *Proceedings of the Berkeley Symposium,* ed. by J. Neyman, University of California Press, Berkeley, 1949, pp. 223–228.

BLACKWELL, D. [1], "On Randomization of Statistical Games with k Terminal Actions," *Contributions to the Theory of Games—II,* Annals of Mathematics Study No. 28, Princeton University Press, Princeton, N.J. (to be published).

BÔCHER, MAXIME [1], *Introduction to Higher Algebra,* The Macmillan Company, New York, 1907.

BOHNENBLUST, H. F., AND S. KARLIN [1], "On a Theorem of Ville," *Contributions to the Theory of Games,* ed. by H. W. Kuhn and A. W. Tucker, Annals of Mathematics Study No. 24, Princeton University Press, Princeton, N.J., 1950.

BOHNENBLUST, H. F., S. KARLIN, AND L. S. SHAPLEY [1], "Solutions of Discrete Two-person Games," *Contributions to the Theory of Games,* ed. by H. W. Kuhn and A. W. Tucker, Annals of Mathematics Study No. 24, Princeton University Press, Princeton, N.J., 1950.

BOHNENBLUST, H. F., S. KARLIN, AND L. S. SHAPLEY [2], "Games with Continuous, Convex Pay-off," *Contributions to the Theory of Games,* ed. by H. W. Kuhn and A. W. Tucker, Annals of Mathematics Study No. 24, Princeton University Press, Princeton, N.J., 1950.

BONNESEN, T., AND W. FENCHEL [1], "Theorie der konvexen Körper," *Ergebnisse der Mathematik und ihrer Grenzgebiete,* Vol. 3, No. 1, Verlag Julius Springer, Berlin, 1934. (Reprint: Chelsea Publishing Company, New York, 1948.)

BOREL, E. [1], "Applications aux jeux de hasard," *Traité du calcul des probabilités et de ses applications,* Vol. 4, f. II, Gauthier-Villars & Cie, Paris, 1938.

BOTT, R. [1], "A Certain Simple Class of Symmetric n-person Games," *Contributions to the Theory of Games—II,* Annals of Mathematics Study No. 28, Princeton University Press, Princton, N.J. (to be published).

BRAY, H. E. [1], "Elementary Properties of the Stieltjes Integral," *Annals of Mathematics,* Vol. 20, 1919, pp. 177–186.

BREMS, H. [1], "Some Notes of the Structure of the Duopoly Problem," *Nordisk Tidsskrift för Teknisk Økonomi,* Nos. 1–4, 1948, pp. 41–74.

BROWN, G. W. [1], "Iterative Solutions of Games by Fictitious Play," *Activity Analysis of Production and Allocation,* Cowles Commission for Research in Economics, Monograph No. 13, John Wiley & Sons, Inc., New York, 1951.

BROWN, G. W., AND T. C. KOOPMANS [1], "Computational Suggestions for Maximizing a Linear Function Subject to Linear Inequalities," *Activity Analysis of Production and Allocation,* Cowles Commission for Research in Economics, Monograph No. 13, John Wiley & Sons, Inc., New York, 1951.

BROWN, G. W., AND J. VON NEUMANN [1], "Solutions of Games by Differential Equations," *Contributions to the Theory of Games,* ed. by H. W. Kuhn and A. W. Tucker, Annals of Mathematics Study No. 24, Princeton University Press, Princeton, N.J., 1950.

CHAMPERNOWNE, D. G. [1], "A Note on J. von Neumann's Article," *Review of Economic Studies,* Vol. 13, No. 1, 1945–1946, pp. 10–18.

CRAMÉR, HARALD [1], *Mathematical Methods of Statistics,* Princeton University Press, Princeton, N.J., 1946.

DALKEY, N. [1], "Equivalence of Information Patterns and Essentially Determinate Games," *Contributions to the Theory of Games—II,* Annals of Mathematics Study No. 28, Princeton University Press, Princeton, N.J. (to be published).

DANTZIG, G. B. [1], "Application of the Simplex Method to a Transportation Problem," *Activity Analysis of Production and Allocation,* Cowles Commission for Research in Economics, Monograph No. 13, John Wiley & Sons, Inc., New York, 1951.

DANTZIG, G. B. [2], "A Proof of the Equivalence of the Programming Problem and the Game Problem," *Activity Analysis of Production and Allocation,* Cowles Commission for Research in Economics, Monograph No. 13, John Wiley & Sons, Inc., New York, 1951.

DANTZIG, G. B. [3], "Maximization of a Linear Function of Variables Subject to Linear Inequalities," *Activity Analysis of Production and Allocation,* Cowles Commission for Research in Economics, Monograph No. 13, John Wiley & Sons, Inc., New York, 1951.

DANTZIG, G. B. [4], "Programming in a Linear Structure," *Econometrica,* Vol. 17, 1949, pp. 73–74.

DANTZIG, G. B., AND M. K. WOOD [1], "Programming of Interdependent Activities: I. General Discussion," *Econometrica,* Vol. 17, 1949, pp. 193–199.

DAVIES, D. W. [1], "A Theory of Chess and Noughts-and-Crosses," *Science News,* No. 16, 1950, pp. 40–64.

DEMARIA, G. [1], "Su una Nuova Logica Economica," *Giornale degli Economisti e Annali di Economia,* Vol. 6 (n.s.), 1947, pp. 661–671.

DINES, L. L. [1], "On a Theorem of von Neumann," *Proceedings, National Academy of Sciences, U.S.A.,* Vol. 33, 1947, pp. 329–331.

DORFMAN, R. [1], "Application of the Simplex Method to a Game Theory Problem," *Activity Analysis of Production and Allocation,* Cowles Commission for Research in Economics, Monograph No. 13, John Wiley & Sons, Inc., New York, 1951.

DRESHER, M. [1], "Methods of Solution in Game Theory," *Econometrica,* Vol. 18, 1950, pp. 179–181.

DRESHER, M. [2], "Games of Strategy," *Mathematics Magazine,* Vol. 25, 1951, pp. 93–99.

DRESHER, M., AND S. KARLIN [1], "Solutions of Convex Games as Fixed-points," *Contributions to the Theory of Games—II,* Annals of Mathematics Study No. 28, Princeton University Press, Princeton, N.J. (to be published).

DRESHER, M., S. KARLIN, AND L. S. SHAPLEY [1], "Polynomial Games," *Contributions to the Theory of Games,* ed. by H. W. Kuhn and A. W. Tucker, Annals of Mathematics Study No. 24, Princeton University Press, Princeton, N.J., 1950.

DVORETZKY, A., A. WALD, AND J. WOLFOWITZ [1], "Elimination of Randomization in Certain Problems of Statistics and of the Theory of Games," *Proceedings, National Academy of Sciences, U.S.A.,* Vol. 36, 1950, pp. 256–260.

FISHER, R. A. [1], "Randomization and an Old Enigma of Card Play," *Mathematical Gazette,* Vol. 18, 1934, pp. 294–297.

FRIEDMAN, M., AND L. J. SAVAGE [1], "The Utility Analysis of Choices Involving Risk," *Journal of Political Economy,* Vol. 56, 1948, pp. 279–304.

GALE, D., H. W. KUHN, AND A. W. TUCKER [1], "On Symmetric Games," *Contributions to the Theory of Games,* ed. by H. W. Kuhn and A. W. Tucker, Annals of Mathematics Study No. 24, Princeton University Press, Princeton, N.J., 1950.

GALE, D., H. W. KUHN, AND A. W. TUCKER [2], "Reductions of Game Matrices," *Contributions to the Theory of Games,* ed. by H. W. Kuhn and A. W. Tucker, Annals of Mathematics Study No. 24, Princeton University Press, Princeton, N.J., 1950.

GALE, D., H. W. KUHN, AND A. W. TUCKER [3], "Linear Programming and the Theory of Games," *Activity Analysis of Production and Allocation,* Cowles Commission for Research in Economics, Monograph No. 13, John Wiley & Sons, Inc., New York, 1951.

GALE, D., AND S. SHERMAN [1], "Solutions of Finite Two-person Games," *Contributions to the Theory of Games,* ed. by H. W. Kuhn and A. W. Tucker, Annals of Mathematics Study No. 24, Princeton University Press, Princeton, N.J., 1950.

GALE, D., AND F. M. STEWART [1], "Infinite Games with Perfect Information," *Contributions to the Theory of Games—II,* Annals of Mathematics Study No. 28, Princeton Unversity Press, Princeton, N.J. (to be published).

GILLIES, D. [1], "Inflated and Bargaining Solutions to the (n, k)-games," *Contributions to the Theory of Games—II,* Annals of Mathematics Study No. 28, Princeton University Press, Princeton, N.J. (to be published).

GILLIES, D., J. MAYBERRY, AND J. VON NEUMANN [1], "Two Variants of Poker," *Contributions to the Theory of Games—II,* Annals of Mathematics Study No. 28, Princeton University Press, Princeton, N.J. (to be published).

GLEZERMAN, M., AND L. PONTRYAGIN [1], *Intersections in Manifolds,* Translation No. 50, American Mathematical Society, New York, 1951.

GLICKSBERG, I., AND O. GROSS [1], "Some Notes on Games on the Unit Square," *Contributions to the Theory of Games—II,* Annals of Mathematics Study No. 28, Princeton University Press, Princeton, N.J. (to be published).

GUILBAUD, G. T. [1], "La théorie des jeux—contributions critiques à la théorie de la valeur," *Économique Appliquée,* Vol. 2, 1949, pp. 275–319.

HELMER, O. [1], "Problems in Game Theory," *Econometrica,* Vol. 20, 1952, p. 90.

HITCHCOCK, F. L. [1], "The Distribution of a Product from Several Sources to Numerous Localities," *Journal of Mathematics and Physics,* Vol. 20, 1941, pp. 224–230.

HURWICZ, L. [1], "The Theory of Economic Behavior," *American Economic Review,* Vol. 35, 1945, pp. 909–925.

JUSTMAN, E. [1], "La théorie des jeux" (Une nouvelle théorie de l'équilibre économique), *Revue d'Économie Politique,* Nos. 5–6, 1949, pp. 616–633.

KAKUTANI, S. [1], "A Generalization of Brouwer's Fixed-point Theorem," *Duke Mathematical Journal,* Vol. 8, 1941, pp. 457–459.

KALMAR, L. [1], "Zur Theorie der abstrakten Spiele," *Acta Szeged,* Vol. 4, 1928–1929, pp. 65–85.

KAPLANSKY, I. [1], "A Contribution to von Neumann's Theory of Games," *Annals of Mathematics,* Vol. 46, 1945, pp. 474–479.

KARLIN, S. [1], "Operator Treatment of Minmax Principle," *Contributions to the Theory of Games,* ed. by H. W. Kuhn and A. W. Tucker, Annals of Mathematics Study No. 24, Princeton University Press, Princeton, N.J., 1950.

KARLIN, S. [2], "The Theory of Infinite Games," *Annals of Mathematics* (to be published).

KARLIN, S. [3], "Reduction of Certain Classes of Games to Integral Equations," *Contributions to the Theory of Games—II,* Annals of Mathematics Study No. 28, Princeton University Press, Princeton, N.J. (to be published).

KARLIN, S. [4], "On a Class of Games," *Contributions to the Theory of Games—II,* Annals of Mathematics Study No. 28, Princeton University Press, Princeton, N.J. (to be published).

KAYSEN, C. [1], "A Revolution in Economic Theory?" *The Review of Economic Studies,* Vol. 14, No. 1, 1946–1947, pp. 1–15.

KRENTEL, W. D., J. C. C. MCKINSEY, AND W. V. QUINE [1], "A Simplification of Games in Extensive Form," *Duke Mathematical Journal,* Vol. 18, 1951, pp. 885–900.

KUHN, H. W. [1], "A Simplified Two-person Poker," *Contributions to the Theory of Games,* ed. by H. W. Kuhn and A. W. Tucker, Annals of Mathematics Study No. 24, Princeton University Press, Princeton, N.J., 1950.

KUHN, H. W. [2], "Extensive Games," *Proceedings, National Academy of Sciences, U.S.A.,* Vol. 36, 1950, pp. 570–576.

KUHN, H. W. [3], "Extensive Games and the Problem of Information," *Contributions to the Theory of Games—II,* Annals of Mathematics Study No. 28, Princeton University Press, Princeton, N.J. (to be published).

KUHN, H. W., AND A. W. TUCKER [1], "Preface," *Contributions to the Theory of Games,* ed. by H. W. Kuhn and A. W. Tucker, Annals of Mathematics Study No. 24, Princeton University Press, Princeton, N.J., 1950.

LEUNBACH, G. [1], "Theory of Games and Economic Behaviour," *Nordisk Tidsskrift för Teknisk Økonomi,* Nos. 1–4, 1948, pp. 175–178.

LOOMIS, L. H. [1], "On a Theorem of von Neumann," *Proceedings, National Academy of Sciences, U.S.A.,* Vol. 32, 1946, pp. 213–215.

MACDUFFEE, C. C. [1], *The Theory of Matrices,* Chelsea Publishing Company, New York, 1946.

MARSCHAK, J. [1], "Neumann's and Morgenstern's New Approach to Static Economics," *Journal of Political Economy,* Vol. 54, 1946, pp. 97–115.

MARSCHAK, J. [2], "Rational Behavior, Uncertain Prospects, and Measurable Utility," *Econometrica,* Vol. 18, 1950, pp. 111–141.

MCDONALD, J. [1], "Poker: An American Game," *Fortune,* Vol. 37, 1948, pp. 128–131 and 181–187.

MCDONALD, J. [2], "The Theory of Strategy," *Fortune,* Vol. 38, 1949, pp. 100–110.

MCDONALD, J. [3], *Strategy in Poker, Business, and War,* W. W. Norton & Company, New York, 1950.

MCKINSEY, J. C. C. [1], "Isomorphism of Games and Strategic Equivalence," *Contributions to the Theory of Games,* ed. by H. W. Kuhn and A. W. Tucker, Annals of Mathematics Study No. 24, Princeton University Press, Princeton, N.J., 1950.

MENDEZ, J. [1], "Progresos en la Teoría Económica de la Conducta Individual," *Rivista Trimestial de Cultura Moderna, Universidad Nacional de Colombia,* Vol. 7, 1946, pp. 259–276.

MILNOR, J. W. [1], "Sums of Positional Games," *Contributions to the Theory of Games—II,* Annals of Mathematics Study No. 28, Princeton University Press, Princeton, N.J. (to be published).

MORGENSTERN, O. [1], "Demand Theory Reconsidered," *Quarterly Journal of Economics,* Vol. 62, 1948, pp. 165–201.

MORGENSTERN, O. [2], "Oligopoly, Monopolistic Competition, and the Theory of Games," *Proceedings, American Economic Review,* Vol. 38, 1948, pp. 10–18.

MORGENSTERN, O. [3], "Theorie des Spiels," *Die Amerikanische Rundschau,* Vol. 5, 1949, pp. 76–87.

MORGENSTERN, O. [4], "The Theory of Games," *Scientific American,* Vol. 180, 1949, pp. 22–25.

MORGENSTERN, O. [5], "Die Theorie der Spiele und des Wirtschaftlichen Verhaltens, Part I," *Jahrbuch für Sozialwissenschaft,* Vol. 1, 1950, pp. 113–139.

MORGENSTERN, O. [6], "Economics and the Theory of Games," *Kyklos,* Vol. 3, 1949, pp. 294–308.

MOTZKIN, T. S., H. RAIFFA, G. L. THOMPSON, AND R. M. THRALL [1], "The Double Description Method," *Contributions to the Theory of Games—II,* Annals of Mathematics Study No. 28, Princeton University Press, Princeton, N.J. (to be published).

NASH, J. F. [1], "Equilibrium Points in *n*-Person Games," *Proceedings, National Academy of Sciences, U.S.A.,* Vol. 36, 1950, pp. 48–49.

NASH, J. F. [2], "The Bargaining Problem," *Econometrica,* Vol. 18, 1950, pp. 155–162.

NASH, J. F. [3], "Non-cooperative Games," *Annals of Mathematics,* Vol. 54, 1951, pp. 286–295.

NASH, J. F., AND L. S. SHAPLEY [1], "A Simple Three-person Poker Game," *Contributions to the Theory of Games,* ed. by H. W. Kuhn and A. W. Tucker, Annals of Mathematics Study No. 24, Princeton University Press, Princeton, N.J., 1950.

NEUMANN, J. VON [1], "Zur Theorie der Gesellshaftsspiele," *Mathematische Annalen,* Vol. 100, 1928, pp. 295–320.

NEUMANN, J. VON [2], "Über ein ökonomisches Gleichungssystem und eine Verallgemeinerung des Brouwer'schen Fixpunktsatzes," *Ergebnisse eines mathematischen Kolloquiums,* Vol. 8, 1937, pp. 73–83.

NEUMANN, J. VON [3], "A Model of General Economic Equilibrium," *Review of Economic Studies,* Vol. 13, 1945–1946, pp. 1–9.

NEUMANN, J. VON [4], "A Certain Zero-sum Two-person Game Equivalent to the Optimal Assignment Problem," *Contributions to the Theory of Games—II,* Annals of Mathematics Study No. 28, Princeton University Press, Princeton, N.J. (to be published).

NEUMANN, J. VON, AND O. MORGENSTERN [1], *Theory of Games and Economic Behavior,* 2d ed., Princeton University Press, Princeton, N.J., 1947.

PAXSON, E. W. [1], "Recent Developments in the Mathematical Theory of Games," *Econometrica,* Vol. 17, 1949, pp. 72–73.

POSSEL, R. DE [1], "Sur la théorie mathématique des jeux de hasard et de reflexion," *Actualités Scientifiques et Industrielles,* No. 436, Hermann & Cie, Paris, 1936.

RAIFFA, H. [1], "Arbitration Schemes for Generalized Two-person Games," *Contributions to the Theory of Games—II,* Annals of Mathematics Study No. 28, Princeton University Press, Princeton, N.J. (to be published).

ROBINSON, J. [1], "An Iterative Method of Solving a Game," *Annals of Mathematics,* Vol. 54, 1951, pp. 296–301.

RUIST, E. [1], "Spelteori och ekonomiska problem," *Economisk Tidsskrift,* Vol. 2, 1949, pp. 112–117.

SHAPLEY, L. S. [1], "Quota Solutions of *n*-person Games," *Contributions to the Theory of Games—II,* Annals of Mathematics Study No. 28, Princeton University Press, Princeton, N.J. (to be published).

SHAPLEY, L. S. [2], "A Value for *n*-person Games," *Contributions to the Theory of Games—II,* Annals of Mathematics Study No. 28, Princeton University Press, Princeton, N.J. (to be published).

SHAPLEY, L. S. [3], "Information and the Formal Solution of Many-moved Games," *Proceedings, International Congress of Mathematicians,* Vol. 1, 1950, pp. 574–575.

SHAPLEY, L. S., AND R. N. SNOW [1], "Basic Solutions of Discrete Games," *Contributions to the Theory of Games,* ed. by H. W. Kuhn and A. W. Tucker, Annals of Mathematics Study No. 24, Princeton University Press, Princeton, N.J., 1950.

SHERMAN, S. [1], "Games and Sub-games," *Proceedings, American Mathematical Society,* Vol. 2, 1951, pp. 186–187.

SHIFFMAN, M. [1], "Games of Timing," *Contributions to the Theory of Games—II,* Annals of Mathematics Study No. 28, Princeton University Press, Princeton, N.J. (to be published).

SINGER, K. [1], "Robot Economics," *Economic Record,* Vol. 25, 1949, pp. 48–73.

STONE, R. [1], "The Theory of Games," *Economic Journal,* Vol. 58, 1948, pp. 185–201.

THOMPSON, G. L. [1], "Signaling Strategies in *n*-person Games," *Contributions to the Theory of Games—II,* Annals of Mathematics Study No. 28, Princeton University Press, Princeton, N.J. (to be published).

THOMPSON, G. L. [2], "Bridge and Signaling," *Contributions to the Theory of Games —II,* Annals of Mathematics Study No. 28, Princeton University Press, Princeton, N.J. (to be published).

TUKEY, J. W. [1], "A Problem in Strategy," *Econometrica,* Vol. 17, 1949, p. 73.

VILLE, J. [1], "Sur la théorie générale des jeux où intervient l'habilité des joueurs," *Traité du calcul des probabilités et de ses applications,* ed. by E. Borel and collaborators, Vol. 2, No. 5, Gauthier-Villars & Cie, Paris, 1938, pp. 105–113.

WALD, A. [1], "Über die eindeutige positive Lösbarkeit der neuen Produktionsgleichungen," *Ergebnisse eines mathematischen Kolloquiums,* Vol. 6, 1935, pp. 12–20 (with discussion by K. Menger).

WALD, A. [2], "Über die Produktionsgleichungen der ökonomischen Wertlehre: II. Mitteilung," *Ergebnisse eines mathematischen Kolloquiums,* Vol. 7, 1936, pp. 1–6.

WALD, A. [3], "Über einige Gleichungssysteme der mathematischen Ökonomie," *Zeitschrift für Nationalökonomie,* Vol. 7, 1936, pp. 637–670.

WALD, A. [4], "Generalization of a Theorem by von Neumann Concerning Zero-sum Two-person Games," *Annals of Mathematics,* Vol. 46, 1945, pp. 281–286.

WALD, A. [5], "Statistical Decision Functions Which Minimize the Maximum Risk," *Annals of Mathematics,* Vol. 46, 1945, pp. 265–280.

WALD, A. [6], "Foundation of a General Theory of Sequential Decision Functions," *Econometrica,* Vol. 15, 1947, pp. 279–313.

WALD, A. [7], "Statistical Decision Functions," *Annals of Mathematical Statistics,* Vol. 20, 1949, pp. 165–205.

WALD, A. [8], *Statistical Decision Functions,* John Wiley & Sons, Inc., New York, 1950.

WALD, A., AND J. WOLFOWITZ [1], "Bayes Solutions of Sequential Decision Problems," *Proceedings, National Academy of Sciences, U.S.A.*, Vol. 35, 1949, pp. 99–102.

WALD, A., AND J. WOLFOWITZ [2], "Bayes Solutions of Sequential Decision Problems," *Annals of Mathematical Statistics*, Vol. 21, 1950, pp. 82–99.

WALD, A., AND J. WOLFOWITZ [3], "Two Methods of Randomization in Statistics and Theory of Games," *Annals of Mathematics*, Vol. 53, 1951, pp. 581–586.

WEYL, H. [1], "Elementare Theorie der konvexen Polyeder," *Commentarii Mathematici Helvetici*, Vol. 7, 1934–1935, pp. 290–306. (English translation, "The Elementary Theory of Convex Polyhedra," in *Contributions to the Theory of Games*, ed. by H. W. Kuhn and A. W. Tucker, Annals of Mathematics Study No. 24, Princeton University Press, Princeton, N.J., 1950.)

WEYL, H. [2], "Elementary Proof of a Minimax Theorem Due to von Neumann," *Contributions to the Theory of Games*, ed. by H. W. Kuhn and A. W. Tucker, Annals of Mathematics Study No. 24, Princeton University Press, Princeton, N.J., 1950.

WIDDER, D. V. [1], *The Laplace Transform*, Princeton University Press, Princeton, N.J., 1941.

WIDDER, D. V. [2], *Advanced Calculus*, Prentice-Hall, Inc., New York, 1947.

Index